The 17th International Conference of the Hellenic Association of Agricultural Economists

The 17th International Conference of the Hellenic Association of Agricultural Economists

Editor

Eleni Theodoropoulou

Basel • Beijing • Wuhan • Barcelona • Belgrade • Novi Sad • Cluj • Manchester

Editor
Eleni Theodoropoulou
Harokopio University of Athens
Athens
Greece

Editorial Office
MDPI
St. Alban-Anlage 66
4052 Basel, Switzerland

This is a reprint of articles from the Proceedings published online in the open access journal *Proceedings* (ISSN 2504-3900) (available at: https://www.mdpi.com/2504-3900/94/1).

For citation purposes, cite each article independently as indicated on the article page online and as indicated below:

Lastname, A.A.; Lastname, B.B. Article Title. *Journal Name* **Year**, *Volume Number*, Page Range.

ISBN 978-3-7258-1197-7 (Hbk)
ISBN 978-3-7258-1198-4 (PDF)
doi.org/10.3390/books978-3-7258-1198-4

© 2024 by the authors. Articles in this book are Open Access and distributed under the Creative Commons Attribution (CC BY) license. The book as a whole is distributed by MDPI under the terms and conditions of the Creative Commons Attribution-NonCommercial-NoDerivs (CC BY-NC-ND) license.

Contents

Eleni Theodoropoulou
Preface of the 17th International Conference of the Hellenic Association of Agricultural Economists (ETAGRO 2023) in Thessaloniki, Greece
Reprinted from: *Proceedings* 2024, 94, 65, doi:10.3390/proceedings2024094065 1

Eleni Theodoropoulou
Statement of Peer Review
Reprinted from: *Proceedings* 2024, 94, 8, doi:10.3390/proceedings2024094008 5

Chrysanthi Charatsari, Anastasios Michailidis, Evagelos D. Lioutas, Thomas Bournaris, Efstratios Loizou, Aikaterini Paltaki and Dimitra Lazaridou
Towards Agricultural Digitalization: Does Higher Agricultural Education Supply Students with Relevant Competencies? [†]
Reprinted from: *Proceedings* 2024, 94, 1, doi:10.3390/proceedings2024094001 6

Stamatis Mantziaris, Irene Tzouramani and Angelos Liontakis
Identifying the External Environment of Greek Fisheries [†]
Reprinted from: *Proceedings* 2024, 94, 2, doi:10.3390/proceedings2024094002 10

Danai Antonopoulou, Ioannis A. Giantsis, George K. Symeon and Melpomeni Avdi
The Association of MTNR1A Gene Alleles with the Response to Estrus Induction Treatments in Improved and Non-Improved Greek Indigenous Sheep Breeds [†]
Reprinted from: *Proceedings* 2023, 94, 3, doi:10.3390/proceedings2024094003 14

Malak Hazimeh, Leonidas Sotirios Kyrgiakos, Georgios Kleftodimos, Christina Kleisiari, Marios Vasileiou and George Vlontzos
Assessing Agroecology Terms for North African Countries: A Literature Review [†]
Reprinted from: *Proceedings* 2024, 94, 4, doi:10.3390/proceedings2024094004 17

Theo Benos, Panagiota Sergaki and Nikos Kalogeras
The Effect of Unfair Trading Practices on the Performance of Agricultural Cooperatives [†]
Reprinted from: *Proceedings* 2024, 94, 5, doi:10.3390/proceedings2024094005 21

Leonidas Sotirios Kyrgiakos, Malak Hazimeh, Marios Vasileiou, Christina Kleisiari, Georgios Kleftodimos and George Vlontzos
The Food Fraud Landscape: A Brief Review of Food Safety and Authenticity [†]
Reprinted from: *Proceedings* 2024, 94, 6, doi:10.3390/proceedings2024094006 25

Aikaterini Paltaki, Efstratios Loizou, Fotios Chatzitheodoridis, Maria Partalidou, Stefanos Nastis and Anastasios Michailidis
Farmers' Knowledge, Training Needs and Skills in the Bioeconomy: Evidence from the Region of Western Macedonia [†]
Reprinted from: *Proceedings* 2024, 94, 7, doi:10.3390/proceedings2024094007 30

Anna Tafidou, Asimina Kouriati, Evgenia Lialia, Angelos Prentzas, Eleni Dimitriadou, Kyriaki Tafidou and Thomas Bournaris
An Electronic Platform for the Integrated Monitoring of Technical and Economic Data of Farms [†]
Reprinted from: *Proceedings* 2024, 94, 9, doi:10.3390/proceedings2024094009 34

Georgios Kountios, Ioannis Chatzis and Georgios Papadavid
Agricultural Knowledge and Innovation Systems and Sustainable Management of Natural Resources [†]
Reprinted from: *Proceedings* **2024**, *94*, 10, doi:10.3390/proceedings2024094010 38

Martina Francescone, Chrysanthi Charatsari, Evagelos D. Lioutas, Luca Bartoli and Marcello De Rosa
Questioning Family Farms' Readiness to Adopt Digital Solutions [†]
Reprinted from: *Proceedings* **2024**, *94*, 11, doi:10.3390/proceedings2024094011 42

Eirini Papadimitriou and Dimitra Lazaridou
Investigating Farmers' Attitudes towards Co-Existence of Agriculture and Renewable Energy Production [†]
Reprinted from: *Proceedings* **2024**, *94*, 12, doi:10.3390/proceedings2024094012 47

Dimitrios Kateris, Anastasios Mitsopoulos, Charalampos Petkoglou and Dionysis Bochtis
Lameness Identification System in Cattle Breeding Units [†]
Reprinted from: *Proceedings* **2024**, *94*, 13, doi:10.3390/proceedings2024094013 51

Gregory Mygdakos, Panagiotis Tournavitis and Emanuel Lekakis
Estimating Farmers' Creditworthiness under a Changing Climate [†]
Reprinted from: *Proceedings* **2024**, *94*, 14, doi:10.3390/proceedings2024094014 55

Penelope Gouta and Christos T. Papadas
Sectoral R&D Activities and Knowledge Production Functions: A Study Using International Data [†]
Reprinted from: *Proceedings* **2024**, *94*, 15, doi:10.3390/proceedings2024094015 60

Giorgos N. Diakoulakis, Konstantinos Tsiboukas and Dimitrios Savvas
Exploring the Financial Viability of Greenhouse Tomato Growers under Climate Change-Induced Multiple Stress [†]
Reprinted from: *Proceedings* **2024**, *94*, 16, doi:10.3390/proceedings2024094016 64

Eleni Pappa and Alex Koutsouris
In Search of the Agronomist as Trusted Advisor: A Farmer-Centric Case Study [†]
Reprinted from: *Proceedings* **2024**, *94*, 17, doi:10.3390/proceedings2024094017 67

Georgia S. Papoutsi and Elena Kourtesi
Factors Influencing Consumer Receptivity to Sustainable Packaging: A Probit Regression Analysis [†]
Reprinted from: *Proceedings* **2024**, *94*, 18, doi:10.3390/proceedings2024094018 70

Georgios Kountios, Ourania Notta, Nikolaos Kratimenos and Ioannis Chatzis
The Implementation of Innovation and the Attitudes of Farmers towards Advisory Services: The Case of Western Macedonia, Greece [†]
Reprinted from: *Proceedings* **2024**, *94*, 19, doi:10.3390/proceedings2024094019 74

Argyrios Georgilas and Zacharoula Andreopoulou
Online Sales Promotion of Geographical Indication Products: The Case of Evia PDO Dried Figs [†]
Reprinted from: *Proceedings* **2024**, *94*, 20, doi:10.3390/proceedings2024094020 78

Athanasios Batzios, Vagis Samathrakis, Alexandros Theodoridis, Georgia Koutouzidou and Alexandros Kakouris
Identifying Veterinary Students' Attitudes on Entrepreneurial Intentions: A Two-Step Cluster Analysis [†]
Reprinted from: *Proceedings* **2024**, *94*, 21, doi:10.3390/proceedings2024094021 82

Dimitrios Kyriazoglou, Vasiliki Makri, Martha Tampaki, Katerina Melfou, Athanasios Ragkos and Ioannis A. Giantsis
Consumers' Trust and Preferences Regarding Local Plant Varieties and Indigenous Farm Animal Breeds in Western Macedonia, Greece [†]
Reprinted from: *Proceedings* **2024**, *94*, 22, doi:10.3390/proceedings2024094022 89

Ekaterini Alexaki, Ioannis Dimitriadis, Efstratios Michalis, Christina-Eleni Giatra and Athanasios Ragkos
Evaluation of the Certification Procedure of Farm Advisors in Greece [†]
Reprinted from: *Proceedings* **2024**, *94*, 23, doi:10.3390/proceedings2024094023 92

Stefania Tselempi, Michalis Frantzis and Panagiota Sergaki
Assessing Economic and Social Security in Agricultural Cooperatives: Two Case Studies from Cooperatives in Northern and Southern Greece [†]
Reprinted from: *Proceedings* **2024**, *94*, 24, doi:10.3390/proceedings2024094024 96

Ranko Gantner, Igor DelVechio, Zvonimir Steiner, Maja Gregić and Vesna Gantner
The Annual Maintenance Costs of Draft Horses as a Part of Fixed Costs in Horse-Powered Agriculture: A Case Study from Požega, Croatia [†]
Reprinted from: *Proceedings* **2024**, *94*, 25, doi:10.3390/proceedings2024094025 102

Vesna Gantner, Ivana Jožef, Ranko Gantner, Maja Gregić and Zvonimir Steiner
The Effect of Farm Size on the Differences in Mastitis Prevalence and Its Consequences on Milk Production in Holstein Cows [†]
Reprinted from: *Proceedings* **2024**, *94*, 26, doi:10.3390/proceedings2024094026 107

Eleni Zafeiriou, Garyfallos Arabatzis, Georgios Tsantopoulos, Spyros Galatsidas and Stavros Tsiantikoudis
Exploring the Impact of the Greening of the Agri-Food Sector on Economic Growth: An Empirical Approach in the BVAR Framework for the EU [†]
Reprinted from: *Proceedings* **2024**, *94*, 27, doi:10.3390/proceedings2024094027 111

Dimitrios K. Papadopoulos, Ioannis Georgoulis, Athanasios Lattos, Konstantinos Feidantsis, Basile Michaelidis and Ioannis A. Giantsis
Partial Substitution of Fresh Microalgae with Baker's Yeast (*Saccharomyces cerevisiae*) Enhances the Growth of Juvenile *Ostrea edulis* and *Ruditapes decussatus* [†]
Reprinted from: *Proceedings* **2024**, *94*, 28, doi:10.3390/proceedings2024094028 116

Epistimi Amerani and Anastasios Michailidis
The Agricultural Knowledge and Innovation System (AKIS) in a Changing Environment in Greece [†]
Reprinted from: *Proceedings* **2024**, *94*, 29, doi:10.3390/proceedings2024094029 120

Andreas Michalitsis, Ferdaous Rezgui, Fatima Lambarraa-Lehnhardt, Paschalis Papakaloudis, Maria Laskari, Efstratios Deligiannis and Christos Dordas
Sustainability Assessment of Highly Biodiversified Farming Systems: Multicriteria Assessment of Greek Arable Crops [†]
Reprinted from: *Proceedings* **2024**, *94*, 30, doi:10.3390/proceedings2024094030 124

Elissavet Ninou, Fokion Papathanasiou, Anthoula Tsipi, Anastasia Kargiotidou, Georgia Vasiligianni, Konstantinos Koutis and Ioannis Mylonas
Important Parameters Connected to Farmers' Networking and Training That Give Added Value to "Fasolia Vanilies Feneou" and "Fava Feneou" Products [†]
Reprinted from: *Proceedings* **2024**, *94*, 31, doi:10.3390/proceedings2024094031 128

Paschalis Papakaloudis, Andreas Michalitsis, Maria Laskari, Efstratios Deligiannis, Fatima Lambarraa-Lehnhardt and Christos Dordas
Co-Design and Co-Evaluation of Traditional and Highly Biodiversity-Based Cropping Systems in the Mediterranean Area [†]
Reprinted from: *Proceedings* 2024, *94*, 32, doi:10.3390/proceedings2024094032 132

Eleni Zarokosta and Alex Koutsouris
Building Advisors and Researchers' Capacity to Support Agricultural Knowledge and Innovation Systems in Europe: The Case of the I2CONNECT Summer School [†]
Reprinted from: *Proceedings* 2024, *94*, 33, doi:10.3390/proceedings2024094033 136

Maja Gregić, Tina Bobić, Ranko Gantner and Vesna Gantner
The Use of Digital Media in Equestrian Clubs in Croatia [†]
Reprinted from: *Proceedings* 2024, *94*, 34, doi:10.3390/proceedings2024094034 140

Konstantinos Demestichas, Antonis Vlandis, Maria Ntaliani and Constantina Costopoulou
An Investigation of the Digital Presence of Agricultural Stores in Greece [†]
Reprinted from: *Proceedings* 2024, *94*, 35, doi:10.3390/proceedings2024094035 145

Ioannis Mylonas, Fokion Papathanasiou, Elissavet Ninou, Anthoula Tsipi, Dimitrios Kostitsis, Iosif Sistanis, et al.
Factors Connected with the Registration of "Sikali Vevis" as a Geographical Indication Protection (PGI) Product [†]
Reprinted from: *Proceedings* 2024, *94*, 36, doi:10.3390/proceedings2024094036 149

Christos P. Pappas and Christos T. Papadas
Agricultural Value Added, Farm Business Cycles and Their Relation to the Non-Farm Economy [†]
Reprinted from: *Proceedings* 2024, *94*, 37, doi:10.3390/proceedings2024094037 153

Konstantinos Ioannou, Evangelia Karasmanaki, Despoina Sfiri, Georgios Tsantopoulos and Kleanthis Xenitidis
Cultivate Crops or Produce Energy? Factors Affecting the Decision of Farmers to Install Photovoltaics on Their Farmland [†]
Reprinted from: *Proceedings* 2024, *94*, 38, doi:10.3390/proceedings2024094038 157

Maria Spilioti, Pavlos Karanikolas, George Papadomichelakis, Konstantinos Tsiboukas and Dimitris Voloudakis
Upgrading Value Chains through Farm Advisory [†]
Reprinted from: *Proceedings* 2024, *94*, 39, doi:10.3390/proceedings2024094039 161

Georgia Koutouzidou, Vagis Samathrakis, Athanasios Batzios and Alexandros Theodoridis
Veterinary Students' Perceptions of Entrepreneurship Education [†]
Reprinted from: *Proceedings* 2024, *94*, 40, doi:10.3390/proceedings2024094040 165

Eleutheria Plousiou, Panagiota Sergaki and Ifigeneia Mylona
The Role of Cooperative Enterprises in the Promotion of Cultural Heritage: A Case Study of the Petrified Forest of Lesvos [†]
Reprinted from: *Proceedings* 2024, *94*, 41, doi:10.3390/proceedings2024094041 169

Asimina Kouriati, Christina Moulogianni, Evgenia Lialia, Angelos Prentzas, Anna Tafidou, Eleni Dimitriadou and Thomas Bournaris
Decision Support Model for Integrating the New Cross-Compliance Rules and Rational Water Management [†]
Reprinted from: *Proceedings* 2024, *94*, 42, doi:10.3390/proceedings2024094042 173

Evgenia Lialia, Anna Tafidou, Asimina Kouriati, Angelos Prentzas, Eleni Dimitriadou, Christina Moulogianni and Thomas Bournaris
Decision Support Model for Input Minimization and the Optimal Economic Efficiency of Agricultural Holdings [†]
Reprinted from: *Proceedings* 2024, 94, 43, doi:10.3390/proceedings2024094043 177

Maria V. Alvanou, Maria Tokamani, Athanasios Toros, Raphael Sandaltzopoulos, Konstantinos Zampakas, Chrysoula Tananaki, et al.
Using Pollen DNA Metabarcoding to Assess the Foraging Preferences of Honeybees in Kastoria Region, Greece [†]
Reprinted from: *Proceedings* 2024, 94, 44, doi:10.3390/proceedings2024094044 181

Alex Koutsouris and Vasiliki Kanaki
Towards a Farmer-Centric Approach to Advise Provision [†]
Reprinted from: *Proceedings* 2024, 94, 45, doi:10.3390/proceedings2024094045 184

Vasiliki Bitsopoulou, Eleni Pastrapa, Eleni Zenakou, Despina Sdrali and Eleni Theodoropoulou
Farmers Vocational Education and Training: The Case of Public Institutes of Vocational Training at ELGO-DIMITRA [†]
Reprinted from: *Proceedings* 2024, 94, 46, doi:10.3390/proceedings2024094046 188

Georgios Papadavid, Georgios Kountios, Diofantos Hadjimitsis and Maria Tsiouni
The Use of Precision Agriculture for Improving the Water Economics of Farms and the Need for Agricultural Advisory [†]
Reprinted from: *Proceedings* 2024, 94, 47, doi:10.3390/proceedings2024094047 193

Efstratios Michalis, Ahmed Yangui, Athanasios Ragkos, Mohamed Kharrat and Dimosthenis Chachalis
Opinions and Perceptions on Sustainable Weed Management: A Comparison between Greek and Tunisian Farmers [†]
Reprinted from: *Proceedings* 2024, 94, 48, doi:10.3390/proceedings2024094048 197

Elissavet Ninou, Fokion Papathanasiou, Christos Alexandris, Elisavet Chatzivassiliou, Garyfallia Economou, Dimitrios Vlachostergios, et al.
Fava Santorinis: Brining Added Value to a Protected Designation of Origin (PDO) Product through the Security of the Traditional Cultivar and Farmers Network [†]
Reprinted from: *Proceedings* 2024, 94, 49, doi:10.3390/proceedings2024094049 201

Georgios Roumeliotis, Elena Raptou, Konstantinos Polymeros and Konstantinos Galanopoulos
An Empirical Investigation of Ethical Food Choices: A Qualitative Research Approach [†]
Reprinted from: *Proceedings* 2024, 94, 50, doi:10.3390/proceedings2024094050 205

Sofia Karampela, Aigli Koliotasi and Konstantinos Kostalis
Unraveling the Research Landscape of Happiness through Agro, Agri, and Rural Tourism for Future Directions [†]
Reprinted from: *Proceedings* 2024, 94, 51, doi:10.3390/proceedings2024094051 209

Elissavet Ninou, Fokion Papathanasiou, Iosif Sistanis, Anastasia Kargiotidou, Sonia Michailidou, Konstantinos Koutis, et al.
The Impact of the Improved Genetic Material to the Economic Value of Plake Fasoli Prespon PGI Product [†]
Reprinted from: *Proceedings* 2024, 94, 52, doi:10.3390/proceedings2024094052 213

Evangelia Stathopoulou, Eleni Theodoropoulou, Antony Rezitis and George Vlahos
Rural Infrastructure Using Dry-Stone Walling, an Asset for Sustainable Development in a Regional and Local Context [†]
Reprinted from: *Proceedings* **2024**, *94*, 53, doi:10.3390/proceedings2024094053 217

Ioannis Sotiriadis, George Sidiropoulos and Maria Partalidou
Social Innovation and Women's Agricultural Cooperatives: Applying Social Change Theory [†]
Reprinted from: *Proceedings* **2024**, *94*, 54, doi:10.3390/proceedings2024094054 221

Lykourgos Chatziioannidis and Maria Partalidou
The Greek Perspective on Foreign Farm Workers and Agricultural Labor [†]
Reprinted from: *Proceedings* **2024**, *94*, 55, doi:10.3390/proceedings2024094055 225

Myrto Paraschou and Panagiota Sergaki
Agricultural Cooperatives as a Vehicle for Small-Scale Farmer's Viability and Sustainable Practices [†]
Reprinted from: *Proceedings* **2024**, *94*, 56, doi:10.3390/proceedings2024094056 229

Aristotelis Batzios and Maria Tsiouni
The Effect of the Regulatory Role of Collective Organizations in Relation to the Consumption of Fruits and Vegetables from Cooperatives [†]
Reprinted from: *Proceedings* **2024**, *94*, 57, doi:10.3390/proceedings2024094057 233

George Bellis, Paris Papaggelos, Evangeli Vlachogianni, Ilias Laleas, Stefanos Moustos, Thanos Patas, et al.
Automated Sorting System for Skeletal Deformities in Cultured Fishes [†]
Reprinted from: *Proceedings* **2024**, *94*, 58, doi:10.3390/proceedings2024094058 237

Panagiota Pantazi
The Role of Cooperatives in the Interconnection of the Agri-Food and Tourism Sectors, Kyllini, 14/09/2023 [†]
Reprinted from: *Proceedings* **2024**, *94*, 59, doi:10.3390/proceedings2024094059 242

Dimitris Alexandridis, Christina Kleisiari and George Vlontzos
Consumers' Behavior toward Plant-Based Milk Alternatives [†]
Reprinted from: *Proceedings* **2024**, *94*, 60, doi:10.3390/proceedings2024094060 246

Maria Tsiouni, Georgios Kountios and Alexandra Pavloudi
How Do Agricultural Education, Advisory, and Financial Factors Affect the Adoption of Precision Farming in Greece? [†]
Reprinted from: *Proceedings* **2024**, *94*, 61, doi:10.3390/proceedings2024094061 250

Sofia Karampela, Thanasis Kizos and Alex Koutsouris
Discovering Innovation, Social Capital and Farm Viability in the Framework of the United Winemaking Agricultural Cooperative of Samos [†]
Reprinted from: *Proceedings* **2024**, *94*, 62, doi:10.3390/proceedings2024094062 254

Aliki Dourountaki, Sofia Karampela and Alex Koutsouris
Gastronomic Tourism and Festivals: Views from Potential Tourists in Greece and South Korea [†]
Reprinted from: *Proceedings* **2024**, *94*, 63, doi:10.3390/proceedings2024094063 258

George Bellis, Paris Papaggelos, Evangeli Vlachogianni, Ilias Laleas, Stefanos Moustos, Thanos Patas, et al.
Development of A Non-Invasive System for the Automatic Detection of Cattle Lameness [†]
Reprinted from: *Proceedings* **2024**, *94*, 64, doi:10.3390/proceedings2024094064 262

Antonello Franca, Marta G. Ferre-Rivera, Feliu Lopez-i-Gelats, Giovanni M. Altana, Dimitrios Skordos, Marisol Dar Ali and Athanasios Ragkos
Pastoral Schools: Diffusing the Italian and Spanish Experience for Sustainable Mediterranean Pastoralism through Co-Creation [†]
Reprinted from: *Proceedings* **2024**, *94*, 66, doi:10.3390/proceedings2024094066 . . . . **266**

Editorial

Preface of the 17th International Conference of the Hellenic Association of Agricultural Economists (ETAGRO 2023) in Thessaloniki, Greece

Eleni Theodoropoulou

Department of Economics and Sustainable Development, Harokopio University of Athens, 17671 Kallithea, Greece; etheodo@hua.gr

1. Conference Overview

The immense political, economic, social, and environmental challenges within which the global agri-food system is operating formulate a complicated context in which the system must discover the innovations and solutions that will ensure its adequate performance and sustainability. Some of these challenges already exist, while others have been reshaped or are brand-new. The primary challenge for most countries worldwide is to achieve food security, as political and economic disturbances cause uncertainty at the production level as well as in the operations of large supply chains. The energy crisis, as a new threat, jeopardizes the viability and competitiveness of farms and the agro-industry. This situation is exacerbated by the effects of the climate crisis, highlighting the need for environmentally sustainable production systems and approaches to reducing the environmental footprint of the agri-food sector—such as a circular economy.

In Europe, the policy framework established by the new CAP 2023-27 and the Farm-to-Fork strategy—in light of the European Green Deal—brings to the forefront, now more than ever, the need to reconcile the often-conflicting goals of strengthening the resilience of the agri-food sector, protecting natural resources, and promoting quality of life in rural areas. At the nexus of these challenges, the depopulation of rural areas is intensifying, leading to the further loosening of social ties and loss of cultural heritage. Farmers operate under the influence of new demands and are in dire need of advisory support and an operational Agricultural Knowledge and Innovation System (AKIS) in order to achieve a balance within the changing structure of the farming sector. On the other hand, as a result of "eco-anxiety" and societal issues, society is increasingly looking for new dietary patterns and foods that meet criteria such as locality, the sustainable use of natural resources, and an interface with a healthy lifestyle. In this complex setting of requirements, challenges, and risks, but also opportunities, the need to connect research and innovation with the agri-food system and social expectations is calling for attention more than ever.

Some of the questions raised in this regard are as follows:

- How will the rural world be able to develop solutions and innovations to meet these challenges?
- How could science contribute to increasing the resilience of the agri-food system?
- What are the roles of new technologies, digitalization, and smart farming?
- What are the necessary synergies between farming and other activities in rural areas aiming to improve sustainability?
- What are the patterns of organization and cooperation among actors in the agri-food system with respect to ensuring sustainable development and social cohesion in rural areas?
- What are the social and economic sustainability aspects of alternative and environmentally friendly production systems, including—but not restricted to—agroecology and hydroponics?

Citation: Theodoropoulou, E. Preface of the 17th International Conference of the Hellenic Association of Agricultural Economists (ETAGRO 2023) in Thessaloniki, Greece. *Proceedings* **2024**, *94*, 65. https://doi.org/10.3390/proceedings2024094065

Published: 2 April 2024

Copyright: © 2024 by the author. Licensee MDPI, Basel, Switzerland. This article is an open access article distributed under the terms and conditions of the Creative Commons Attribution (CC BY) license (https://creativecommons.org/licenses/by/4.0/).

- What are the adoption patterns of innovative production methods, practices, and systems, and which policies and strategies could improve them?
- What are the evolving consumer profiles? Which agri-food production patterns and alternative products will fulfil their expectations, and what is the role of certification and labelling?
- What are the non-market values associated with environmentally friendly food production?

The Hellenic Association of Agricultural Economists (ETAGRO) invited contributors across the globe to attend the 17th International Conference of the Hellenic Association of Agricultural Economists (ETAGRO 2023) held in Thessaloniki from 2 to 3 November 2023. The main topic was "The agri-food system facing complex challenges: Responses towards economic, social, environmental sustainability". It was organized under the auspices of the School of Agriculture of the Aristotle University of Thessaloniki and co-organized by the University of Western Macedonia and the International Hellenic University. The Conference was sponsored by Ergoplanning Ltd., Qlab, the Geotechnical Chamber of Greece, Harokopio University of Athens, AgroApps, the Agricultural University of Athens, and Ecodevelopment.

ETAGRO 2023 invited leaders, policymakers, academics, scientists, producers, and political bodies to attend and share their work experience and thoughts regarding various subjects related to the bio-economy, such as sustainable agri-food systems, sustainable food security, climate change mitigation, new technologies, and others.

The specific objective of this meeting was to serve as a bridge in terms of communication and collaboration between science and research as well as the bodies and the stakeholders in agriculture, thereby facilitating the transfer of research data and expertise. All interested scholars or representatives from governmental entities, institutions, industries, NGOs, etc., were invited to contribute up-to-date approaches to the meeting. Young scholars and highly motivated students were strongly encouraged to participate, presenting their most recent theoretical and empirical research.

During the two days of the on-site event, nearly 250 participants had a chance to listen to 43 talks from experts in their fields, as well as from early-career researchers. In addition, 51 posters were presented in electronic form during the conference.

The conference started with a welcome event and an award ceremony for Prof. Stamatis Aggelopoulos, Dean of the International Hellenic University, and a keynote speech by Prof. Karl Behrendt, Elizabeth Creak Chair of Agri-Tech Economic Modelling at Harper Adams University (UK), the founding Director of the Global Institute for Agri-Tech Economics (HAU), and Co-Director of the Centre for Effective Innovation in Agriculture (UK). Prof. Behrendt presented a talk titled "The potential for agri-tech and digitalisation to improve the sustainability of food production".

The conference was divided in two parallel sessions per session, one in English and the other in Greek. There were five sessions in English and three sessions in Greek. The sessions were divided into categories such as "Farm advisory and agricultural higher education", "Current trends in consumer habits and value chains", "Innovation, digitalization and Cooperatives", "Agricultural Industry and Cooperatives", "Consumer behaviour", "Agricultural economics and trade", "Environment, bioeconomy and agroecology", and "Agricultural counselling and training".

The conference included three special sessions, two of which were round tables titled "Common Agricultural Policy Strategic Plan 2023-2027: Opportunities and challenges for the country's rural economy" and "The future of Greek agriculture in a complex international environment: Challenges and prospects", and one was an industry session titled "Innovation and entrepreneurship in Precision Agriculture in Greece".

2. Committees

Each one of the following contributors played a role in the success of ETAGRO 2023; therefore, we would like to list them here for future reference.

2.1. Organizing Committee
Thomas Bournaris (Chair), Aristotle University of Thessaloniki, Greece
Athanasios Batzios, University of Western Macedonia, Greece
Giannis Gousios, Agricultural University of Athens, Greece
Pavlos Karanikolas, Agricultural University of Athens, Greece
George Kountios, International Hellenic University, Greece
Stavriani Koutsou, International Hellenic University, Greece
Dimitra Lazaridou, Agricultural University of Athens, Greece
Katerina Melfou, University of Western Macedonia, Greece
Stefanos Nastis, Aristotle University of Thessaloniki, Greece
Dimitrios Natos, Aristotle University of Thessaloniki, Greece
Dimitrios Paparas, Harper Adams University, UK
Eleni Pastrapa, Harokopio University of Athens, Greece
Athanasios Ragkos, Agricultural Economic Research Institute, Greece
Despina Sdrali, Harokopio University of Athens, Greece
Panagiota Sergaki, Aristotle University of Thessaloniki, Greece
Maria Tsiouni, International Hellenic University, Greece
Elias Tsourapas, European Association for Viewers Interest, Belgium

2.2. Secretarial Support
Maria Balamoti, International Hellenic University, Greece
Lykourgos Chatziioannidis, Aristotle University of Thessaloniki, Greece
Evgenia Lialia, Aristotle University of Thessaloniki, Greece
Alexandra Niniraki, International Hellenic University, Greece
Georgios Omouridis, International Hellenic University, Greece
Aggelos Prentzas, Aristotle University of Thessaloniki, Greece

2.3. Scientific Committee
Eleni Theodoropoulou (Chair), Harokopio University of Athens, Greece
Konstadinos Abeliotis, Harokopio University of Athens, Greece
Mikael Akimowicz, University of Toulouse III, France
Maria Alebaki, Agricultural Economics Research Institute, ELGO-DIMITRA, Greece
Garyffalos Arabatzis, Democritus University of Thrace, Greece
Athanasios Batzios, University of Western Macedonia, Greece
Karl Behrendt, Harper Adams University, UK
Julio Berbel, University of Cordoba, Spain
Thomas Bournaris, Aristotle University of Thessaloniki, Greece
Christopher Brewster, Maastricht University, Netherlands
Kostas Chatzimichail, Agricultural University of Athens, Greece
Fotios Chatzitheodoridis, University of Western Macedonia, Greece
Elias Giannakis, Energy Environment and Water Research Center, The Cyprus Institute, Cyprus
Constantine Iliopoulos, Agricultural Economics Research Institute, ELGO-DIMITRA, Greece
Eleni Kaimakoudi, Agricultural Economics Research Institute, ELGO-DIMITRA, Greece
Irini Kamenidou, International Hellenic University, Greece
Efstathios Klonaris, Agricultural University of Athens, Greece
Ioannis Kostakis, Harokopio University of Athens, Greece
Georgios Kountios, International Hellenic University, Greece
Stavriani Koutsou, International Hellenic University, Greece
Alexandros Koutsouris, Agricultural University of Athens, Greece
Aggelos Liontakis, Agricultural University of Athens, Greece
Jaume Lloveras, Universitat de Lleida, Spain
George Malindretos, Harokopio University of Athens, Greece
Spiridon Mamalis, International Hellenic University, Greece

Daniel E. May, Harper Adams University, UK
Romain Melot, INRAE, SADAPT—Université Paris Saclay, AgroParisTech, France
Dimitrios Moutopoulos, University of Patras, Greece
Claude Napoleon, INRAE-Avignon, France
Dimitrios Natos, Aristotle University of Thessaloniki, Greece
Ourania Notta, International Hellenic University, Greece
Jean Christophe Paoli, INRAE-Corse, France
George Papadavid, Agricultural Research Institute of Cyprus, Cyprus
Georgia Papoutsi, Agricultural Economics Research Institute, ELGO-DIMITRA, Greece
Maria Partalidou, Aristotle University of Thessaloniki, Greece
Merri Raggi, University of Bologna, Italy
Elena Raptou, Democritus University of Thrace, Greece
Anthony Rezitis, Agricultural University of Athens, Greece
Vagis Samathrakis, International Hellenic University, Greece
Panagiota Sergaki, Aristotle University of Thessaloniki, Greece
Sofoklis Skoultsos, Harokopio University of Athens, Greece
Ledia Sula, University College of "LOGOS", Albania
Alexandros Theodoridis, Aristotle University of Thessaloniki, Greece
George Tsekouropoulos, International Hellenic University, Greece
Irini Tzouramani, Agricultural Economics Research Institute, ELGO-DIMITRA, Greece
Michael Vassalos, Clemson University, USA
Davide Viaggi, University of Bologna, Italy
Marco Vieri, University of Firenze, Italy
Georgios Vlachos, Agricultural University of Athens, Greece
Aspasia Vlachvei, University of Western Macedonia, Greece
George Vlontzos, University of Thessaly, Greece

Conflicts of Interest: The author declares no conflicts of interest.

Disclaimer/Publisher's Note: The statements, opinions and data contained in all publications are solely those of the individual author(s) and contributor(s) and not of MDPI and/or the editor(s). MDPI and/or the editor(s) disclaim responsibility for any injury to people or property resulting from any ideas, methods, instructions or products referred to in the content.

Editorial
Statement of Peer Review

Eleni Theodoropoulou

Department of Economics and Sustainable Development, Harokopio University of Athens, 17671 Kallithea, Greece; etheodo@hua.gr

In submitting conference proceedings to *Proceedings*, the volume editors of the proceedings certify to the publisher that all papers published in this volume have been subjected to peer review administered by the volume editors. Reviews were conducted by expert referees to the professional and scientific standards expected of the *Proceedings* journal.

- Type of peer review: double-blind.
- Conference submission management system: EasyChair.
- Number of submissions sent for review: 102.
- Number of submissions originally accepted: 101.
- Number of submissions withdrawn after the reviewing process: 7.
- Number of submissions opted-out from publication: 27.
- Number of submissions accepted for publication: 67.
- Acceptance rate (number of submissions accepted/number of submissions received): 99%.
- Average number of reviews per paper: 2.
- Total number of reviewers involved: 52.
- Any additional information on the review process: N/A.

Conflicts of Interest: The author declares no conflicts of interest.

Disclaimer/Publisher's Note: The statements, opinions and data contained in all publications are solely those of the individual author(s) and contributor(s) and not of MDPI and/or the editor(s). MDPI and/or the editor(s) disclaim responsibility for any injury to people or property resulting from any ideas, methods, instructions or products referred to in the content.

Citation: Theodoropoulou, E. Statement of Peer Review. *Proceedings* **2024**, *94*, 8. https://doi.org/10.3390/proceedings2024094008

Published: 22 January 2024

Copyright: © 2024 by the author. Licensee MDPI, Basel, Switzerland. This article is an open access article distributed under the terms and conditions of the Creative Commons Attribution (CC BY) license (https://creativecommons.org/licenses/by/4.0/).

Proceeding Paper

Towards Agricultural Digitalization: Does Higher Agricultural Education Supply Students with Relevant Competencies? †

Chrysanthi Charatsari [1,*], Anastasios Michailidis [1], Evagelos D. Lioutas [2], Thomas Bournaris [1], Efstratios Loizou [3], Aikaterini Paltaki [1] and Dimitra Lazaridou [4]

1. Department of Agricultural Economics, School of Agriculture, Aristotle University of Thessaloniki, 54124 Thessaloniki, Greece; tassosm@auth.gr (A.M.); tbournar@agro.auth.gr (T.B.); apaltaki@agro.auth.gr (A.P.)
2. Department of Supply Chain Management, International Hellenic University, 60100 Katerini, Greece; evagelos@agro.auth.gr
3. Department of Regional and Cross Border Development, University of Western Macedonia, 50100 Kozani, Greece; lstratos@agro.auth.gr
4. Department of Forestry and Natural Environment Management, Agricultural University of Athens, 36100 Karpenisi, Greece; dimitral@for.auth.gr
* Correspondence: chcharat@agro.auth.gr
† Presented at the 17th International Conference of the Hellenic Association of Agricultural Economists, Thessaloniki, Greece, 2–3 November 2023.

Abstract: Agricultural digitalization is gaining momentum, urging a transition from process-driven to technology-enhanced and data-driven agriculture. To support such a transition and help farmers derive benefits from digital technologies, also avoiding the potential threats associated with digitalization, future advisors need a variety of competencies, ranging from pure technocentric skills to more complex capabilities, such as impact forecasting and transition facilitation. Do Greek students who study to become advisors have these competencies? In this study, we attempted to answer this question following a quantitative approach. The results indicate that participants possess low levels in all the examined sets of competencies and, as a result, have limited overall competency in dealing with digital agriculture. These findings suggest the need for agricultural universities to reset competence-related targets and design strategies to supply future farm advisors with the competencies needed to act as facilitators of agricultural digitalization.

Keywords: agricultural digitalization; advisors; students; competencies; competence development; smart farming; precision agriculture; skills; advisory services; farming

Citation: Charatsari, C.; Michailidis, A.; Lioutas, E.D.; Bournaris, T.; Loizou, E.; Paltaki, A.; Lazaridou, D. Towards Agricultural Digitalization: Does Higher Agricultural Education Supply Students with Relevant Competencies? *Proceedings* **2024**, *94*, 1. https://doi.org/10.3390/proceedings2024094001

Academic Editor: Eleni Theodoropoulou

Published: 18 January 2024

Copyright: © 2024 by the authors. Licensee MDPI, Basel, Switzerland. This article is an open access article distributed under the terms and conditions of the Creative Commons Attribution (CC BY) license (https://creativecommons.org/licenses/by/4.0/).

1. Introduction

Digitalization of agriculture refers to the introduction of technologies belonging to the so-called fourth industrial revolution to the agricultural sector. These tools are expected to have a positive transformative potential for farming and the wider agrifood sector [1], without, however, being free from negative impacts [2,3]. To unfold their potential and bring about the desired outcomes, digital technologies depend on the adopters' aptitude to exploit them [4]. Since farmers do not always possess the skills needed to autonomously use these technologies [5] and cope with the new complexities that digitalization creates [6], advisors are called to undertake the role of digital transition facilitators, helping adopters extract value from these technologies [7–9].

Such a role is demanding, requiring a broad array of competencies. Research indicates that advisors working in the field develop digitalization-related competencies through their participatory engagement in the digital transition process [8]. However, to date, there is no evidence of the extent to which agricultural universities supply students who, after graduating, will act as advisors with the competencies needed to deal with digitalization.

Our study aims to offer some preliminary insights into this topic by examining the levels of Greek agronomy students' competencies in nine different areas, including competencies related to the interactions between human actors and technology, involving both advisor–technology interaction and the mediation of the farmer–technology relationship; understanding the use and potential of technologies; integrating technologies into farms; effectively and responsibly exploiting technologies; anticipating the impacts of digitalization; and managing digitalization-related risks. We also focused on competencies associated with the abilities of future advisors to guide the digital transition of farms, namely adaptation to the new conditions that digitalization creates, facilitation of the transition process, empathy towards adopters, and the ability to orient the self toward the future. As a set, these competencies allow the future advisors to adapt themselves to external changes, paving the way for the transition process, understanding the adopters' needs and difficulties, and developing and attaining goals for the future.

2. Methods

Our analysis draws on data from a sample of 108 students (55.6% women; mean age = 23.5 years; S.D. = 4.1) who study agronomy at a large Greek university. To measure students' competencies, we developed nine scales referring to basic technology understanding (example item: "Understanding the potential of technologies"), technology integration competencies (example item: "Solving problems associated with newly introduced technologies on the farms"), technology exploitation skills (example item: "Transforming technologies to productive resources"), impact forecasting competencies (example item: "Predicting how technologies will transform farming systems"), risk reduction competencies (example item: "Minimizing the social risks associated with technologies"), adaptation competencies (example item: "Adapting to profound change when innovative technologies emerge"), transition facilitation competencies (example item: "Facilitating through my collaboration with farmers the technology-enabled transition of farm enterprises"), empathy (example item: "Understanding how farmers feel about technologies and resolving potential conflicts"), and future orientation (example item: "Anticipating the potential futures that technologies create"). To generate items for the first four scales, we lean upon conceptual and empirical literature on the technology-related competencies of farm advisors [8,10–12]. For the remaining scales, we formulated items based on social science research that refers to farmer–advisor interaction during the digitalization process [8,13–16].

For all the items, students were instructed to indicate their competency level on a scale ranging from 1 (not at all) to 5 (very much). Principal axis factor analyses confirmed that items loaded on the theoretically expected factors. Cronbach's alpha values were satisfactory for all the scales, ranging from 0.80 to 0.94.

To assess students' overall digital agriculture-related competency, we used a single item measured on a ten-point scale, where higher values correspond to a higher level of competency.

To analyze data, beyond descriptive statistics, we built a simultaneous regression model to examine what types of competencies are associated with participants' overall level of competency in dealing with digital agriculture.

3. Results

The summary statistics of the variables are presented in Table 1. Interestingly, students' overall digital agriculture-related competency was low (M = 4.12; S.D. = 1.94). However, it is worth mentioning that considerable differences exist between participants, given that the overall competency scores ranged from 1 to 8. The mean scores for the nine sets of competencies were moderate, ranging from 2.92 to 3.53. Notably, only three competency sets yielded mean scores higher than the baseline level of 3.0; two involving a high degree of self-direction (future orientation and empathy) and the capacity to understand technologies.

Table 1. Summary statistics of the study variables.

Variable	Mean Score	Standard Deviation
Basic technology understanding competencies	3.22	1.05
Technology integration competencies	2.99	1.00
Technology exploitation competencies	2.92	0.98
Impact forecasting competencies	2.98	0.93
Risk reduction competencies	2.85	1.03
Adaptation competencies	3.10	0.90
Transition facilitation competency	2.93	0.92
Empathy competency	3.31	1.00
Future orientation competencies	3.53	0.86
Overall competency	4.13	1.94

When all the sets of competencies were entered in a simultaneous regression model ($R^2 = 0.20$, $F = 2.76$, $p = 0.006$), only the technology integration capacity ($\beta = 0.54$, $p = 0.021$) and transition facilitation competencies ($\beta = 0.46$, $p = 0.048$) received significant beta coefficients. In both cases, higher scores led to an increase in the future advisors' overall competency.

4. Discussion and Conclusions

The present study uncovered that future advisors in Greece have low levels of agricultural digitalization-related competencies. Some of the examined sets, like future orientation and empathy, had higher—yet questionably sufficient—scores. Our regression analysis revealed the pivotal role of technology integration and transition facilitation competencies in shaping the overall competency. Since both these variables had scores below the baseline level, it is not surprising that the overall competency in dealing with digital agriculture was also low.

These results point out the need to rethink the ability of the offered curricula to supply students with the competencies needed to cope with the challenges posed by digitalization in the agrifood sector. Modern approaches and a redefinition of the priorities set by their designers can help future farm advisors effectively support the transition to digital agriculture.

Author Contributions: Conceptualization, C.C. and A.M.; methodology, C.C., A.M. and E.D.L.; validation, T.B., E.L., A.P. and D.L.; formal analysis, C.C., A.M., E.D.L., T.B., E.L., A.P. and D.L.; investigation, A.P. and D.L.; writing—original draft preparation, C.C., A.M. and E.D.L.; writing—review and editing, C.C., A.M., E.D.L., T.B., E.L., A.P. and D.L.; project administration, A.M.; funding acquisition, A.M. All authors have read and agreed to the published version of the manuscript.

Funding: This research was funded by the European Union, grant number 101056291.

Institutional Review Board Statement: This study was conducted in accordance with the Declaration of Helsinki. Ethical review and approval were waived for this study since the study was conducted in accordance with the Declaration of Helsinki and the EU General Data Protection Regulation.

Informed Consent Statement: Informed consent was obtained from all the subjects involved in the study.

Data Availability Statement: Data will be made available upon request by the first author after the completion of the BOOST project.

Acknowledgments: This study is part of an ongoing project titled "BOOSTing agribusiness acceleration and digital hub networking by an advanced training program on sustainable Precision Agriculture". The research project is co-funded by the European Union. Project number: 101056291.

Conflicts of Interest: The authors declare no conflicts of interest. The funders had no role in the design of the study; in the collection, analyses, or interpretation of data; in the writing of the manuscript; or in the decision to publish the results.

References

1. Iaksch, J.; Fernandes, E.; Borsato, M. Digitalization and big data in smart farming—A review. *J. Manag. Anal.* **2021**, *8*, 333–349. [CrossRef]
2. McGrath, K.; Brown, C.; Regan, Á.; Russell, T. Investigating narratives and trends in digital agriculture: A scoping study of social and behavioural science studies. *Agric. Syst.* **2023**, *207*, 103616. [CrossRef]
3. Rozenstein, O.; Cohen, Y.; Alchanatis, V.; Behrendt, K.; Bonfil, D.J.; Eshel, G.; Harari, A.; Harris, W.E.; Klapp, I.; Laor, Y.; et al. Data-driven agriculture and sustainable farming: Friends or foes? *Precis. Agric.* **2023**, in press. [CrossRef]
4. Carmela Annosi, M.; Brunetta, F.; Capo, F.; Heideveld, L. Digitalization in the agri-food industry: The relationship between technology and sustainable development. *Manag. Decis.* **2020**, *58*, 1737–1757. [CrossRef]
5. Schnebelin, É.; Labarthe, P.; Touzard, J.M. How digitalisation interacts with ecologisation? Perspectives from actors of the French Agricultural Innovation System. *J. Rural Stud.* **2021**, *86*, 599–610. [CrossRef]
6. da Silveira, F.; da Silva, S.L.C.; Machado, F.M.; Barbedo, J.G.A.; Amaral, F.G. Farmers' perception of barriers that hinder the implementation of agriculture 4.0. *Agric. Syst.* **2023**, *208*, 103656. [CrossRef]
7. Higgins, V.; van der Velden, D.; Bechtet, N.; Bryant, M.; Battersby, J.; Belle, M.; Klerkx, L. Deliberative assembling: Tinkering and farmer agency in precision agriculture implementation. *J. Rural Stud.* **2023**, *100*, 103023. [CrossRef]
8. Charatsari, C.; Lioutas, E.D.; Papadaki-Klavdianou, A.; Michailidis, A.; Partalidou, M. Farm advisors amid the transition to Agriculture 4.0: Professional identity, conceptions of the future and future-specific competencies. *Sociol. Rural.* **2022**, *62*, 335–362. [CrossRef]
9. Eastwood, C.; Ayre, M.; Nettle, R.; Rue, B.D. Making sense in the cloud: Farm advisory services in a smart farming future. *NJAS-Wagening. J. Life Sci.* **2019**, *90*, 100298. [CrossRef]
10. Cook, S.; Jackson, E.L.; Fisher, M.J.; Baker, D.; Diepeveen, D. Embedding digital agriculture into sustainable Australian food systems: Pathways and pitfalls to value creation. *Int. J. Agric. Sustain.* **2022**, *20*, 346–367. [CrossRef]
11. Shepherd, M.; Turner, J.A.; Small, B.; Wheeler, D. Priorities for science to overcome hurdles thwarting the full promise of the 'digital agriculture' revolution. *J. Sci. Food Agric.* **2020**, *100*, 5083–5092. [CrossRef] [PubMed]
12. Ayre, M.; Mc Collum, V.; Waters, W.; Samson, P.; Curro, A.; Nettle, R.; Paschen, J.-A.; King, B.; Reichelt, N. Supporting and practising digital innovation with advisers in smart farming. *NJAS-Wagening. J. Life Sci.* **2019**, *90*, 100302. [CrossRef]
13. Lioutas, E.D.; Charatsari, C. Innovating digitally: The new texture of practices in agriculture 4.0. *Sociol. Rural.* **2022**, *62*, 250–278. [CrossRef]
14. Ogunyiola, A.; Gardezi, M. Restoring sense out of disorder? Farmers' changing social identities under big data and algorithms. *Agric. Hum. Values* **2022**, *39*, 1451–1464. [CrossRef]
15. Charatsari, C.; Lioutas, E.D.; De Rosa, M.; Papadaki-Klavdianou, A. Extension and advisory organizations on the road to the digitalization of animal farming: An organizational learning perspective. *Animals* **2020**, *10*, 2056. [CrossRef] [PubMed]
16. Newton, J.E.; Nettle, R.; Pryce, J.E. Farming smarter with big data: Insights from the case of Australia's national dairy herd milk recording scheme. *Agric. Syst.* **2020**, *181*, 102811. [CrossRef]

Disclaimer/Publisher's Note: The statements, opinions and data contained in all publications are solely those of the individual author(s) and contributor(s) and not of MDPI and/or the editor(s). MDPI and/or the editor(s) disclaim responsibility for any injury to people or property resulting from any ideas, methods, instructions or products referred to in the content.

Proceeding Paper

Identifying the External Environment of Greek Fisheries [†]

Stamatis Mantziaris [1],*, Irene Tzouramani [1] and Angelos Liontakis [2]

[1] Agricultural Economics Research Institute (AGR.E.R.I.), Hellenic Agricultural Organization-DIMITRA, 11145 Athens, Greece; tzouramani@agreri.gr or tzouramani@elgo.gr

[2] Department of Agribusiness and Supply Chain Management, Agricultural University of Athens, 32200 Thebes, Greece; aliontakis@aua.gr

* Correspondence: sta.athens@hotmail.com or smantziaris@elgo.gr

[†] Presented at the 17th International Conference of the Hellenic Association of Agricultural Economists, Thessaloniki, Greece, 2–3 November 2023.

Abstract: The Greek fishing sector faces various challenges which can threaten its long-term sustainability. The PESTLE analysis is used to assess the impact of the external environment on the Greek fishing sector. According to our analysis, appropriate strategic planning should emphasize promoting the integration of innovation and technology transfer from the laboratory to the fisheries sector to address the challenges and capitalize on the opportunities. Future research can be conducted on the prioritization of external factors by sector experts and the coupling with other strategic planning tools.

Keywords: external environment; PESTLE analysis; Greek fishing sector; long-term sustainability; fisheries policy

Citation: Mantziaris, S.; Tzouramani, I.; Liontakis, A. Identifying the External Environment of Greek Fisheries. *Proceedings* **2024**, *94*, 2. https://doi.org/10.3390/proceedings2024094002

Academic Editor: Eleni Theodoropoulou

Published: 19 January 2024

Copyright: © 2024 by the authors. Licensee MDPI, Basel, Switzerland. This article is an open access article distributed under the terms and conditions of the Creative Commons Attribution (CC BY) license (https://creativecommons.org/licenses/by/4.0/).

1. Introduction

Fishing is a critical sector for the national economy but, above all, for the social cohesion of disadvantaged and remote areas [1]. However, the sector faces various challenges, affecting its long-term sustainability.

For example, political factors, such as government policies and regulations, can significantly impact the sector's operations and profitability. Economic factors, such as market trends and the macroeconomic environment, can influence the sector's financial performance. Social factors, such as changing consumer preferences and attitudes towards sustainability, can affect the fisheries sector. Technological advancements and innovations can bring new opportunities and challenges to the sector. Legal factors, such as international regulations and environmental laws, can significantly impact the sector's operations and sustainability. Finally, environmental factors, such as climate change, can threaten the sector's future.

The analysis of the above external factors presupposes a holistic and multidisciplinary approach such as PESTLE analysis [2,3]. PESTLE analysis of the fisheries sector can help stakeholders gain a comprehensive understanding of the sector's external environment and develop strategies to address the challenges and capitalize on the opportunities [4]. This can lead to the development of sustainable practices and policies that promote the long-term viability of the fisheries sector.

PESTLE analysis has been widely used as a strategic planning tool in fisheries in various regions worldwide [4–7]. Nevertheless, applying the PESTLE strategic planning tool to the analysis of the fisheries sector of Greece constitutes a contribution to the existing bibliographic background.

2. Materials and Methods

PESTLE analysis is a strategic planning tool to identify the external factors affecting a particular industry or sector. The acronym PESTLE stands for Political, Economic, Social,

Technological, Legal, and Environmental factors [8,9]. This analytical framework can provide valuable insights into the external factors that may impact the fisheries sector's growth and sustainability [10]. In particular, conducting a PESTLE analysis of the fisheries sector can help stakeholders to identify the opportunities and challenges that arise from these external factors.

3. Results and Discussion

Table 1 outlines the profile of exogenous factors that can affect the long-term sustainability of Greek fisheries. In this context, we highlight the primary challenges in the sector that need to be addressed. In particular, although the positive role of support measures in the common fisheries policy (CFP) framework is found, a deficit is observed in promoting the integration of the innovative component and transferring technology from the laboratory to the sector. The specific challenges can be addressed, given the flexibility provided through the CFP to member states to develop policy tools adapted to the current economic, social, and technological needs. The existing high-level scientific staff serving the fisheries sector can help in this effort by promoting the development of multi-level pilot actions. Moreover, cases of successful collaboration between scientific institutions and the private sector can be the "pilot" for future collaborations between stakeholders.

Table 1. External factors affecting Greek fisheries, according to PESTLE analysis.

P Political factors	- Policy measures to promote the modernization of the fishing fleet; - Policy measures to support fishermen due to the pandemic; - Design of policy measures with a more national orientation - Designing policies to mitigate climate change; - Funding of training actions and introduction of new fishermen.
E Economic factors	- Economic efficiency of small-scale fisheries due to diverse distribution channel; - Significant degree of dependence of small-scale fishing on tourist flows; - Low bargaining power of large-scale fishing; - Limited number of small-scale processing units (run by fishermen and their families) that produce high-value-added fishery products; - Adverse macroeconomic environment.
S Social factors	- Positive effect of the role of women in the development of the sector; - High-level scientific staff serving the fisheries sector; - Reduced level of social sustainability in disadvantaged and remote areas; - Reduction in tourist flows due to the pandemic; - Consumers are turning to long-lasting products due to the pandemic.
T Technological factors	- Collaboration of scientific institutions and the private sector for the construction of innovative fishing technologies; - Collaboration of scientific institutions, organizations, fishermen, and the private sector for the implementation of innovative fisheries management systems; - Limited actions integrating innovation and transfer technology from the laboratory to the sector.
L Legal factors	- Special fishing licenses; - Advanced tracking systems of fishing activity; - International governance efforts in the Mediterranean region.
E Environmental factors	- The rich biodiversity of the Greek seas; - Seasonality of fishing species; - Increase in competitive foreign fishing species due to climate change.

4. Conclusions

In conclusion, the deficits in the promotion of innovation integration and technology transfer from the laboratory to the sector can be reduced, given that there are three main elements: (i) the critical number of high-level scientific staff serving the fisheries sector; (ii) the interest of the private sector in collaborating with scientific institutions; and (iii) the possibility of co-financing from policy measures (under the CFP) and the private sector.

Therefore, coupling these elements is necessary, which could be implemented and achieved by creating research and innovation units (Innovation Hubs). The successful operation of such structures can promote, for example, the development of innovative small-scale processing units (run by fishermen and their families) that produce high-value-added fishery products [1], which is judged to be another severe challenge of the sector. By extension, developing such innovative small-scale processing units can cure the tough challenge of reducing social sustainability in the country's disadvantaged and remote areas [1]. Future research can be conducted on the prioritization of external factors by sector experts utilizing multi-criteria decision analysis methods (e.g., analytic hierarchy process (AHP) method) [11,12]. The PESTLE analysis should also be used with other tools to support strategy (e.g., coupling with SWOT analysis) [11–14].

Author Contributions: Conceptualization, S.M., I.T. and A.L.; methodology, S.M.; formal analysis, S.M., I.T. and A.L.; investigation, S.M., I.T. and A.L.; resources, I.T.; writing—original draft preparation, S.M.; writing—review and editing, S.M., I.T. and A.L.; supervision, I.T.; project administration, I.T.; funding acquisition, I.T. All authors have read and agreed to the published version of the manuscript.

Funding: This research was funded by the Greek Work Plan for data collection in the fisheries and aquaculture sectors 2022–2023, co-ordinated by the General Directorate of Sustainable Fisheries, Greek Ministry of Rural Development, and Food.

Institutional Review Board Statement: Not applicable.

Informed Consent Statement: Not applicable.

Data Availability Statement: No new data were created or analyzed in this study. Data sharing is not applicable to this article.

Conflicts of Interest: The authors declare no conflicts of interest.

References

1. Mantziaris, S.; Liontakis, A.; Valakas, G.; Tzouramani, I. Family-run or business-oriented fisheries? Integrating socioeconomic and environmental aspects to assess the societal impact. *Mar. Policy* **2021**, *131*, 104591. [CrossRef]
2. Zalengera, C.; Blanchard, R.E.; Eames, P.C.; Juma, A.M.; Chitawo, M.L.; Gondwe, K.T. Overview of the Malawi Energy Situation and a PESTLE Analysis for Sustainable Development of Renewable Energy. *Renew. Sustain. Energy Rev.* **2014**, *38*, 335–347. [CrossRef]
3. Eichhorn, T.; Schaller, L.; Hamunen, K.; Runge, T. Exploring macro-environmental factors influencing adoption of result-based and collective agri-environmental measures: A PESTLE approach based on stakeholder statements. *Bio-Based Appl. Econ.* **2023**. [CrossRef]
4. De Silva, D. *Value Chain of Fish and Fishery Products: Origin, Functions and Application in Developed and Developing Country Markets*; Food and Agriculture Organization (FAO): Rome, Italy, 2011. Available online: https://www.fao.org/fileadmin/user_upload/fisheries/docs/De_Silva_report_with_summary_doc (accessed on 12 September 2023).
5. Qatan, S.; Knútsson, Ö.; Gestsson, H. *Operating a Wholesale Fish Market in the Sultanate of Oman Analyses of External Factors*; UNU-Fisheries Training Programme: Reykjavik, Iceland, 2010. Available online: https://www.grocentre.is/static/gro/publication/234/document/salim2010prf.pdf (accessed on 12 September 2023).
6. Ahmadzai, B. *Fish Value Chain Analysis and Fisheries Sector Development Opportunities: Afghanistan*; Technical Report Kabul; Afghanistan, Ministry of Agriculture Irrigation and Livestock (MAIL): Kabul, Afghanistan, 2017. [CrossRef]
7. Fillie, M.T. Socioeconomic Impacts of Illegal Unreported and Unregulated (IUU) Fishing on Sierra Leone. Master's Thesis, World Maritime University, Malmö, Sweden, 2019. Available online: https://commons.wmu.se/all_dissertations/1198 (accessed on 12 September 2023).
8. UNICEF. *SWOT and PESTEL-Understanding Your External and Internal Context for Better Planning and Decision-Making*; UNICEF KE Toolbox; UNICEF: New York, NY, USA, 2015. Available online: https://www.unicef.org/knowledge-exchange/files/SWOT_and_PESTEL_production.pdf (accessed on 12 September 2023).
9. USYD. *Marketing: PESTLE Analysis*; The University of Sydney: Sydney, Australia, 2023. Available online: https://libguides.library.usyd.edu.au/c.php?g=508107&p=5994242 (accessed on 12 September 2023).
10. DataBio. *D7.3–PESTLE Analysis*; Data-Driven Bioeconomy: Brussels, Belgium, 2017. Available online: https://www.databio.eu/wp-content/uploads/2017/05/DataBio_D7.3-PESTLE-Analysis_v1.0_2017-12-29_VTT.pdf (accessed on 12 September 2023).
11. Tsangas, M.; Jeguirim, M.; Limousy, L.; Zorpas, A. The Application of Analytical Hierarchy Process in Combination with PESTEL-SWOT Analysis to Assess the Hydrocarbons Sector in Cyprus. *Energies* **2019**, *12*, 791. [CrossRef]

12. Vardopoulos, I.; Tsilika, E.; Sarantakou, E.; Zorpas, A.A.; Salvati, L.; Tsartas, P. An Integrated SWOT-PESTLE-AHP Model Assessing Sustainability in Adaptive Reuse Projects. *Appl. Sci.* **2021**, *11*, 7134. [CrossRef]
13. Zhu, L.; Hiltunen, E.; Antila, E.; Huang, F.; Song, L. Investigation of China's bio-energy industry development modes based on a SWOT–PEST model. *Int. J. Sustain. Energy* **2014**, *34*, 552–559. [CrossRef]
14. Christodoulou, A.; Cullinane, K. Identifying the Main Opportunities and Challenges from the Implementation of a Port Energy Management System: A SWOT/PESTLE Analysis. *Sustainability* **2019**, *11*, 6046. [CrossRef]

Disclaimer/Publisher's Note: The statements, opinions and data contained in all publications are solely those of the individual author(s) and contributor(s) and not of MDPI and/or the editor(s). MDPI and/or the editor(s) disclaim responsibility for any injury to people or property resulting from any ideas, methods, instructions or products referred to in the content.

Proceeding Paper

The Association of MTNR1A Gene Alleles with the Response to Estrus Induction Treatments in Improved and Non-Improved Greek Indigenous Sheep Breeds †

Danai Antonopoulou [1,2], Ioannis A. Giantsis [1,*], George K. Symeon [3] and Melpomeni Avdi [2]

1. Division of Animal Science, Faculty of Agricultural Sciences, University of Western Macedonia, 53100 Florina, Greece; dsantono@agro.auth.gr
2. Department of Animal Production, Faculty of Agriculture, Aristotle University of Thessaloniki, 54124 Thessaloniki, Greece; avdimel@agro.auth.gr
3. Research Institute of Animal Science, HAO-Demeter, 58100 Giannitsa, Greece; gsymeon@elgo.gr
* Correspondence: igiantsis@uowm.gr
† Presented at the 17th International Conference of the Hellenic Association of Agricultural Economists, Thessaloniki, Greece, 2–3 November 2023.

Abstract: Seasonality in sheep reproduction and related limitations make milk production challenging throughout the year. In the present study, we investigated the response to estrus induction treatments in three indigenous breeds, Florina, Chios, and Karagouniko, as well as the melatonin receptor 1A gene variants in relation to this response. The three distinct synchronization methods were A: intravaginal sponges, B: GNRH use, and C: male effect. In group A, fertility was 85%, and Florina ewes expressed estrus at 90% in July. Ewes from Karagouniko and Chios had fecundity rates of 95% and 99%, respectively, and 100% estrus expression. The Florina ewes in group B expressed estrus at a percentage of 60%, with a fecundity rate of 57%, the Karagouniko ewes at a percentage of 65%, with a fecundity rate of 54%, and the Chios breed animals at a percentage of 87%, with a fecundity rate of 85%. Twenty to twenty-five days after ram induction, 68% of the Florina breed in group C showed signs of estrus, compared to 84% and 94% of Karagouniko and Chios breeds, respectively. In both Florina and Karagouniko breeds, all treatments showed a substantial difference in the frequency of the four identified SNPs in the MTNR1A gene between ewes who expressed estrus and ewes who did not. The genetic improvement based on the alleles analyzed in the current study is expected to decrease seasonality rates in indigenous sheep breeds.

Keywords: reproduction; sheep; MTNR1A gene

1. Introduction

In sheep, milk production is often not feasible throughout the year due to the seasonality of reproduction. The goal of the current study was to compare how indigenous Greek sheep breeds (Florina, Chios, and Karagouniko) respond to various estrus synchronization treatments, as well as to associate this response with their genetic composition. This was accomplished by molecularly analyzing the melatonin receptor 1A (MTNR1A) gene in order to determine the alleles that are associated with those treatments. Exon 2 of the MTNR1A gene influences the seasonality of reproduction in small ruminants, with particular alleles linked with long anestrus periods [1–4]. A total of 450 ewes from three different indigenous breeds were examined, with each breed being divided into three different groups, where each group was treated with a different synchronization treatment.

2. Materials and Methods

The experimental procedures were carried out using 450 ewes (150 of Chios breed, 150 of Florina breed, and 150 of Karagouniko breed). Three groups of 50 ewes each were created for each breed. The intravaginal progesterone sponges in Group A (sponges) contained 20 mg of flurogestone acetate FGA and were given for 14 days. At the conclusion of the treatment, 300 IU of chorionic gonadotrophin was then infused. In Group B (GnRH), 0.0084 mg buserelin acetate was given twice between D0 and D9, and 0.263 mg of prostaglandin was given seven days later. The male (or ram) effect, which states that sexually active males should be introduced to females that have been isolated for three months via visual contact from males at a geographic distance of more than 500 m, was used in group C. Blood samples were collected from each ewe. Using the PureLink Genomic DNA Mini Kit, DNA was extracted from 150 mL of collected blood. A partial segment of the MTNR1A gene located in the exon 2 was amplified as described in Giantsis et al. [3] and was sequenced using the Sanger methodology.

3. Results and Discussion

After the implementation of the first synchronization treatment (group A), Chios ewes had a fecundity rate of 99%, Karagouniko ewes 95%, and Florina ewes 85%. In group B, Chios ewes had a fecundity rate of 85%, Karagouniko ewes 54%, and Florina ewes 57%. In the third synchronization treatment (group C), Chios ewes had a fecundity rate of 94%, Karagouniko ewes 84%, and Florina ewes 68%.

The estrus expression rate of the sponge-treated Florina ewes (group A) was 90% in July, and their fecundity rate was 85% when they gave birth in December. Following the removal of the intravaginal sponges, Karagouniko and Chios ewes who were given the same treatment displayed 100% estrus expression, with fecundity rates of 95% and 99%, respectively. Ewes of the Florina breed expressed estrus at a percentage of 60%, with fecundity at 57%, Karagouniko ewes expressed estrus at a percentage of 65%, with fecundity at 54%, and Chios breed animals expressed estrus at a percentage of 87%, with a fecundity rate of 85% with regard to GnRH–PGF2a–GnRH (GnRH Protocol). Last but not least, 68% of the Florina breed displayed estrus 20–25 days after ram induction following the implementation of the male effect technique (group C), compared to 84% and 94% for the Karagouniko and Chios breeds, respectively. Chios ewes had a multiplicity (prolificacy) of 1.9, Karagouniko ewes had 1.3, and Florina ewes had 1.4.

The MTNR1A gene's amplified aligned sequence was 824 bp long and corresponded to the bases 285–1108 of the reference melatonin receptor mRNA haplotype in Ovis aries with the GenBank accession number U14109. In the Florina and Karagouniko breeds, all frequencies of the four identified alleles were statistically significantly different in ewes that expressed estrus compared to ewes who did not express estrus for all treatments. Only few animals from the groups that belonged to the Chios breed showed any statistical relationship with the allele frequencies. Therefore, in the two seasonal indigenous breeds, but not in the Chios breed, the scored alleles were statistically significantly related with the response to the treatments for out-of-season reproduction.

4. Conclusions

In conclusion, estrus synchronization can be achieved in indigenous sheep breeds from temperate latitudes using all three treatments. However, it has been demonstrated that some breeds are more receptive than others, and this trait is directly linked to their genetic composition. An efficient method to achieve the best estrus synchronization for milk production throughout the year is marker-assisted selection based on the MTNR1A gene.

Author Contributions: Conceptualization, I.A.G.; methodology, D.A. and G.K.S.; software, D.A. and I.A.G.; validation, M.A.; formal analysis, M.A.; investigation, D.A. and G.K.S.; resources, G.K.S.; data curation, I.A.G.; writing—original draft preparation, D.A.; writing—review and editing, G.K.S. and I.A.G.; visualization, M.A.; supervision, I.A.G.; project administration, M.A.; funding acquisition, M.A. All authors have read and agreed to the published version of the manuscript.

Funding: This research work was supported by the Hellenic Foundation for Research and Innovation (H.F.R.I.) under the 'First Call for H.F.R.I. Research Projects to support Faculty members and Researchers and the procurement of high-cost research equipment grant' (Project Number: HFRI-FM17-2987).

Institutional Review Board Statement: All animal manipulations were carried out according to the EU Directive on the protection of animals' usage for scientific purposes (2010/63/EU). No other permission was needed, since only blood samples were collected.

Informed Consent Statement: Not applicable.

Data Availability Statement: All data from this research are available after communication with the corresponding author.

Conflicts of Interest: The authors declare no conflicts of interest.

References

1. Mura, M.C.; Luridiana, S.; Vacca, G.M.; Bini, P.P.; Carcangiu, V. Effect of genotype at the MTNR1A locus and melatonin treatment on first conception in Sarda ewe lambs. *Theriogenology* **2010**, *74*, 1579–1586. [CrossRef] [PubMed]
2. Mura, M.C.; Luridiana, S.; Bodano, S.; Daga, C.; Cosso, G.; Diaz, M.L.; Bini, P.P.; Carcangiu, V. Influence of melatonin receptor 1A gene polymorphisms on seasonal reproduction in Sarda ewes with different body condition scores and ages. *Anim. Reprod. Sci.* **2014**, *149*, 173–177. [CrossRef] [PubMed]
3. Giantsis, I.A.; Laliotis, G.; Stoupa, O.; Avdi, M. Polymorphism of the melatonin receptor 1A (MNTR1A) gene and association with seasonality of reproductive activity in a local Greek sheep breed. *J. Biol. Res.-Thessalon.* **2016**, *23*, 9. [CrossRef] [PubMed]
4. Starič, J.; Farci, F.; Luridiana, S.; Mura, M.C.; Pulinas, L.; Cosso, G.; Carcangiu, V. Reproductive performance in three Slovenian sheep breeds with different alleles for the MTNR1A gene. *Anim. Reprod. Sci.* **2020**, *216*, 106352. [CrossRef] [PubMed]

Disclaimer/Publisher's Note: The statements, opinions and data contained in all publications are solely those of the individual author(s) and contributor(s) and not of MDPI and/or the editor(s). MDPI and/or the editor(s) disclaim responsibility for any injury to people or property resulting from any ideas, methods, instructions or products referred to in the content.

Proceeding Paper

Assessing Agroecology Terms for North African Countries: A Literature Review [†]

Malak Hazimeh [1], Leonidas Sotirios Kyrgiakos [2,*], Georgios Kleftodimos [1,3], Christina Kleisiari [2], Marios Vasileiou [2] and George Vlontzos [2]

[1] Mediterranean Agronomic Institute of Montpellier (CIHEAM-IAMM), University of Montpellier, 34090 Montpellier, France; malakhzm24@gmail.com (M.H.); kleftodimos@iamm.fr (G.K.)
[2] Department of Agriculture Crop Production and Rural Environment, University of Thessaly, 38446 Volos, Greece; chkleisiari@uth.gr (C.K.); mariosvasileiou@uth.gr (M.V.); gvlontzos@uth.gr (G.V.)
[3] Montpellier Interdisciplinary Center on Sustainable Agri-Food Systems (MoISA), Univ Montpellier, CIHEAM-IAMM, CIRAD, INRAE, Institut Agro, IRD, 34060 Montpellier, France
* Correspondence: lkyrgiakos@uth.gr; Tel.: +30-24210-93013
[†] Presented at the 17th International Conference of the Hellenic Association of Agricultural Economists, Thessaloniki, Greece, 2–3 November 2023.

Abstract: Conventional agricultural techniques cannot fulfill the requirements of a sustainable food value chain. Agroecology can be a great alternative practice for transforming the current agricultural systems. This approach combines ecology and agriculture, considering different stakeholders' opinions. An assessment of the current literature about "agroecology practices" using the Web of Science database was made, and 1235 results were collected and unified into a bibtex file using R studio. The final results were extracted through the bibliometix library. The acquired results show that annual scientific production on the aforementioned term was limited between the 1990s and 2010s and has recently increased due to increased interest in the topic. Additionally, the terms "agriculture", "management biodiversity", and "conservation" are frequently correlated with agroecology, covering all three dimensions of sustainability. Agroecology as a trending topic has great potential to serve North African countries, increasing food security levels while assuring sustainability standards.

Keywords: agroecology; agroecological practices; sustainability; North African; countries; literature review

1. Introduction

Conventional agricultural management techniques focus on monoculture, heavy use of mechanized production, genetic manipulation, and mass production. However, these techniques do not provide feasible solutions to the current needs of sustainable food value chains, especially for countries with lower incomes [1]. It is essential for actions to be taken in this regard, as small farmers are the most affected and they usually have traditional farming systems [2]. A new perspective is needed for a transformation in the agricultural systems, in order to deal with food security and environmental issues, as well as diminishing the gap between small and large farmers in terms of yield and eco-friendliness.

Agroecology can be defined as the combination of agriculture and ecology. According to FAO (2020) [3], it is a holistic approach that covers ecological and social aspects in order to create and manage agricultural and food systems in a sustainable way. Agroecology involves various stakeholders utilizing inter-disciplinary methods that promote accountable and transparent resource governance. Implementation of agroecological techniques has a positive impact on seven Sustainable Development Goals: 2—zero hunger; 1—poverty alleviation; 13—climate change resilience; 15—biodiversity; 8—youth engagement; 5—gender determination; 10—human rights.

North African (NA) countries are less resilient to climate change due to their location, low technological level, and their lack of knowledge about the implementation of agroecological practices [4]. The NATAE project involves five NA countries, namely Tunisia, Morocco, Algeria, Egypt, and Mauritania, which are facing water scarcity problems as well as food safety issues. Water is a crucial source for the agricultural sector as irrigation is a key factor for food production. Due to this problem, small farmers' production is affected, as it is highly dependent on extreme weather events [5]. Considering that consumer standards have increased all around the globe, it is of paramount importance for these countries to produce agricultural products that fulfill all sustainability dimensions (economy, environment, and society). This can be achieved by implementing new approaches based on agroecological practices [6].

Furthermore, agroecology is a multidimensional approach that considers different stakeholders' opinions, and that is the reason why five different living labs in total have been set up in order to record their opinions and deliver their needs for the overall optimization process. A part of the NATAE project will evaluate the differences between the conventional and agroecology practices for the NA countries in local value chains. However, it is considered appropriate to highlight the main results of the existing literature in order to identify the existing gaps prior to the implementation of agroecological practices.

2. Methodology

In order to extract the literature review results, an assessment of Web of Science results was performed, regarding the existence of the term "Agroecology practices" in abstracts, titles, and keywords, which led to the extraction of the main factors influencing the dynamics of the field. More precisely, partnerships and trending topics were assessed, and 1235 results were collected through the web of science database. After this stage, results were transformed into a unified Bibtex file, as the web of science database permits the extraction of only 500 results at the time. The unification process has been achieved through the R studio program and the use of R version 4.2.3. Moreover, the bibliometrix library was used to extract the figures and data to be analyzed.

3. Results and Discussion

The acquired results of the literature review analysis include the following: (1) annual scientific production; (2) most relevant authors; (3) most relevant affiliations; (4) most frequently used terms; (5) most relevant resources; (6) source production over time; (7) affiliation production over time; (8) most cited countries; (9) co-citations network; (10) collaboration network; (11) conceptual structure map; (12) dendrogram. However, due to the limitations of this abstract, only two titles will be discussed below.

Agroecology is a new sector of agriculture that started to be considered in the late 1980s and early 1990s as a solution and method to be applied on industrialized food systems and the sustainability social movement. It came as a solution to climate change and its consequences, particularly food security and the environmental risks [1,4]. As seen in Figure 1 below, scientific publications on agricultural practices were low and stagnant between the 1990s and 2010s. With the rise in the concept of food sovereignty in the 1990s, agroecological approaches gained attention. However, literature regarding agroecological transition is considered limited; thus, a slight increase in scientific production has been recorded lately [7]. More precisely, a significant increase can be observed after 2016, reflecting the high interest in agroecological practices facing climate change and food security issues.

In Figure 2, the most frequently used terms emphasize the significance of the terms and keywords used in publications related to agroecological practices. The terms most abundantly used include management, agriculture, systems, and biodiversity as the focal heart of agroecology. Ecosystem services, diversity, sustainability, conservation, food, and soil impact are the second most important cluster of terms. The combination of these terms reflects how agroecology covers different dimensions, not only environment and agriculture. Akakpo et al. (2021) [4] have highlighted that agroecology is a necessary agricultural model that is capable of mitigating the effects of climate change, conserving biodiversity, securing sustainable food production, and preserving local ecosystems while valorizing them.

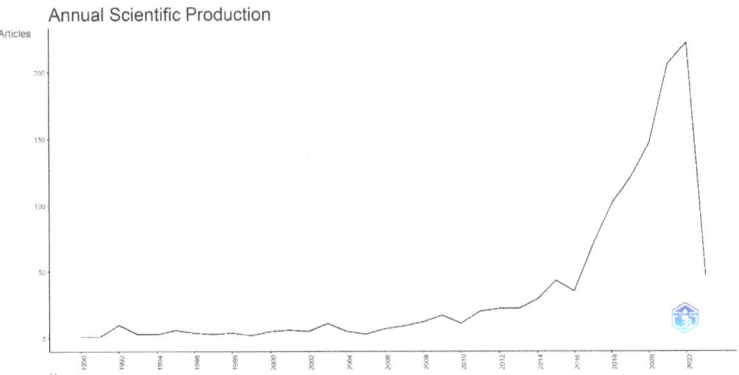

Figure 1. Annual scientific production trend.

Figure 2. Representation of the most frequent terms.

4. Conclusions

Agroecology is part of a transformative trajectory towards a more sustainable food supply chain covering economic, social, and environmental aspects. Through this approach, feasible solutions can be provided tackling food security issues for the NA countries. The aforementioned literature review highlighted that agroecology is a trending topic with significant potential for NA countries.

Author Contributions: Conceptualization, G.K. and L.S.K.; methodology, G.V.; software, L.S.K. and M.V.; validation, C.K. and M.V.; formal analysis, L.S.K. and C.K.; investigation, M.V.; resources, M.V.; data curation, M.V.; writing—original draft preparation, M.H.; writing—review and editing, M.H. and L.S.K.; visualization, L.S.K.; supervision, G.V. and G.K.; project administration, C.K.; funding acquisition, G.V. and G.K. All authors have read and agreed to the published version of the manuscript.

Funding: This project has received funding from the European Union's HE research and innovation program under grant agreement No 101084647 and official acronym NATAE. Views and opinions expressed are, however, those of the authors only and do not necessarily reflect those of the European Union or the European Research Executive Agency (REA). Neither the European Union nor the granting authority can be held responsible for any use of the information contained in the document.

Institutional Review Board Statement: Not applicable.

Informed Consent Statement: Not applicable.

Data Availability Statement: All data used for this study are widely accessible, through Web of Science, by searching "Agroecology practices" as described in the Methodology section.

Conflicts of Interest: The authors declare no conflicts of interest.

References

1. Amoak, D.; Luginaah, I.; McBean, G. Climate Change, Food Security, and Health: Harnessing Agroecology to Build Climate-Resilient Communities. *Sustainability* **2022**, *14*, 13954. [CrossRef]
2. Vasileiou, M.; Kyrgiakos, L.S.; Kleisiari, C.; Kleftodimos, G.; Vlontzos, G.; Belhouchette, H.; Pardalos, P.M. Transforming Weed Management in Sustainable Agriculture with Artificial Intelligence: A Systematic Literature Review towards Weed Identification and Deep Learning. *Crop Prot.* **2024**, *176*, 106522. [CrossRef]
3. FAO Overview | Agroecology Knowledge Hub | Food and Agriculture Organization of the United Nations. Available online: https://www.fao.org/agroecology/overview/en/ (accessed on 5 June 2023).
4. Akakpo, K.; Bouarfa, S.; Benoît, M.; Leauthaud, C. Challenging Agroecology through the Characterization of Farming Practices' Diversity in Mediterranean Irrigated Areas. *Eur. J. Agron.* **2021**, *128*, 126284. [CrossRef]
5. Fusco, G. Climate Change and Food Security in the Northern and Eastern African Regions: A Panel Data Analysis. *Sustainability* **2022**, *14*, 12664. [CrossRef]
6. Francis, C.A.; Wezel, A. Agroecology and Agricultural Change. In *International Encyclopedia of the Social & Behavioral Sciences*, 2nd ed.; Elsevier: Amsterdam, The Netherlands, 2015; pp. 484–487. [CrossRef]
7. López-García, D.; Cuéllar-Padilla, M.; de Azevedo Olival, A.; Laranjeira, N.P.; Méndez, V.E.; Peredo y Parada, S.; Barbosa, C.A.; Barrera Salas, C.; Caswell, M.; Cohen, R.; et al. Building Agroecology with People. Challenges of Participatory Methods to Deepen on the Agroecological Transition in Different Contexts. *J. Rural Stud.* **2021**, *83*, 257–267. [CrossRef]

Disclaimer/Publisher's Note: The statements, opinions and data contained in all publications are solely those of the individual author(s) and contributor(s) and not of MDPI and/or the editor(s). MDPI and/or the editor(s) disclaim responsibility for any injury to people or property resulting from any ideas, methods, instructions or products referred to in the content.

Proceeding Paper

The Effect of Unfair Trading Practices on the Performance of Agricultural Cooperatives †

Theo Benos [1],*, Panagiota Sergaki [2] and Nikos Kalogeras [1]

[1] Sustainable International Business (SIB) Research Centre, Zuyd University of Applied Sciences, Brusselseweg 150, 6217 HB Maastricht, The Netherlands; nikos.kalogeras@zuyd.nl

[2] Department of Agricultural Economics, Aristotle University of Thessaloniki, 54124 Thessaloniki, Greece; gsergaki@agro.auth.gr

* Correspondence: theo.benos@zuyd.nl; Tel.: +31-4333466275

† Presented at the 17th International Conference of the Hellenic Association of Agricultural Economists, Thessaloniki, Greece, 2–3 November 2023.

Abstract: In the European agri-food sector, operators with substantial bargaining power often engage in unfair trading practices (UTPs). Our paper aims to empirically examine the occurrence of UTPs and their influence on the performance of cooperatives. To fulfill the goal of our paper, we collected responses from 109 cooperatives in Greece after the transposition of a specialized EU Directive (i.e., Directive (EU) 2019/633). We found that, on average, the sampled cooperatives encountered three prohibited ("black") UTPs, while all reported at least one prohibited UTP. Moreover, the two most commonly reported practices (i.e., "unduly late payments" and "buyers' demand that suppliers pay for the deterioration or loss of products that occurred after ownership transfer") exerted a significant negative influence on cooperative performance, even in the presence of a proficient Board of Directors. Consequently, policymakers may need to pay more attention to UTPs and ensure that the national enforcement authorities are well-equipped to act rapidly and effectively against offenders.

Keywords: unfair trading practices; agricultural cooperatives; performance

Citation: Benos, T.; Sergaki, P.; Kalogeras, N. The Effect of Unfair Trading Practices on the Performance of Agricultural Cooperatives. *Proceedings* **2024**, *94*, 5. https://doi.org/10.3390/proceedings2024094005

Academic Editor: Eleni Theodoropoulou

Published: 19 January 2024

Copyright: © 2024 by the authors. Licensee MDPI, Basel, Switzerland. This article is an open access article distributed under the terms and conditions of the Creative Commons Attribution (CC BY) license (https:// creativecommons.org/licenses/by/ 4.0/).

1. Introduction

The power imbalance between actors in the food chain, which is closely associated with the increasing concentration of markets [1], repeatedly results in unfair behaviors, particularly to the detriment of the chain stakeholders with the lowest bargaining power (i.e., agricultural producers) [2]. In the European Union (EU), the Common Agricultural Policy (CAP) contains measures that aim to strengthen farmers' position in the food supply chain, including start-up funding for producer groups and regulatory exemptions from competition law for producer organizations [3]. Still, operators with substantial bargaining power (e.g., traders, retailers) continue to impose pressure on the weaker actors in the supply chain, giving rise to various unfair trading practices (e.g., short-notice order cancellations, unduly late payments) [4]. In response to these concerns, the EU issued a Directive (i.e., [5]) on unfair trading practices (UTPs) aiming at protecting weaker suppliers (primarily farmers) and their organizations (e.g., cooperatives) against their buyers.

Despite the renewed interest at a policy-making level (i.e., the EU Directive) and the recent surge of specialized policy reports on UTPs (e.g., [2,6]), only a few studies have empirically investigated the incidence of UTPs in the agri-food sector and their consequences (e.g., [4,7]). Interestingly, the effects of UTPs on cooperatives have been largely overlooked. To the best of our knowledge, only Di Marcantonio et al. [4,8] and Russo et al. [9] have studied the impact of UTPs on producer-owned groups. More specifically, Di Marcantonio et al. [4] performed a farm survey with 1258 dairy producers in five EU regions and found some weak evidence about producer organizations' role in helping farmers to set fairer contractual arrangements. Similarly, Di Marcantonio et al. [8]

conducted a farm survey with 1061 dairy producers in four EU regions and showed that membership in producer organizations makes farmers less likely to report UTPs. Finally, Russo et al. [9] measured fairness perceptions using a sample of 85 Italian kiwi fruit producers. They concluded that membership in producer-owned groups raises the probability that a farmer perceives a transaction as fair, but the countervailing power bestowed by collective action does not offset all unfair practices in the same way.

Consequently, little is still known about the actual presence of UTPs and what they induce, especially in producer-owned organizations like agricultural cooperatives, which farmers form to help them deal with power imbalances and unfair market behaviors, among others. Our paper aims to fill this knowledge gap and empirically examine the occurrence of UTPs and their influence on the performance of cooperatives. On top of studying the effect of UTPs on cooperative performance, we set out to explore the influence of the quality of the Board of Directors (BoD) as a cooperative's BoD customarily deals with its buyers.

2. Materials and Methods

To fulfill the goal of our paper, we administered a survey among cooperatives in Greece. We drew a sample from the official national registry of cooperatives in Greece (i.e., [10]). We targeted cooperatives from two of the most productive regions, namely Central Macedonia and Thessaly. From the 400 cooperatives officially registered in these regions at the time of the study, we randomly selected 200 and contacted two types of key informants (i.e., general managers and commercial managers). Those who agreed to participate were emailed a link to an online survey. After removing 13 questionnaires with incomplete responses, our sample size was 109, with an effective response rate of 54.5%.

To collect the responses, we used a structured questionnaire with sections on background characteristics (i.e., region, key informant type, and whether the cooperative offered perishable products), the presence of UTPs (i.e., whether the cooperatives experienced the prohibited "black" practices stipulated in the EU Directive; see Appendix A), the BoD quality (i.e., general satisfaction with the BoD, trust in BoD members, the competence of BoD members, experienced BoD members, and BoD vision to develop the coop), and the perception of cooperative performance (i.e., sales volume, profitability, market share, and new market entry). The constructs we used for BoD quality and cooperative performance proved to be sufficiently reliable and valid.

3. Results

Interestingly, while some of the UTPs did not occur at all (i.e., "payments requested but not related to a specific transaction" and "misuse of trade secrets"), all cooperatives reported at least one UTP. We also found that the three most common ones (see Table 1) were experienced by the vast majority of cooperatives (>60%). Perhaps it should not be surprising that "unduly late payments" and "unilateral changes in supply agreements" are so common. However, the high occurrence of "the risk of loss and deterioration transferred to suppliers" warrants special attention. Moreover, the practice that was added by the Greek transposition law was reported by 1/5 of the respondents (i.e., "the buyer demands from the seller(s), in writing or orally, to sell a certain quantity of their products without at the same time committing themselves to the purchase price"; we titled this as UTP "one-way commitment" in Table 1). This suggests that the decision of the policymakers to include this UTP in the transposition document (i.e., [11]) was right. Finally, on average, the participating cooperatives were subject to about three UTPs.

Table 1. Results of OLS regression analysis predicting cooperative performance.

Variables	Cooperative Performance (Standardized β)
Control variables	
Region	0.02
Role	0.01
Perishable products offered	0.08
Independent variables-UTPs	
Unduly late payments (86%)	−0.19 **
Short-notice order cancellations (33%)	−0.08
Unilateral changes in supply agreements (62%)	−0.01
Paying for loss or deterioration of products that occurred after ownership transfer (67%)	−0.21 **
Refusal of a written confirmation (8%)	−0.03
Commercial retaliation (11%)	0.04
Transferring the costs of examining customer complaints (4%)	0.10
One-way commitment (22%)	0.01
Independent variables-other	
BoD quality	0.77 **
R^2	0.68
F	20.40 **

Notes: β values are standardized coefficients; ** $p < 0.01$; The percentages for UTPs represent the share of cooperatives that experienced the respective UTPs in their usual transactions at least once in the previous year.

Using an OLS regression model, we then tested the UTPs' influence on cooperative performance, but we also explored the influence of "BoD quality". In addition, we entered the three background characteristics as control variables. Table 1 presents the regression results. We found that only the two most common UTPs significantly and adversely affected performance. That is, "unduly late payments" and "risk of loss and deterioration" were significantly and negatively associated with cooperative performance (β = −0.19, $p < 0.01$, and β = −0.21, $p < 0.01$, respectively). Furthermore, "BoD quality" had a strong positive effect on cooperative performance (β = 0.77, $p < 0.01$). As for the control variables, none of them exhibited any significant effect.

4. Discussion and Conclusions

Taken together, our survey results suggest that UTPs are widespread, and some are experienced by most of the cooperatives in the sample. Moreover, the two most commonly reported "black" practices exert a significant negative influence on cooperative performance, even in the presence of a proficient BoD. If such UTPs undermine producer groups' capacity to perform well, one may wonder how individual producers may cope with the ever-increasing competition in the food chain.

Overall, this paper offers fresh evidence of UTPs' occurrence in the agri-food sector and is among the few to empirically document UTPs' detrimental effects on cooperative organizations. The results contribute to the nascent UTPs-related literature, providing novel insights into the mark of UTPs on the weaker chain actors' organizations. They also advance cooperative literature, improving our understanding of an external peril that harms cooperative performance. Finally, the findings of this paper have important policy implications. That is, policymakers may need to pay more attention to UTPs, particularly to the impactful ones, and ensure that the national enforcement authorities are well-equipped to act rapidly and effectively against offenders.

Author Contributions: Conceptualization, T.B., P.S. and N.K.; methodology, T.B., P.S. and N.K.; investigation, T.B. and P.S.; data curation, T.B.; formal analysis, T.B.; writing—original draft preparation, T.B.; writing—review and editing, T.B., P.S. and N.K.; project administration, T.B. and P.S.; funding acquisition, T.B. and P.S. All authors have read and agreed to the published version of the manuscript.

Funding: The research project was supported by the Hellenic Foundation for Research and Innovation (H.F.R.I.) under the "3rd Call for H.F.R.I. Research Projects to support Post-Doctoral Researchers" (Project Number: 7606).

Institutional Review Board Statement: The corresponding author sought and received approval from the privacy officer of Zuyd University of Applied Sciences. As an official approval number was not issued, the contact details of the privacy officer who can confirm that approval was granted are available upon request.

Informed Consent Statement: Informed consent was obtained from all subjects involved in the study.

Data Availability Statement: The data presented in this study are available upon request from the corresponding author. The data are publicly unavailable due to privacy restrictions.

Conflicts of Interest: The authors declare no conflicts of interest. The funders had no role in the design of the study; in the collection, analyses, or interpretation of data; in the writing of the manuscript; or in the decision to publish the results.

Appendix A

The EU Directive (i.e., [5]) required Member States to prohibit a specific set of unfair practices, splitting them into two lists. The first list contained practices that are regarded as unfair per se (the "black" practices), while the second list consisted of practices that are deemed unfair if not explicitly agreed upon in the supply agreement (the "grey" practices). Member States could add other practices to the lists, extend the scope of listed prohibitions, make the prohibitions stricter, and even move practices from the "grey list" to the "black list". We concentrated on the "black" practices because they constitute unconditional prohibitions.

References

1. Bonanno, A.; Russo, C.; Menapace, L. Market power and bargaining in agrifood markets: A review of emerging topics and tools. *Agribusiness* **2018**, *34*, 6–23. [CrossRef]
2. Agricultural Markets Task Force (AMTF). Improving Market Outcomes. Enhancing the Position of Farmers in the Supply Chain (November 2016). Available online: https://agriculture.ec.europa.eu/common-agricultural-policy/agri-food-supply-chain/agricultural-markets-task-force_en (accessed on 20 March 2022).
3. European Commission. *Commission Staff Working Document. Impact Assessment. Initiative to Improve the Food Supply Chain (Unfair Trading Practices). Accompanying the Document Proposal for a Directive of the European Parliament and of the Council on Unfair Trading Practices in Business-to-Business Relationships in the Food Supply Chain*; European Commission: Brussels, Belgium, 2018.
4. Di Marcantonio, F.; Ciaian, P.; Fałkowski, J. Contracting and farmers' perception of unfair trading practices in the EU dairy sector. *J. Agric. Econ.* **2020**, *71*, 877–903. [CrossRef]
5. European Union. Directive (EU) 2019/633 of the European Parliament and of the Council of 17 April 2019 on unfair trading practices in business-to-business relationships in the agricultural and food supply chain. *Off. J. Eur. Union* **2019**, *L111*, 59–72.
6. Di Marcantonio, F.; Ciaian, P.; Castellanos, V. *Unfair Trading Practices in the Dairy Farm Sector: Evidence from Selected EU Regions*; Publications Office of the European Union: Luxembourg, 2018. [CrossRef]
7. Markou, M.; Stylianou, A.; Giannakopoulou, M.; Adamides, G. Identifying business-to-business unfair trading practices in the food supply chain: The case of Cyprus. *New Medit.* **2020**, *1*, 19–34. [CrossRef]
8. Di Marcantonio, F.; Havari, E.; Colen, L.; Ciaian, P. Do producer organizations improve trading practices and negotiation power for dairy farms? Evidence from selected EU countries. *Agric. Econ.* **2022**, *53*, 121–137. [CrossRef]
9. Russo, C.; Di Marcantonio, F.; Cacchiarelli, L.; Menapace, L.; Sorrentino, A. Unfair trading practices and countervailing power. *Food Policy* **2023**, *119*, 102521. [CrossRef]
10. National Registry of Agricultural Cooperatives. Available online: http://www.minagric.gr/index.php/el/for-farmer-2/sillogikes-agrotikes-organoseis (accessed on 20 March 2022).
11. Greek Law 4792/2021. Incorporation of Directive (EU) 2019/633 of the European Parliament and of the Council of 17 April 2019 on unfair commercial practices in business-to-business relations in the agricultural and food supply chain. *Gov. Gaz.* **2021**, 7075–7082. Available online: https://www.fao.org/faolex/results/details/en/c/LEX-FAOC213005/ (accessed on 20 March 2022).

Disclaimer/Publisher's Note: The statements, opinions and data contained in all publications are solely those of the individual author(s) and contributor(s) and not of MDPI and/or the editor(s). MDPI and/or the editor(s) disclaim responsibility for any injury to people or property resulting from any ideas, methods, instructions or products referred to in the content.

Proceeding Paper

The Food Fraud Landscape: A Brief Review of Food Safety and Authenticity [†]

Leonidas Sotirios Kyrgiakos [1], Malak Hazimeh [2], Marios Vasileiou [1,*], Christina Kleisiari [1], Georgios Kleftodimos [2,3] and George Vlontzos [1]

[1] Department of Agriculture Crop Production and Rural Environment, University of Thessaly, Fytoko, 38446 Volos, Greece; lkyrgiakos@uth.gr (L.S.K.); chkleisiari@uth.gr (C.K.); gvlontzos@uth.gr (G.V.)
[2] CIHEAM-IAMM (Institut Agronomique Méditerranéen de Montpellier), 34090 Montpellier, France; malakhzm24@gmail.com (M.H.); kleftodimos@iamm.fr (G.K.)
[3] Montpellier Interdisciplinary Center on Sustainable Agri-Food Systems (MoISA), Univ Montpellier, CI-HEAM-IAMM, CIRAD, INRAE, Institut Agro, IRD, 34060 Montpellier, France
* Correspondence: mariosvasileiou@uth.gr; Tel.: +30-24210-93013
[†] Presented at the 17th International Conference of the Hellenic Association of Agricultural Economists, Thessaloniki, Greece, 2–3 November 2023.

Abstract: Food fraud poses a significant challenge within the global food supply chain, with apprehensions regarding safety, authenticity, and efficiency. This study conducts a brief review of the literature by utilizing the Web of Science database, analyzing 2331 outcomes pertaining to the subject of food fraud. The analysis results demonstrated a noteworthy surge in scientific publications after 2013, which was propelled by events such as the horsemeat scandal and the formation of the European Food Safety Authority. Utilizing Multiple Correspondence Analysis (MCA), the study identified significant clusters pertaining to food transformation, safety, traceability, and distinct meat sources. In addition, trending topics shifted towards a holistic approach to food safety and the implementation of technologies like Blockchain (BC), Internet of Things (IoT), Artificial Intelligence (AI), and Big Data (BD). These technologies offer enhanced traceability, authentication, automation, and decision-making capabilities. The present research offers valuable perspectives on the evolving landscape of food fraud research and the potential of nascent technologies to tackle these issues.

Keywords: food fraud; food safety; food authenticity; food supply chain; review; Industry 4.0; sustainability; blockchain; food traceability; Multiple Correspondence Analysis (MCA)

Citation: Kyrgiakos, L.S.; Hazimeh, M.; Vasileiou, M.; Kleisiari, C.; Kleftodimos, G.; Vlontzos, G. The Food Fraud Landscape: A Brief Review of Food Safety and Authenticity. *Proceedings* **2024**, *94*, 6. https://doi.org/10.3390/proceedings2024094006

Academic Editor: Eleni Theodoropoulou

Published: 19 January 2024

Copyright: © 2024 by the authors. Licensee MDPI, Basel, Switzerland. This article is an open access article distributed under the terms and conditions of the Creative Commons Attribution (CC BY) license (https://creativecommons.org/licenses/by/4.0/).

1. Introduction

Over the last decade, the Food Supply chain (FSC) has been facing one of the most emerging challenges and issues on a global scale, specifically "Food Fraud". Food fraud is considered an intentional act of misrepresentation of food for economic gain that is intended to remain undetected by the consumer, and often includes food modification or false documentation [1]. Food products are heterogeneous, as they come in various proportions from different geographical sources and comply with different legislation and norms depending on their origin, destination, and manufacturing [2,3]. Thus, food commodities are prone to fraudulent acts. In addition, FSCs have several interconnected and intercorrelated elements and phases that should be considered for assuring elimination of food fraud along the supply chain [4].

2. Materials and Methods

This literature review is mainly focused on the assessment of Web of Science (WoS) database results, regarding the term "food fraud" in abstracts, titles and keywords, leading to the extraction of the factors influencing this specific field. More precisely, partnerships and trending topics were assessed, focusing on the technological, social and economic dimensions. In this study, 2331 results have been collected through the WoS database, and have been transformed into a unified Bibtex file. Moreover, the Bibliometrix library was used to extract the figures and data presented in the subsequent sections [5].

3. Results and Discussion

The literature review assessment covered the period from 2003 to 2023, coinciding with the establishment of the European Food Safety Authority (EFSA) and extending up to February 2023 (Figure 1). This timeframe reflects the European Union's transition towards producing safer food products for consumers, and it can be divided into three sub-timeframes. From 2003 to 2013 (first time frame), scientific production was limited and low, resulting on the annual production of 25 papers on average for this period regarding food fraud. For the second time frame (2013–2018), the scientific production increased significantly, leading to an annual production of 200 articles in 2018. Post-2018 (third time frame), the annual scientific production had a straight increase reaching up to 400 articles in 2021 on an annual basis. This gap between the different time frames is due to raised awareness regarding food fraud issues. Both the EFSA's report on pesticides and the horsemeat scandal that broke in 2013 indicate that these two incidents were the catalysts for the European strategy to eliminate food fraud [6,7].

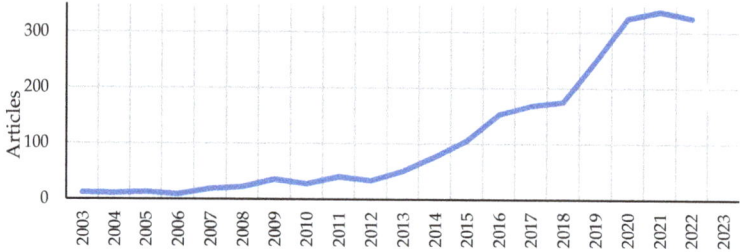

Figure 1. Trend of annual scientific production on food fraud.

3.1. Conceptual Structure Map

For a deeper understanding of the dynamics of the terms used in the scientific production, a conceptual structure map was obtained through the Multiple Correspondence Analysis (MCA) method. Two main groups were recognized, as shown in Figure 2. The first group, which is highlighted in red, contains the majority of the keywords regarding food transformation and science, as well as safety and traceability. Meanwhile, the second group contains seafood and substitution. Two subgroups can be identified within the red group. The first one refers to the applied methodologies (e.g., chemometrics, metabolomics and markers), and the second one refers to the different meat sources (e.g., meat, beef, pork). An important finding is that fish meat is an independent cluster, meaning that there is a special treatment towards this sensitive product. Overall, the MCA model can explain 69.3% of the involved keyword variability, which is considered representative of the whole sample of 2331 papers being incorporated into this literature review.

3.2. Trend Topic

Over the last decade, trend topics have been changing, leading to the creation of new directions of the scientific orientation regarding food fraud and its assessment in the FSC. Figure 3 presents the food fraud trend topics over the years. Up until 2017, the terms *quality, authenticity, food safety,* and *supply chain monitoring* were absent. Prior to 2017, almost all keywords and trend topics were focused more on the food science and biochemistry domains, rather than ensuring the quality and the elimination of food fraud in the FSC. It was no later than 2019 that there was a shift towards a holistic approach for increasing food safety standards and providing more insights about the implementation of new technologies for monitoring.

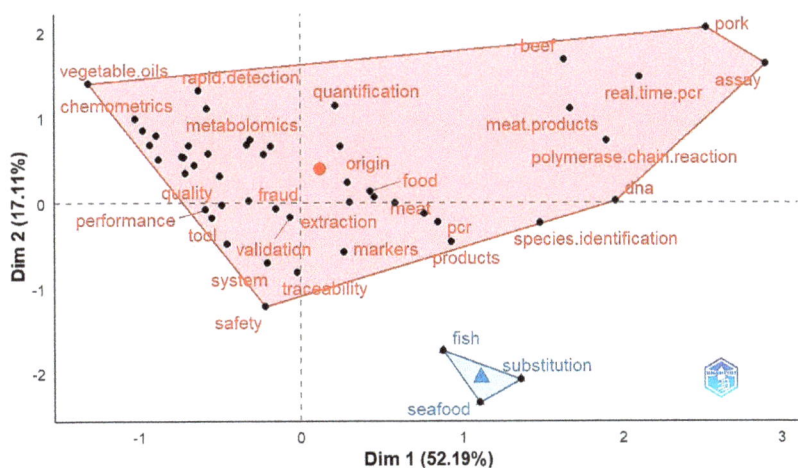

Figure 2. Analysis of food fraud most relevant keywords. Conceptual structure map using MCA.

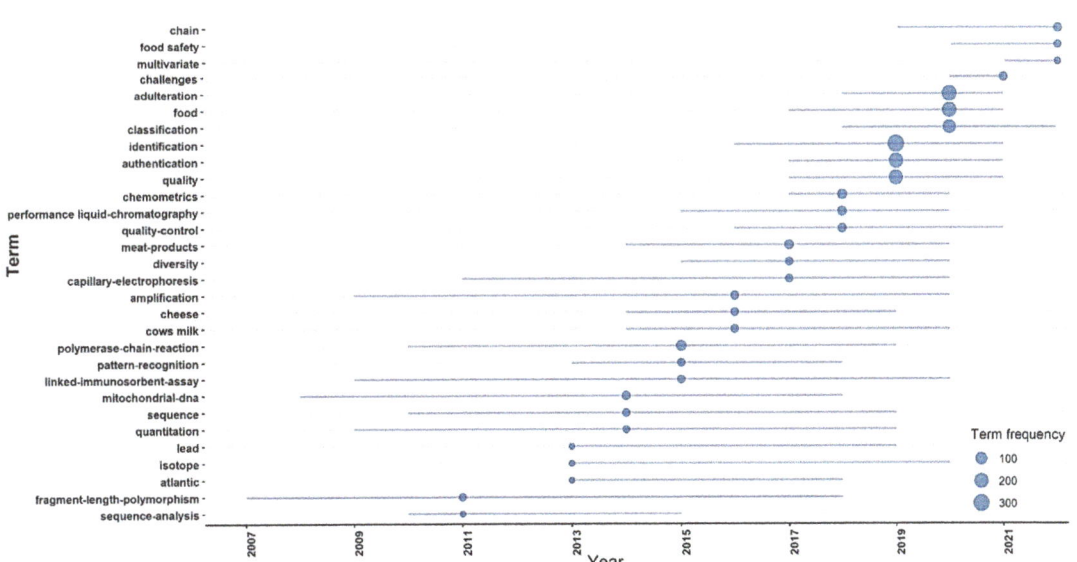

Figure 3. Representation of food fraud trend topics over the years.

Furthermore, a subsequent review was conducted regarding trending technologies on FSC for food safety and authenticity. The dominant technology is Blockchain (BC), followed by IoT, Artificial Intelligence (AI) and Big Data (BD). BC and IoT provide reliable traceability systems, and offer the assurance of food authenticity and safety, given the input of the data is reliable. AI, on the other hand, enhances automation and digitalization, and can provide predictions for food fraud, while Big Data supports the prementioned technologies while improving decision-making [8]. Table 1 quotes selected publications classified based on these technologies.

Table 1. Selected publications classified based on these technologies.

Source	Category	AI	BC	BD	IoT
2019 [9]	Food quality and Authenticity	•			
2022 [10]	Food Safety		•		
2017 [11]	Food Safety			•	
2022 [12]	Food Authenticity		•		
2019 [13]	Food Fraud Detection	•	•		
2020 [14]	Food Provenance and Authenticity	•	•	•	•
2018 [15]	Food Quality & Safety				•
2021 [16]	Food Authenticity		•		

4. Conclusions

The issue of food fraud poses a continuous and dynamic challenges in FSC. The literature review of this study highlighted the growing recognition and scholarly inquiry into comprehending and addressing the issue of food fraud. The identification of pivotal clusters pertaining to safety, authenticity, and meat origins yields valuable insights into the research's focal points. The shift towards a holistic approach and the adoption of technologies like Blockchain, IoT, AI, and Big Data demonstrate promising solutions for ensuring food authenticity and safety. However, further research and collaboration are required to bridge gaps and inconsistencies in FSC, ultimately safeguarding consumers and upholding the integrity of the industry. The mitigation of food fraud necessitates an ongoing level of vigilance, advancements in the field of technology, and comprehensive strategies focused on enhancing the transparency and efficacy of the FSC.

Author Contributions: Conceptualization, L.S.K., M.H., M.V. and G.V.; methodology, L.S.K., C.K., M.V. and G.V.; validation, G.K. and G.V.; formal analysis, L.S.K., M.V. and C.K.; resources, G.V.; data curation, L.S.K., M.H., M.V. and C.K.; writing—original draft preparation, M.V., L.S.K. and M.H.; writing—review and editing, M.V., C.K., G.K. and G.V.; visualization, L.S.K.; supervision, G.V.; project administration, G.V. All authors have read and agreed to the published version of the manuscript.

Funding: This research has received funding from the European Union's HE research and innovation program under grant agreement No. 101084188. Views and opinions expressed are, however, those of the author only and do not necessarily reflect those of the European Union or the European Research Executive Agency (REA). Neither the European Union nor the granting authority can be held responsible for any use that may be made of the information the document contains.

Institutional Review Board Statement: Not applicable.

Informed Consent Statement: Not applicable.

Data Availability Statement: The datasets generated and supporting the findings of this article are obtainable from the corresponding author upon reasonable request.

Conflicts of Interest: The authors declare no conflicts of interest.

References

1. Visciano, P.; Schirone, M. Food Frauds: Global Incidents and Misleading Situations. *Trends Food Sci. Technol.* **2021**, *114*, 424–442. [CrossRef]
2. Vasileiou, M.; Kyriakos, L.S.; Kleisiari, C.; Kleftodimos, G.; Vlontzos, G.; Belhouchette, H.; Pardalos, P.M. Transforming weed management in sustainable agriculture with artificial intelligence: A systematic literature review towards weed identification and deep learning. *Crop Prot.* **2024**, *176*, 106522. [CrossRef]
3. Brooks, C.; Parr, L.; Smith, J.M.; Buchanan, D.; Snioch, D.; Hebishy, E. A Review of Food Fraud and Food Authenticity across the Food Supply Chain, with an Examination of the Impact of the COVID-19 Pandemic and Brexit on Food Industry. *Food Control* **2021**, *130*, 108171. [CrossRef]
4. Tanveer, U.; Kremantzis, M.D.; Roussinos, N.; Ishaq, S.; Kyrgiakos, L.S.; Vlontzos, G. A Fuzzy TOPSIS Model for Selecting Digital Technologies in Circular Supply Chains. *Supply Chain. Anal.* **2023**, *4*, 100038. [CrossRef]
5. Bibliometrix. Available online: https://www.bibliometrix.org/home/ (accessed on 28 April 2023).
6. European Food Safety Authority. The 2013 European Union Report on Pesticide Residues in Food. *EFSA J.* **2015**, *13*, 4038. [CrossRef]

7. European Food Safety Authority, Horsemeat in the EU Food Chain | EFSA. 2013. Available online: https://www.efsa.europa.eu/en/press/news/130211 (accessed on 10 March 2023).
8. Vasileiou, M. *Industry 4.0 Technologies in Supply Chain Management: A Systematic Literature Review and Classification of Technologies*; University of the Aegean: Chios, Greece, 2022.
9. Jiménez-Carvelo, A.M.; González-Casado, A.; Bagur-González, M.G.; Cuadros-Rodríguez, L. Alternative Data Mining/Machine Learning Methods for the Analytical Evaluation of Food Quality and Authenticity—A Review. *Food Res. Int.* **2019**, *122*, 25–39. [CrossRef] [PubMed]
10. Xu, Y.; Li, X.; Zeng, X.; Cao, J.; Jiang, W. Application of Blockchain Technology in Food Safety Control: Current Trends and Future Prospects. *Crit. Rev. Food Sci. Nutr.* **2022**, *62*, 2800–2819. [CrossRef] [PubMed]
11. Marvin, H.J.P.; Janssen, E.M.; Bouzembrak, Y.; Hendriksen, P.J.M.; Staats, M. Big Data in Food Safety: An Overview. *Crit. Rev. Food Sci. Nutr.* **2017**, *57*, 2286–2295. [CrossRef] [PubMed]
12. Patro, P.K.; Jayaraman, R.; Salah, K.; Yaqoob, I. Blockchain-Based Traceability for the Fishery Supply Chain. *IEEE Access* **2022**, *10*, 81134–81154. [CrossRef]
13. Lo, S.K.; Xu, X.; Wang, C.; Weber, I.; Rimba, P.; Lu, Q.; Staples, M. *Digital-Physical Parity for Food Fraud Detection*; Springer: Cham, Switzerland; San Diego, CA, USA, 2019; pp. 65–79. [CrossRef]
14. Khan, P.W.; Byun, Y.-C.; Park, N. IoT-Blockchain Enabled Optimized Provenance System for Food Industry 4.0 Using Advanced Deep Learning. *Sensors* **2020**, *20*, 2990. [CrossRef] [PubMed]
15. Ping, H.; Wang, J.; Ma, Z.; Du, Y. Mini-Review of Application of IoT Technology in Monitoring Agricultural Products Quality and Safety. *Int. J. Agric. Biol. Eng.* **2018**, *11*, 35–45. [CrossRef]
16. Katsikouli, P.; Wilde, A.S.; Dragoni, N.; Høgh-Jensen, H. On the Benefits and Challenges of Blockchains for Managing Food Supply Chains. *J. Sci. Food Agric.* **2021**, *101*, 2175–2181. [CrossRef] [PubMed]

Disclaimer/Publisher's Note: The statements, opinions and data contained in all publications are solely those of the individual author(s) and contributor(s) and not of MDPI and/or the editor(s). MDPI and/or the editor(s) disclaim responsibility for any injury to people or property resulting from any ideas, methods, instructions or products referred to in the content.

Proceeding Paper

Farmers' Knowledge, Training Needs and Skills in the Bioeconomy: Evidence from the Region of Western Macedonia [†]

Aikaterini Paltaki [1,*], Efstratios Loizou [2], Fotios Chatzitheodoridis [2], Maria Partalidou [1], Stefanos Nastis [1] and Anastasios Michailidis [1]

[1] Department of Agricultural Economics, School of Agriculture, Aristotle University of Thessaloniki, 54124 Thessaloniki, Greece; parmar@agro.auth.gr (M.P.); snastis@agro.auth.gr (S.N.); tassosm@auth.gr (A.M.)
[2] Department of Regional Development and Cross Border Studies, School of Economic Sciences, University of Western Macedonia, 50100 Kozani, Greece; eloizou@uowm.gr (E.L.); fxtheodoridis@uowm.gr (F.C.)
* Correspondence: apaltaki@agro.auth.gr; Tel.: +30-6973860469
[†] Presented at the 17th International Conference of the Hellenic Association of Agricultural Economists, Thessaloniki, Greece, 2–3 November 2023.

Abstract: The aim of this paper is to explore farmers' training needs, their lack of knowledge and skills, and their willingness to participate in related training programs in the Western Macedonia Region. Summary statistics and multivariate analyses were performed for the data analysis. The results indicate a low level of knowledge about the bioeconomy and its practices. Furthermore, the findings revealed the high willingness of farmers for future adoption of the bioeconomy, and the need to create bioeconomy training programs.

Keywords: bioeconomy; multivariate statistical analysis; sustainability; training needs assessment; Western Macedonia

1. Introduction

The reduced availability of fossil fuels, climate change, resource conversion, food security, and population growth are some challenges that rural areas and agriculture are facing [1]. The transition to the bioeconomy contributes to the economic development of rural areas, as it refers to the shift of society towards sustainability [2]. There are several definitions available in the literature, and the most representative is the one that defines bioeconomy as "the production of renewable biological resources and their conversion into food, feed, bio-based products, and bioenergy via innovative, efficient technologies. In this regard, bioeconomy is the biological motor of a future circular economy, which is based on the optimal use of resources and the production of primary raw materials from renewably sourced feedstock [3]". To achieve sustainability in the agricultural sector, farmers and workers in agriculture must have the knowledge and skills to implement new practices and technologies [4].

The aim of this paper is to provide input on farmers' training needs, their lack of knowledge and skills, as well as their willingness to participate in related training programs. These outcomes would be useful for understanding the several training dimensions of the bioeconomy in the agricultural sector of the Western Macedonia Region (WMR) and future research on this subject.

2. Materials and Methods

Quantitative research was conducted between 1 January and 10 March 2023 using a structured questionnaire. Most questions were formulated on the typical five-point Likert scale of agreement. The questionnaire was completed by 331 farmers, from the four Regional Units of the WMR (Grevena, Kastoria, Kozani, and Florina).

Validity and reliability tests were performed prior to multivariate statistical analysis, using the statistical program SPSS (version 28). Validity tests for the structure of the questionnaire were conducted by five experts in questionnaire research before it was distributed to farmers. Then, the *a-Cronbach* test was used to ensure the reliability of this research and determine the consistency, accuracy, and objectivity of the research instruments. In total, 116 variables were included in the analysis. The *a-Cronbach* coefficient value was found equal to 0.920, showing a reliable scale. Two-Step Cluster Analysis (TSCA) was performed in order to classify the farmers based on common characteristics. Furthermore, a Categorical Regression model (CATREG) was used to determine the factors that influence the farmers' choices to implement bioeconomy practices.

3. Results

3.1. Summary Statistics

Results indicated a low level of knowledge of bioeconomy and its practices (M = 2.67). The level of bioeconomy practices implementation is low (M = 2.46). The main barriers to bioeconomy practices adoption are: (a) a lack of related financial resources (M = 4.81), (b) a lack of incentive to invest (M = 4.43), (c) the high cost of the bioeconomy (M = 4.42), (d) the high technological level of the bioeconomy (M = 4.35), and (e) the lack of training, and unqualified research and labor staff (M = 4.29). It is worth mentioning that responders' willingness to adopt bioeconomy practices in the future is high (M = 3.33).

To promote bioeconomy in the WMR, efforts should be mainly focused on developing bioeconomy training programs (M = 3.91). Actually, the majority of the responders mentioned their interest in participating in a training program in the future. More specifically, their interest is higher in issues related to water conservation and irrigation management (M = 3.62), national and EU funding and programs for bioeconomy (M = 3.37), waste management (M = 3.09), rational use of natural resources (M = 3.00), transition to the post-lignite era incorporating the bioeconomy (M = 2.97), utilization of biomass and liquid manure for energy production (M = 2.95) and, finally, application of sustainable agriculture-livestock technologies such as Precision Agriculture (M = 2.94).

3.2. TSCA

TSCA was implemented to segment the population into groups of farmers with common characteristics in terms of "Willingness to apply bioeconomy practices". Five clusters were created using 11 variables. According to the Silhouette measure of cohesion and separation, the clustering process is satisfactory. The first cluster consists of 81 farmers (24.5%), and the second cluster consists of 41 farmers (12.4%). The third cluster, which is the smallest, has 35 farmers (10.6%). In the fourth cluster, 59 farmers (17.8%) are classified.

Finally, the fifth cluster is the most numerous, as there are 115 farmers (34.7%). Table 1 lists the mean values of the variables of each cluster.

Table 1. Characteristics of each cluster.

Variable	Clusters				
	1	2	3	4	5
Knowledge in bioeconomy [1]	2.77	2.83	2.06	2.61	2.75
Advantages of bioeconomy on the farm [1]	3.22	3.15	2.65	3.54	3.09
High cost of bioeconomy [2]	3.98	3.71	4.51	4.95	4.86
Unqualified research and labor staff [2]	4.40	2.63	4.66	4.27	4.74
Lack of incentive to invest [2]	4.12	3.15	4.54	4.81	4.90
Lack of financial resources and financing [2]	4.62	4.59	4.71	4.98	4.99
High technological level and lack of know-how [2]	4.12	2.98	4.51	4.66	4.80
Application of bioeconomy practices [1]	2.42	3.12	3.34	3.19	1.61
Interest in applying bioeconomy practices [1]	3.77	2.80	2.20	4.03	3.16
Promoting bioeconomy through training programs [2]	3.38	4.20	4.83	4.24	3.49
Interest in adopting innovations [1]	3.70	3.80	3.43	3.98	3.15

[1] (1 = very low, 5 = very high), [2] (1 = strongly disagree, 5 = strongly agree).

3.3. CATREG

Then, to further analyze the variable that was created from TSCA "Willingness to apply bioeconomy practices" (dependent variable), CATREG was performed for the total sample (331 questionnaires) to identify the factors that influence farmers' choice to implement bioeconomy practices. The 13 independent variables were gender, age, marital status, occupation, educational level, income, municipality, distance from the nearest city, current adoption of bioeconomy (1 = very low, 5 = very high), barriers of bioeconomy adoption (1 = strongly disagree, 5 = strongly agree), advantages of bioeconomy adoption (1 = strongly disagree, 5 = strongly agree), participation in training programs regarding the application of bioeconomy practices (1 = strongly disagree, 5 = strongly agree), and interest in innovation adoption (1 = very low, 5 = very high). Additionally, it yielded an R2 value equal to 0.800, indicating a significant relationship between the "Willingness to apply bioeconomy practices" and the group of selected predictors (80.0% of the variance in the "Willingness to apply bioeconomy practices" rankings is explained by the regression of the optimally transformed variables used). The F statistic value 4.075, with $\alpha = 0.00$, indicated a consistently well-performing model.

The relative-importance measures of the independent variables show that the most important predictors are: (a) current adoption of bioeconomy (31.1%); (b) barriers to bioeconomy adoption (16.9%); and (c) age (10.3%). The additional significance of the independent variables is estimated at 58.30%.

A better prediction of "Willingness to apply bioeconomy practices" can be derived from the transformed plots (Figure 1) of the main independent variables that present the higher relative importance measures (more than 0.100). The most influential factors predicting the "Willingness to apply bioeconomy practices" are "current adoption of bioeconomy" (1 = very low, 2 = low, 3 = neutral, 4 = high, 5 = very high), "barriers of bioeconomy adoption" (1 = strongly disagree, 2 = disagree, 3 = nor disagree/nor agree, 4 = agree, 5 = strongly agree), and "age" (1 $\leq$ 20, 2 = 21–30, 3 = 31–40, 4 = 41–50, 5 = 51–60, 6 $\geq$ 61). This means that farmers have very low levels of bioeconomy adoption, agree that bioeconomy's application has many barriers, are 41–50 years old, and are more willing to adopt bioeconomy practices in their farms.

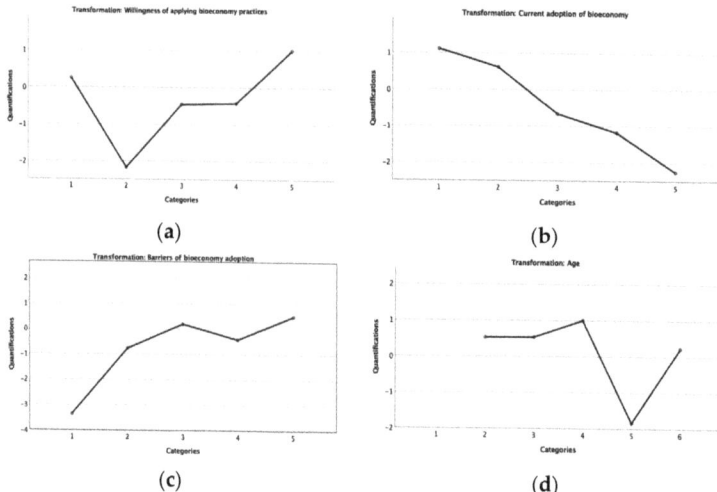

Figure 1. Transformed plots: (**a**) willingness to apply bioeconomy practices; (**b**) current adoption of bioeconomy; (**c**) barriers of bioeconomy adoption; (**d**) age.

4. Discussion

The results showed that farmers' low level of knowledge of the bioeconomy is one of the main barriers to the bioeconomy transition, which is also supported by the bibliography [5]. However, their high willingness for future adoption has to be the key to promoting the bioeconomy. Training is necessary for turning toward new sustainable practices [6]. Based on these results, separate training programs for each cluster should be created, focusing on the specific needs of each group. In addition, CATREG revealed the three variables that influence farmers' willingness to adopt bioeconomy practices in their farms.

5. Conclusions

The findings of this paper highlight the importance of constant and relevant training. The segmentation of the farmers into several discrete clusters with common characteristics is a great opportunity to improve the already-existing training programs. Moreover, the outcomes showed a remarkable interest in bioeconomy training, so they would be useful for understanding the current development of the bioeconomy in Greece and future research on this subject.

Author Contributions: Conceptualization, A.P. and A.M.; methodology, S.N.; validation, M.P., E.L. and F.C.; analysis, A.P. and A.M.; writing—original draft preparation, A.P.; writing—review and editing, E.L., F.C., M.P. and S.N.; visualization, A.M.; supervision, A.M. All authors have read and agreed to the published version of the manuscript.

Funding: This research received funding by the project "80601" (MIS 5047196), which is implemented under the Action "Reinforcement of the Research and Innovation Infrastructure", funded by the operational program "Competitiveness, Entrepreneurship and Innovation" (NSRF 2014–2020) and co-financed by Greece and the European Union (European Regional Development Fund).

Institutional Review Board Statement: Not applicable.

Informed Consent Statement: Not applicable.

Data Availability Statement: Data available upon request.

Acknowledgments: The study protocol was approved by the Research Ethics and Deontology Committee of the Aristotle University of Thessaloniki (#300543/2022) before its application.

Conflicts of Interest: The authors declare no conflicts of interest.

References

1. Wei, X.; Liu, Q.; Pu, A.; Wang, S.; Chen, F.; Zhang, L.; Zhang, Y.; Dong, Z.; Wan, X. Knowledge Mapping of bioeconomy: A bibliometric analysis. *J. Clean. Prod.* **2022**, *373*, 133824. [CrossRef]
2. Shcherbak, A.; Tishkov, S.; Karginova-Gubinova, V. Bioeconomy in Arctic regions of Russia: Problems and prospects. *E3S Web Conf.* **2019**, *135*, 03005. [CrossRef]
3. Kardung, M.; Cingiz, K.; Costenoble, O.; Delahaye, R.; Heijman, W.; Lovrić, M.; van Leeuwen, M.; M'barek, R.; van Meijl, H.; Piotrowski, S.; et al. Development of the Circular Bioeconomy: Drivers and Indicators. *Sustainability* **2021**, *13*, 413. [CrossRef]
4. Baptista, F.; Lourenço, P.; Fitas da Cruz, V.; Silva, L.L.; Silva, J.R.; Correia, M.; Picuno, P.; Dimitriou, E.; Papadakis, G. Which are the best practices for MSc programmes in sustainable agriculture? *J. Clean. Prod.* **2021**, *303*, 126914. [CrossRef]
5. Fytili, D.; Zabaniotou, A. Organizational, societal, knowledge and skills capacity for a low carbon energy transition in a Circular Waste Bioeconomy (CWBE): Observational evidence of the Thessaly region in Greece. *Sci. Total Environ.* **2021**, *813*, 151870. [CrossRef] [PubMed]
6. Urmetzer, S.; Lask, J.; Vargas-Carpintero, R.; Pyka, A. Learning to change: Transformative knowledge for building a sustainable bioeconomy. *Ecol. Econ.* **2020**, *167*, 106435. [CrossRef]

Disclaimer/Publisher's Note: The statements, opinions and data contained in all publications are solely those of the individual author(s) and contributor(s) and not of MDPI and/or the editor(s). MDPI and/or the editor(s) disclaim responsibility for any injury to people or property resulting from any ideas, methods, instructions or products referred to in the content.

Proceeding Paper

An Electronic Platform for the Integrated Monitoring of Technical and Economic Data of Farms [†]

Anna Tafidou [1], Asimina Kouriati [2], Evgenia Lialia [2], Angelos Prentzas [2], Eleni Dimitriadou [2], Kyriaki Tafidou [3] and Thomas Bournaris [2,*]

[1] Department of Mathematics, Aristotle University of Thessaloniki, 54124 Thessaloniki, Greece; annatafidou@gmail.com
[2] Department of Agricultural Economics, Aristotle University of Thessaloniki, 54124 Thessaloniki, Greece; kouriata@agro.auth.gr (A.K.); evlialia@agro.auth.gr (E.L.); aprentzas@agro.auth.gr (A.P.); edimitri@agro.auth.gr (E.D.)
[3] Department of Architectural Engineering, Democritus University of Thrace, 67100 Xanthi, Greece; ktafidou@yahoo.gr
* Correspondence: tbournar@agro.auth.gr
[†] Presented at the 17th International Conference of the Hellenic Association of Agricultural Economists, Thessaloniki, Greece, 2–3 November 2023.

Abstract: The digitalization of farming is considered the fourth revolution in agriculture. The necessity of providing decision support tools and electronic platforms to help Greek farmers in their work is becoming increasingly evident. For this reason, this article presents the electronic platform called "FarmEconomicMonitoring" to monitor the operations of farms to control production costs and improve efficiency. With the use of the electronic platform by the farmer–entrepreneurs, their easy adaptation to the new technologies concerning decision-making and farm management systems becomes achieved.

Keywords: digitalization; technical and economic analysis; management

Citation: Tafidou, A.; Kouriati, A.; Lialia, E.; Prentzas, A.; Dimitriadou, E.; Tafidou, K.; Bournaris, T. An Electronic Platform for the Integrated Monitoring of Technical and Economic Data of Farms. *Proceedings* **2024**, *94*, 9. https://doi.org/10.3390/proceedings2024094009

Academic Editor: Eleni Theodoropoulou

Published: 22 January 2024

Copyright: © 2024 by the authors. Licensee MDPI, Basel, Switzerland. This article is an open access article distributed under the terms and conditions of the Creative Commons Attribution (CC BY) license (https://creativecommons.org/licenses/by/4.0/).

1. Introduction

Today, in the context of income support for farmers from the EU, their support is decoupled from the quantity produced and is based on the size of the farms. To increase their profits, farmers are incentivized to connect and adapt to market demands (demand) while promoting sustainable agriculture, which requires adaptation to EU rules on the environment, plant and animal health, and their management [1].

To achieve farm adaptation, farmers should closely monitor the operation of their farms. Thus, they will be able to make decisions quickly and immediately, adapting the production plan and the requirements of the crops adapted to the new conditions created. After all, the detailed recording of the financial data of agricultural holdings is necessary for immediate and correct decision making, future planning, and dealing with emergencies [2]. For the detailed monitoring of agricultural undertakings, it is essential for farmers to adapt to new technologies, such as decision-making systems and electronic management platforms.

This article presents the electronic platform "FarmEconomicMonitoring" for sustainable management. It concerns farm management software, which aims to monitor the operations of the farm through its crops. It aims to help farmers thoroughly monitor the management of the agricultural inputs they use during the growing season, along with the financial results from implementing the farm plan. The farmer is allowed to know in detail the requirements of each crop of production he adopts, in seeds, fertilizers, medicines, water, energy, labor (human and mechanical), as well as the returns in economic data, to decide whether to continue following the cultivation of the specific crops or not, adapting to the demands of the market and the EU to enhance his income and reduce production costs.

Digital technologies in agriculture provide a variety of data-based services that improve different applications on farms [3]. However, a key challenge is compatibility between technologies and protocols [3]. Conveniently, the specific platform helps through daily monitoring so that basic allowances that strengthen each crop are not lost.

This article is structured as follows: The first part is the project description. The second part is the presentation of the flow of the "FarmEconomicMonitoring" platform, and, finally, the conclusions regarding how the use of the service affects the management of the farm, with specific regard to the increase in its income and the reduction of production costs.

2. Materials and Methods

The digitalization of agriculture is hailed as the fourth revolution in agriculture [3], as it offers new opportunities for agriculture [4]. Digital agriculture platforms are a crucial aspect of agricultural innovation building on the broader agricultural innovation landscape [5,6]. The "FarmEconomicMonitoring" platform is based on an existing structure of the Laboratory of Informatics in Agriculture and Agricultural Economic Research Laboratory of Aristotle University of Thessaloniki and the farm accounting theory regarding production crops [7]. The chief priority is that it is easy to use by everyone and on any device. The primary users of this will be the producers themselves, who will also help with its optimization. Each producer will have a personal login code to enter the data whenever they wish. Through direct registration, an organized input–output management framework is created that facilitates producers in reducing the use of inputs, along with receiving CAP subsidies.

The originality of this idea is manifold. This attempt is an innovation in the agricultural sector, as only a few research studies, according to the existing literature [8–11], have dealt with monitoring farms by production crop. Through the connection of farmers with Information and Communication Technologies (ICT), the reduction of the movements of farmers from remote areas is achieved. Farmers are allowed to register their farm data and monitor market conditions without need visiting an accounting office. Also, the efficiency of the farms is enhanced as the inputs and outputs are controlled directly by them. More generally, the project refers to the correct and more cost-effective management of inputs and outputs of farms. Its purpose is to, through deploying the direct participation of the farmer in the monitoring and financial management of the farm's data, reduce production costs, improve competitiveness, and better adapt to the requirements of the CAP. By using the platform, each producer can directly check the technical and economic data of their farm, their obligations, and their reserves.

The application of the "FarmEconomicMonitoring" platform follows a standard procedure (Figure 1) where, as follows:

1. The data of the farm are documented, which concerns its assets, the available production factors, and details regarding the requirements (cultivation care, working hours, quantity and expenditure of medicines, fertilizers, fuel, et cetera) and performance of the production crops it adopts in the production plan of the specific growing season. Achieving this is through the inventory and completion of the farm's production crop diaries.
2. Data entered are processed (automatically) following the rules of farm accounting and estimation [9].
3. The information regarding the economic results of the operation, the production costs, and the level of employment obtained, while at the same time, there is a series of additional information transferred to the decision-making centers.
4. The new information helps to make decisions, the adoption of which leads to actions that affect the farm's business activities and the creation of new data for it.

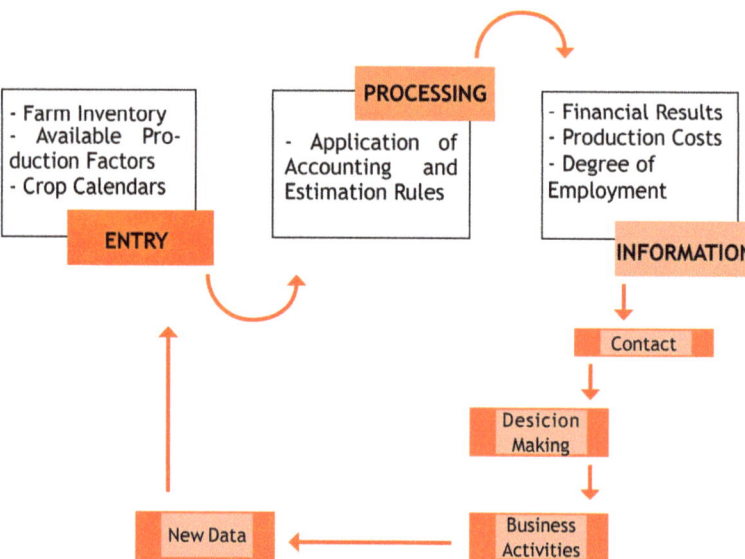

Figure 1. The flow of actions and results of the "FarmEconomicMonitoring" platform.

3. Conclusions

The "FarmEconomicMonitoring" online platform was designed keeping in mind the farmer–entrepreneurs who have little time to manage their farms and lack specialized knowledge in technology. Thus, the philosophy of the service and its operation is simple, and its use leads to the creation of information that helps the farmer–entrepreneur make immediate and quick decisions, which in turn leads to the improvement of the position of the farm. More specifically, "FarmEconomicMonitoring" enables the average farmer–entrepreneur to monitor the operation of their farm by recording in detail all the changes that take place in their assets and in the production crops they cultivate. Additionally, estimates the economic results of the growing season and evaluates the implementation of the production plan adopted in the specific period. Gives the opportunity to compare the current state of the farm with that of previous years or similar farms in the area, and to proceed with the restructuring of the production plan guided by the current state and its data. More generally, the collection of data, the processing, and the results resulting from the use of "FarmEconomicMonitoring" help to make decisions related to the improvement of the economic position of the farm, along with the advancement of the use of the production factors it has. At the same time, with the detailed monitoring of the operation of the farm, it is easier to comply with the rules of the CAP, which also allows earnings to increase through the linked aids and subsidies. For the successful operation of the platform, the user should be trained and familiar with the use and interpretation of the results to turn them into functional information for making decisions that will lead the user to adopt new actions and adjust the production plan of the farm to increase their income.

The weakness of this platform lies in the fact that technology evolves quickly, which means that it should be adapted at regular intervals and have its databases updated so that the information the farmer receives is up to date and in line with market conditions. However, the fact that it can adapt to modern circumstances allows for an evolutionary dynamic and confirms that it is a decision-making and management tool for the modern farmer.

Author Contributions: Conceptualization, A.K. and A.T.; methodology, A.K. and A.T.; validation, E.L. and A.P.; formal analysis, A.K.; investigation, A.T.; resources, A.K., E.L. and A.T.; data curation, E.D. and A.P.; writing—original draft preparation, A.K., K.T. and A.P.; writing—review and editing, E.D. and T.B.; supervision, T.B. All authors have read and agreed to the published version of the manuscript.

Funding: This research was funded by the Rural Development Program (RDP) and is co-financed by the European Agricultural Fund for Rural Development (EAFRD) and Greece, grant number M16ΣYN2-00225.

Institutional Review Board Statement: Not applicable.

Informed Consent Statement: Informed consent was obtained from all subjects involved in the study.

Data Availability Statement: Data sharing is not applicable to this article.

Conflicts of Interest: The authors declare no conflicts of interest.

References

1. European Commission. Agriculture and Rural Development. 2022. Available online: https://agriculture.ec.europa.eu/common-agricultural-policy/income-support/cross-compliance_el (accessed on 30 March 2023).
2. Kouriati, A.; Dimitriadou, E.; Bournaris, T. Farm accounting for farm decision making: A case study in Greece. *Int. J. Sustain. Agric. Manag. Inform.* **2021**, *7*, 77. [CrossRef]
3. Amiri-Zarandi, M.; Hazrati Fard, M.; Yousefinaghani, S.; Kaviani, M.; Dara, R. A Platform Approach to Smart Farm Information Processing. *Agriculture* **2022**, *12*, 838. [CrossRef]
4. Borrero, J.D.; Mariscal, J. A Case Study of a Digital Data Platform for the Agricultural Sector: A Valuable Decision Support System for Small Farmers. *Agriculture* **2022**, *12*, 767. [CrossRef]
5. Runck, B.C.; Joglekar, A.; Silverstein, K.; Chan-Kang, C.; Pardey, P.; Wilgenbusch, J.C. Digital agriculture platforms: Driving data-enabled agricultural innovation in a world fraught with privacy and security concerns. *Agron. J.* **2022**, *114*, 2635–2643. [CrossRef]
6. Gustafson, A.; Erdmann, J.; Milligan, M.; Onsongo, G.; Pardey, P.; Prather, T.; Zhang, Y. A Platform for Computationally Advanced Collaborative AgroInformatics Data Discovery and Analysis. In Proceedings of the Practice and Experience in Advanced Research Computing 2017 on Sustainability, Success and Impact, New Orleans, LA, USA, 9–13 July 2017; ACM: New York, NY, USA, 2017; Volume 1287, pp. 1–4. [CrossRef]
7. Martika-Vakirtzi, M.; Dimitriadou, E. *Accounting in Types of Agricultural Holdings*; Grafima: Thessaloniki, Greece, 2007.
8. Mpoutakidis, D.; Pavloudi, A.; Aggelopoulos, S.; Rapti, M. Development of Software for the Farms Accounting. In Proceedings of the 7th International Conference on Information and Communication Technologies in Agriculture, Food and Environment (HAICTA 2015), Kavala, Greece, 17–20 September 2015.
9. Bournaris, T.; Manos, B.; Vlachopoulou, M.; Manthou, V. AgroMANAGER, a web application for farm management. *Int. J. Bus. Inf. Syst.* **2011**, *8*, 440–455. [CrossRef]
10. Bournaris, T. Evaluation of e-Government Web Portals: The Case of Agricultural e-Government Services in Greece. *Agronomy* **2020**, *10*, 932. [CrossRef]
11. Bournaris, T.; Manos, B.; Vlachopoulou, M.; Manthou, V. E-government and farm management agricultural services in Greece. *Int. J. Bus. Innov. Res.* **2011**, *5*, 325–337. [CrossRef]

Disclaimer/Publisher's Note: The statements, opinions and data contained in all publications are solely those of the individual author(s) and contributor(s) and not of MDPI and/or the editor(s). MDPI and/or the editor(s) disclaim responsibility for any injury to people or property resulting from any ideas, methods, instructions or products referred to in the content.

Proceeding Paper

Agricultural Knowledge and Innovation Systems and Sustainable Management of Natural Resources †

Georgios Kountios [1,*], Ioannis Chatzis [2] and Georgios Papadavid [3]

1. Department of Agriculture, International Hellenic University, Sindos, 574 00 Thessaloniki, Greece
2. Payment and Control Agency for Guidance and Guarantee Community Aid, Domokou 5, 104 45 Athens, Greece; ioannis.chatzis@opekepe.gr
3. Agricultural Research Institute, Athalassa, Nicosia 1516, Cyprus; gpapadavid@ari.moa.gov.cy

* Correspondence: gkountios@ihu.gr
† Presented at the 17th International Conference of the Hellenic Association of Agricultural Economists, Thessaloniki, Greece, 2–3 November 2023.

Abstract: The question of how agricultural knowledge and innovation systems (AKISs) can address the issue of sustainable management of natural resources (SMNR) is presented in this conference paper. This literature review, which collected published research from the Scopus electronic database, aimed to explore the value of AKISs in enhancing the sustainability of natural resources. Therefore, it examined and evaluated the roles of AKISs as either positive or negative overall. Moreover, it analyzed whether the use of AKISs supports the goal of creating a sustainable system that links agriculture with natural resources. Among its findings, this review presents the positive and negative outcomes of each element and potential future scenarios/suggestions if the current trends persist.

Keywords: agricultural knowledge and innovation systems; sustainable management of natural resources; advisory; agricultural extension; innovation

1. Introduction

Agricultural knowledge and innovation systems (AKISs) are meant to foster collaboration among all of the actors involved in the development, dissemination, and adoption of the current knowledge and technology in agriculture [1]. Ref. [2] stated that this includes the research process, the extension of agricultural knowledge, and the provision of effective education for farmers. Additionally, Ref. [3] noted that the AKIS also encompasses other organizations and institutions that have an interest in advancing agricultural technology and knowledge, such as governments. The key role of AKISs is integrated in the new Common Agricultural Policy (CAP) 2021-27 of the European Union (EU) as one of the ten goals for this period, along with environmental, social, and economic objectives that relate closely to SMNR. Therefore, SMNR seems to be closely interrelated with AKISs. This paper presents the results of a systematic literature review that focuses on the contribution of agricultural knowledge and innovation systems to the achievement of the sustainability objectives in EU countries.

2. Materials and Methods

This review was based on extensive research of the available studies cited in the Scopus database on the topic of AKISs and sustainable management of natural resources (SMNR). The initial search with specific keywords for papers from 2009 to 2023 resulted in 616 articles, out of which 114 were removed as duplicates and another 63 were removed for reasons such as being older than the minimum publication date (2009).

We also conducted an individualized screening of the papers to attain a more remote data acquisition procedure and reduce the risk of bias altogether. The terms that were used for searching in the above-mentioned database aligned with this review's objectives.

Therefore, 399 articles passed an initial screening process based on their content and relevance, which led to eliminating an additional 186 articles. As a result, 213 articles were selected for retrieval to continue with the quality assessment. After thorough research, 200 articles were excluded because they were published in a language other than English, their content was irrelevant to the subject matter, and they had unclear methodology for data acquisition and processing. Therefore, only thirteen (13) were approved as eligible for a systematic review after passing all of the stipulated quality filtration procedures.

3. Results

AKISs seem to play a great role in maintaining the management of natural resources. This role is evident in the transition from conventional agricultural systems to agroecological systems [4]. According to Ref. [5], agroecological agricultural systems also established transitions to sustainable soil management.

Although some countries are behind on the AKIS concept [6], one main finding is that the EU has been active in promoting AKISs and SMNR in most of its countries [7].

Additionally, some of the strategies developed by the EU need to be more complete and conclusive. Thus, Ref. [8] revealed the gaps in the EU Farm to Fork (F2F) strategy of transforming a large part of the food system to a more sustainable form. In particular, they pointed out that many of the F2F targets were unrealistic as the EU focused on technical aspects and less on the social pillar that would ensure the durability of the outcomes.

Another main finding is that nearly all farmers from the involved countries depended on the knowledge flow from researchers to advisors who conveyed this information to the farmers who practiced it directly [9,10]. Ref. [11] conducted also research to determine whether the advisory services of EU countries could perform the activities of knowledge flow to farmers who ensured proper SMNR. The significant reliance on their peers (e.g., other farmers) and social media farming influencers is seen to result from the tendency of most farmers from EU countries to trust professional soil researchers and the government for information on reasonable soil maintenance practices, while [12] discovered that most Hungarian and UK farmers depend heavily on online sources for soil practices and knowledge.

In general, EU countries seem to be making efforts to enhance agricultural extension services and empower the structure of AKISs. Additionally, AKISs seem to be adopted in most EU countries and are expected to grow to higher levels [13], enhancing the assimilation of private and public interests, such as Belgium, France, Ireland, Germany, etc. [14], while countries like Bulgaria have experienced a deteriorating level of AKIS incorporation into the agricultural processes [15].

4. Discussion

After reviewing the relevant literature, AKISs seem to receive positive feedback in advancing the agricultural production sector toward sustainability. The relationship between the two is such that a sufficient flow of information from researchers to farmers and proper governance in the sector through credible institutions to oversee the whole process leads to positive outcomes in enabling and maintaining sustainability in the agricultural sector within the EU community. Most countries studied within the EU community were receptive of the AKIS model in their agriculture and tended to encourage innovation and sustainability in the agri-food sector by facilitating many policies and enhancing their coherence [16]. The effort to foster agricultural innovation in rural areas has led to the EU developing targeted rural development in specific locations. Rural development has also been supported by EU initiatives such as (EIP-AGRI) and the Program of Operational Groups (OGs) [17]. Overall, there is still a lot of work to be done as far as the integration of AKISs in many EU countries goes, as in many cases, like Greece, it is fragmented [18,19].

5. Conclusions

The European Union (EU) has been actively promoting agricultural knowledge and innovation systems (AKISs) across its member countries to foster sustainability in the agricultural sector. This research paper specifically focuses on addressing cutting-edge issues in policy debates, namely, water, soil, and pest management.

In this study, most of the countries examined displayed significant efforts in integrating sustainable natural resource management (SMNR) into their AKIS, with the notable exception of Bulgaria, which showed a lagging trend, emphasizing the urgency of modernization in their agricultural practices. Conversely, Portugal emerged as well prepared, possessing the essential knowledge required for effective AKIS implementation.

For many countries, the necessity of introducing a modern innovation model was underscored as a crucial step. However, one of the most crucial findings of this study is the insufficient exploration of the relationship between SMNR and AKISs. This highlights the pressing need for more substantial research and development efforts. There is an urgent requirement to collect and systemize existing knowledge related to SMNR to ensure its effective dissemination to farmers. Accompanying this, there is a critical need for intensive training of advisors, representing the two primary priorities that will enable the AKIS to fulfill its essential role not only in achieving sustainable natural resource management but also in advancing all objectives of the Common Agricultural Policy (CAP).

Author Contributions: Conceptualization and methodology, G.P. and I.C.; investigation, data curation, writing—original draft preparation, writing—review and editing, G.K.; visualization, I.C.; supervision, G.K. All authors have read and agreed to the published version of the manuscript.

Funding: This research received no external funding.

Data Availability Statement: Data are contained within the article.

Conflicts of Interest: Author Ioannis Chatzis was employed by the Payment and Control Agency for Guidance and Guarantee Community Aid. The remaining authors declare that the research was conducted in the absence of any commercial or financial relationships that could be construed as a potential conflict of interest.

References

1. Busse, M.; Doernberg, A.; Siebert, R.; Kuntosch, A.; Schwerdtner, W.; König, B.; Bokelmann, W. Innovation mechanisms in German precision farming. *Precis. Agric.* **2013**, *15*, 403–426. [CrossRef]
2. Lindblom, J.; Lundström, C.; Ljung, M.; Jonsson, A. Promoting sustainable intensification in precision agriculture: Review of decision support systems development and strategies. *Precis. Agric.* **2016**, *18*, 309–331. [CrossRef]
3. Ahuja, L.R.; Ma, L.; Howell, T.A. *Agricultural System Models in Field Research and Technology Transfer*; CRC Press: Boca Raton, FL, USA, 2016.
4. Giagnocavo, C.; de Cara-García, M.; González, M.; Juan, M.; Marín-Guirao, J.I.; Mehrabi, S.; Rodríguez, E.; van der Blom, J.; Crisol-Martínez, E. Reconnecting Farmers with Nature through Agroecological Transitions: Interacting Niches and Ex-perimentation and the Role of Agricultural Knowledge and Innovation Systems. *Agriculture* **2022**, *12*, 137. [CrossRef]
5. Wezel, A.; Herren, B.G.; Kerr, R.B.; Barrios, E.; Gonçalves, A.L.R.; Sinclair, F. Agroecological principles and elements and their implications for transitioning to sustainable food systems. A review. *Agron. Sustain. Dev.* **2020**, *40*, 40. [CrossRef]
6. Terziev, V.; Arabska, E. Enhancing Competitiveness and Sustainability of Agri-Food Sector through Mar-ket-Oriented Technology Development in Agricultural Knowledge and Innovation System in Bulgaria. 22 June 2015. Available online: https://papers.ssrn.com/abstract=3039595 (accessed on 7 August 2023).
7. Ingram, J.; Mills, J.; Black, J.E.; Chivers, C.-A.; Aznar-Sánchez, J.A.; Elsen, A.; Frac, M.; López-Felices, B.; Mayer-Gruner, P.; Skaalsveen, K. Do Agricultural Advisory Services in Europe Have the Capacity to Support the Transition to Healthy Soils? *Land* **2022**, *11*, 599. [CrossRef]
8. Moschitz, H.; Muller, A.; Kretzschmar, U.; Haller, L.; Porras, M.; Pfeifer, C.; Oehen, B.; Willer, H.; Stolz, H. How can the EU Farm to Fork strategy deliver on its organic promises? Some critical reflections. *EuroChoices* **2021**, *20*, 30–36. [CrossRef]
9. Keesstra, S.; Mol, G.; de Leeuw, J.; Okx, J.; Molenaar, C.; de Cleen, M.; Visser, S. Soil-Related Sustainable Development Goals: Four Concepts to Make Land Degradation Neutrality and Restoration Work. *Land* **2018**, *7*, 133. [CrossRef]
10. Kountios, G. The role of agricultural consultants and precision agriculture in the adoption of good agricultural practices and sustainable water management. *Int. J. Sustain. Agric. Manag. Inform.* **2022**, *8*, 144–155. [CrossRef]

11. Ingram, J.; Mills, J. Are advisory services "fit for purpose" to support sustainable soil management? An assessment of advice in Europe. *Soil Use Manag.* **2019**, *35*, 21–31. [CrossRef]
12. Rust, N.A.; Stankovics, P.; Jarvis, R.M.; Morris-Trainor, Z.; de Vries, J.R.; Ingram, J.; Mills, J.; Glikman, J.A.; Parkinson, J.; Toth, Z. Have farmers had enough of experts? *Environ. Manag.* **2021**, *69*, 31–44. [CrossRef] [PubMed]
13. Knickel, K.; Brunori, G.; Rand, S.; Proost, J. Towards a Better Conceptual Framework for Innovation Processes in Agri-culture and Rural Development: From Linear Models to Systemic Approaches. *J. Agric. Educ. Ext.* **2009**, *15*, 131–146. [CrossRef]
14. Knierim, A.; Kernecker, M.; Erdle, K.; Kraus, T.; Borges, F.; Wurbs, A. Smart farming technology innovations—Insights and reflections from the German Smart-AKIS hub. *NJAS-Wagening. J. Life Sci.* **2019**, *90–91*, 100314. [CrossRef]
15. Bachev, H. Governance of Agricultural Knowledge and Innovation System (AKIS) in Bulgaria. *SSRN Electron. J.* **2022**, *2022*, 4050617. [CrossRef]
16. Moreddu, C.; Poppe, K.J. Agricultural Research and Innovation Systems in Transition. *EuroChoices* **2013**, *12*, 15–20. [CrossRef]
17. Oliveira, M.d.F.; Gomes da Silva, F.; Ferreira, S.; Teixeira, M.; Damásio, H.; Ferreira, A.D.; Gonçalves, J.M. Innovations in Sustainable Agriculture: Case Study of Lis Valley Irrigation District, Portugal. *Sustainability* **2019**, *11*, 331. [CrossRef]
18. Koutsouris, A. AKIS and Advisory Services in Greece. Report for the AKIS Inventory (WP3) of the PRO AKIS Project. Online Resource. 2014. Available online: www.proakis.eu/publicationsandevents/pubs (accessed on 13 August 2023).
19. Birke, F.; Bae, S.; Schober, A.; Wolf, S.; Gerster-Bentaya, M.; Knierim, A. *AKIS in European Countries: Cross Analysis of AKIS Country Reports from the I2connect Project*; I2Connect: Paris, France, 2022.

Disclaimer/Publisher's Note: The statements, opinions and data contained in all publications are solely those of the individual author(s) and contributor(s) and not of MDPI and/or the editor(s). MDPI and/or the editor(s) disclaim responsibility for any injury to people or property resulting from any ideas, methods, instructions or products referred to in the content.

Proceeding Paper

Questioning Family Farms' Readiness to Adopt Digital Solutions [†]

Martina Francescone [1,*], Chrysanthi Charatsari [2], Evagelos D. Lioutas [3], Luca Bartoli [1] and Marcello De Rosa [1]

[1] Department of Economics and Law, University of Cassino and Southern Lazio, 03043 Cassino, Italy; bartoli@unicas.it (L.B.); mderosa@unicas.it (M.D.R.)
[2] Department of Agricultural Economics, School of Agriculture, Aristotle University of Thessaloniki, 54124 Thessaloniki, Greece; chcharat@agro.auth.gr
[3] Department of Supply Chain Management, International Hellenic University, 60100 Katerini, Greece; evagelos@agro.auth.gr
* Correspondence: martina.francescone@unicas.it
[†] Presented at the 17th International Conference of the Hellenic Association of Agricultural Economists, Thessaloniki, Greece, 2–3 November 2023.

Abstract: This paper explores the adoption of digital solutions by Italian farmers. The hypothesis is that digital technology adoption relies on an articulated set of socioeconomic variables that deserve attention. To test this hypothesis, we analyzed data from the last census of Italian agriculture. The analysis showed significant differences in the adoption of digital technologies, which can be viewed from territorial, structural, and sociodemographic points of view. This casts some doubt on the fairness of the digital transition in rural areas, calling for the strengthening of rural policies at the beginning of the new programming period in 2023–2027.

Keywords: digital agriculture; Italian farms; context-related analysis; smart farming; technology adoption; innovation

1. Introduction

This paper analyzes the readiness of farmers to adopt digital technologies in Italian farms. More precisely, we aim to explore how multiple contexts (business, personal/social, spatial) interplay and affect farmers' readiness to adopt digital solutions. This aspect is investigated through the lens of "omnibus context", by evidencing the relevance of multiple contexts in innovation adoption [1].

Innovation adoption is a complex process involving various dimensions. Concerning digitalization, a growing body of literature has analyzed these factors by emphasizing the disruptive character of digitalization and the related risks of non-neutrality [2–4], which call for more responsible digital innovations [5]. Set against the background of potential beneficiaries, the concept of readiness plays a significant role in the adoption process of digital solutions in agricultural practices in that it emerges as a prerequisite for digitalization, emphasizing the broad gap between the current state of farmers and the ideal farmer 4.0 [4]. As pointed out in Heiman et al. [6] thresholds model of diffusion, microeconomic behavior, heterogeneity, and dynamics are the critical variables that explain innovation adoption. This paper focuses on the second dimension by analyzing heterogeneity in the readiness to adopt digital solutions among Italian farms. To do that, we performed a context-related analysis, taking into account the various dimensions of the "omnibus" context emphasized in the seminal works of Welter and Baker: business, social, and spatial [1,7].

2. Methodology

To test heterogeneity and neutrality issues, we refer to the readiness of farmers in adopting an innovation. This concept is analyzed with reference to digital solutions and is grounded on secondary data provided by the Italian Census of Agriculture of the National

Citation: Francescone, M.; Charatsari, C.; Lioutas, E.D.; Bartoli, L.; De Rosa, M. Questioning Family Farms' Readiness to Adopt Digital Solutions. *Proceedings* **2024**, *94*, 11. https://doi.org/10.3390/proceedings2024094011

Academic Editor: Eleni Theodoropoulou

Published: 22 January 2024

Copyright: © 2024 by the authors. Licensee MDPI, Basel, Switzerland. This article is an open access article distributed under the terms and conditions of the Creative Commons Attribution (CC BY) license (https://creativecommons.org/licenses/by/4.0/).

Institute of Statistics. More precisely, data were extracted from section F of the questionnaire (use of information or digital technologies by farms). A "looking behind the data" approach was adopted through an "omnibus" lens of analysis referring to the broad perspective of the context [7], which focuses on context-related variables: business (farm size, standard output, commercialization), social (farmer's age, level of education) and spatial (regional level). To verify the eventual relationships between the context variables and digitalization, a χ^2 test was carried out to test the following hypotheses:

H0: $\chi^2 = 0$ (no association between context variables and digital technologies).

H1: $\chi^2 \neq 0$ (there is an association between context variables and the adoption of digital technologies).

The Chi-squared test was calculated through the following formula:

$$\chi^2 = \sum \frac{(Observed\ frequencies - Expected\ frequencies)^2}{Expected\ frequencies}$$

3. Results

Data from the Italian Agricultural Census demonstrate, on the one hand, the increase in the percentage of farms that have adopted digital technologies (from 3.8% to 25.8%) and, on the other hand, the relevance of context in shaping different trajectories of digitalization.

3.1. Business Context

As far as the business context is concerned, the utilized agricultural area, standard output, and marketing channels are considered. Table 1 allows us to reject the null hypothesis: as a matter of fact, large and economically viable farms are more digitalized than small-scale farms. Regarding marketing channels, farms directly selling to other farms or producers' organizations are more equipped with digital technologies.

Table 1. Business context: contingencies.

UAA *		Standard Output (€)		Marketing Channels	
	Digitalized		Digitalized		Digitalized
<1 ha	−23,439.26109	<4000	−53,212.28313	direct selling	18,405.31721
1–4.99	−31,060.08915	4000–15,000	−18,421.39692	other farms	380.0263797
5–9.99	6243.864275	15,000–50,000	15,856.57355	food industries	−2662.394058
10–19.99	13,098.33523	50,000–500,000	46,163.80925	trade companies	−19,229.80497
20–49.99	17,973.52573	>500,000	9613.297256	producers' organ.	3106.855436
50–99.99	9954.075728				
>100 ha	7229.549277				
χ^2	118,320.64	χ^2	204,114.61	χ^2	13,710.74
normalized χ^2	0.104429163	normalized χ^2	0.180150458	normalized χ^2	0.015388227
Cramer's V	0.323155013	Cramer's V	0.424441349	Cramer's V	0.124049294

* UAA: Utilized Agricultural Area.

3.2. Social Context

For this category, we took into account farmers' age and their level of education. Table 2 shows the results, confirming the connection between the selected variables, more precisely, by indicating the following: (a) younger farmers are more equipped with digital technologies than their elderly counterparts; (b) the higher the level of education, the higher the rate of digitalization in the farms.

Table 2. Social context: contingencies.

Farmer's Age		Level of Education	
	digitalized	No education	−0.1261882
<40	18,612.745	Primary school certificate	−0.1076158
41–64	23,308.333	Secondary school certificate	−0.0316419
≥65	−41,921.078	Professional (agricultural) qualification diploma	0.1816126
		Professional (not agricultural) qualification diploma	0.0645273
		High school specialized (agriculture) diploma	0.2048297
		High school Diploma	0.0384361
		University degree in agriculture	0.3292959
		University degree	0.0860448
χ^2	58575.165	χ^2	76,320.052
normalized χ^2	0.0518122	normalized χ^2	0.0675083
Cramer's V	0.227623	Cramer's V	0.2598236

3.3. Spatial Context

The spatial context was here analyzed through the regional distribution of innovation in the Italian regions, which allows a digital gap among regions to emerge. As evident from Table 3, we can reject the null hypothesis and confirm the association between the selected variables. The digital divide between northern and central-southern Italy (with the exception of Tuscany) emerges.

Table 3. Spatial context: contingencies.

Northern Italy		Central Italy		Southern Italy	
	Digitalized		Digitalized		Digitalized
Piedmont	8196.55208	Tuscany	3805.572	Abruzzo	−3511.128
Valle D'Aosta	385.60472	Umbria	−124.20022	Molise	−1536.2406
Lombardy	10,357.3808	Marche	−384.33698	Campania	−5605.2783
Veneto	8298.92492	Lazio	−2809.738	Apulia	−19,642.919
Friuli Venezia Giulia	2237.31579			Basilicata	−2835.9181
Liguria	501.470861			Calabria	−9059.9993
Emilia Romagna	8353.71655			Sicily	−11,795.249
Bolzano	8715.99573			Sardinia	1305.3146
Trento	5147.15973				
		χ^2	128,435.44		
		normalized χ^2	0.1133564		
		Cramer's V	0.3366845		

Table 4 provides a synthesis of the empirical analysis through the lens of heterogeneity and neutrality in digital technology adoption. High levels of heterogeneity mark the endowment of digital technologies, which mostly penalize smaller farms, conducted by aged farmers prevailingly located in central and southern Italy.

Table 4. Synthesis of the results.

	Heterogeneity	Neutrality	Who Is Excluded?
Business context	Yes	No	Small-size farms
Social context	Yes	No	Aged and low educated farmers
Spatial context	Yes	No	Central (but Tuscany) and southern regions

4. Discussion and Conclusions

The analysis represents a first step towards a broader investigation concerning digitalization in the Italian farming sector and, as such, has limitations due to the lack of available data from the Italian census of agriculture. That is why we have conducted a study based on an "omnibus" context while waiting for more detailed data supporting a discrete context work. Moreover, an analysis of the other two elements of Heiman et al.'s [6] threshold model ("microlevel behavior" and "dynamics") is required to excavate the complex process of innovation adoption. Despite these limits, we can agree with the recent literature emphasizing that heterogeneity and non-neutrality issues characterize digital technology adoption [2]. This paper confirms that the digital gap mainly affects smallholder, aged farmers located in southern and (surprisingly) central Italy. The results of the analysis cast some doubts on the potential capacity to fill the ambitious objectives expected in the long-term vision for rural areas 2040, where is the following is posited: the further development of rural areas is dependent on them being well connected between each other and to peri-urban and urban areas [8]. Based on our results, it seems difficult to answer the following question: how long is expected to be the "Long-Term Vision"?

Author Contributions: Conceptualization, M.F., C.C., E.D.L., L.B. and M.D.R.; methodology, M.F., C.C., E.D.L., L.B. and M.D.R.; software, M.F., C.C., E.D.L., L.B. and M.D.R.; validation, M.F., C.C., E.D.L., L.B. and M.D.R.; formal analysis, M.F., C.C., E.D.L., L.B. and M.D.R.; investigation, M.F., C.C., E.D.L., L.B. and M.D.R.; resources, M.F., C.C., E.D.L., L.B. and M.D.R.; data curation, M.F., C.C., E.D.L., L.B. and M.D.R.; writing—original draft preparation, M.F, C.C., E.D.L., L.B. and M.D.R.; writing—review and editing, M.F., C.C., E.D.L., L.B. and M.D.R.; visualization, M.F., C.C., E.D.L., L.B. and M.D.R.; supervision, M.F., C.C., E.D.L., L.B. and M.D.R.; project administration, M.F., C.C., E.D.L., L.B. and M.D.R.; funding acquisition, M.F., C.C., E.D.L., L.B. and M.D.R. All authors have read and agreed to the published version of the manuscript.

Funding: This research received no external funding.

Institutional Review Board Statement: This study was conducted in accordance with the Declaration of Helsinki. Ethical review and approval were waived for this study since the study was conducted in accordance with the Declaration of Helsinki and the EU General Data Protection Regulation.

Informed Consent Statement: Informed consent was obtained from all subjects involved in the study.

Data Availability Statement: The data presented in this study are available on request from the corresponding author (status: local repository).

Conflicts of Interest: The authors declare no conflicts of interest.

References

1. Baker, T.; Welter, F. *Contextualizing Entrepreneurship Theory*; Routledge: London, UK, 2020.
2. Fielke, S.J.; Taylor, B.M.; Jakku, E.; Mooij, M.; Stitzlein, C.; Fleming, A.; Thorburn, P.J.; Webster, A.J.; Davis, A.; Vilas, M.P. Grasping at digitalisation: Turning imagination into fact in the sugarcane farming community. *Sustain. Sci.* **2021**, *16*, 677–690. [CrossRef] [PubMed]
3. Schnebelin, E.; Labarthe, P.; Touzard, J.-M. How digitalisation interacts with ecologisation? Perspectives from actors of the French Agricultural Innovation System. *J. Rural. Stud.* **2021**, *86*, 599–610. [CrossRef]
4. Klerkx, L.; Rose, D. Dealing with the game-changing technologies of Agriculture 4.0: How do we manage diversity and responsibility in food system transition pathways? *Glob. Food Sec.* **2020**, *24*, 100347.

5. Lioutas, E.D.; Charatsari, C. Innovating digitally: The new texture of practices in Agriculture 4.0. *Sociol. Rural.* **2022**, *62*, 250–278. [CrossRef]
6. Heiman, A.; Ferguson, J.; Zilberman, D. Marketing and technology adoption and diffusion. *Appl. Econ. Perspect. Policy* **2020**, *42*, 21–30. [CrossRef]
7. Welter, F. Contextualizing entrepreneurship—Conceptual Challenges and Ways Forward. *Entrep. Theory Pract.* **2011**, *35*, 165–184. [CrossRef]
8. European Commission. Communication from the Commission to the European Parliament, the Council, the European economic and social committee and the committee of the regions, A long-term Vision for the EU's Rural Areas—Towards stronger, connected, resilient and prosperous rural areas by 2040. COM(2021) 345 final. Available online: https://eur-lex.europa.eu/legal-content/EN/TXT/?uri=CELEX:52021DC0345 (accessed on 18 January 2024).

Disclaimer/Publisher's Note: The statements, opinions and data contained in all publications are solely those of the individual author(s) and contributor(s) and not of MDPI and/or the editor(s). MDPI and/or the editor(s) disclaim responsibility for any injury to people or property resulting from any ideas, methods, instructions or products referred to in the content.

Proceeding Paper

Investigating Farmers' Attitudes towards Co-Existence of Agriculture and Renewable Energy Production [†]

Eirini Papadimitriou [1] and Dimitra Lazaridou [1,2,*]

1 School of Forestry and Natural Environment, Aristotle University of Thessaloniki, 54124 Thessaloniki, Greece; eirinipd@for.auth.gr
2 Department of Forestry and Natural Environment Management, Agricultural University of Athens, 36100 Karpenisi, Greece
* Correspondence: dlazaridou@aua.gr
† Presented at the 17th International Conference of the Hellenic Association of Agricultural Economists, Thessaloniki, Greece, 2–3 November 2023.

Abstract: Agri-voltaics (AVs) refer to combining agricultural activities and photovoltaic power generation. This dual use of the land has been identified as an important measure to address some of the main current and future social and environmental challenges. AVs constitute an upward trend at a global level. However, a limited number of studies have been carried out to identify the views of the interested parties, farmers, regarding the adoption of AVs on their agricultural lands. This paper reports research findings of the investigation of farmers' views and attitudes towards the adoption of photovoltaics in agricultural lands. The non-parametric Mann–Whitney U Test was used in order to make comparisons between the group of participants that were willing to adopt AVs and those who were not. Chi-square (χ^2) test of independence was performed to identify statistically significant relationships between farmers' willingness to adopt AVs and their socioeconomic characteristics or variables that represent knowledge about agro-energy. The results reveal that educational level and age had a significant role on accepting the installation of PV agriculture. Farmers' knowledge concerning agro-energy and their participation in farmers' associations are positively related to their willingness to adopt AV as well.

Keywords: renewable energy; agri-voltaics; farmers; attitudes; adoption

1. Introduction

Agri-voltaics (AVs) are a new approach that ensure the production of renewable energy, alongside the possibility of growing agricultural products on the same land. AV systems combining solar photovoltaic panels and food crops can optimize land use and increase overall productivity [1]. This new approach has been identified as a promising way to deal with some of the main current and future social and environmental challenges, such as climate change [2].

The majority of studies on the adoption of PV systems have focused on the adoption of solar PV systems among householders. To our knowledge, to date, a restricted number of studies have examined the key factors that influence the diffusion of PV power generation among farmers. Frantal and Prousek [3] explored why and how Czech farmers become renewable energy producers and concluded that the main reason for this is their intention for economic diversification and stabilization of their farms. Li et al. [4] investigated the variables affecting the adoption willingness of farmers regarding photovoltaic agriculture in China. According to their findings, usefulness perception and technical training positively influenced the adoption willingness of the farmers, whereas PV investment cost had a negative impact.

The present study is an attempt to investigate farmers' attitudes towards the co-existence of agriculture and renewable energy production and to examine the factors influencing farmers' adoption of AVs.

2. Methods

The questionnaire survey took place between November 2022 and February 2023 in Western Macedonia, Greece. This specific region was selected for the survey because of the high percentage of photovoltaic installations. Convenience sampling was conducted and, at the end of the collection process, 287 questionnaires had been gathered. Chi-square tests for independence were conducted between the variable that represents the question "are you willing to adopt AVs in your agricultural land?" and variables that represent the characteristics of the farmers in order to see if any of those influenced respondents' intention. Significant associations in Chi-square tests were examined by standardized residuals (stand. res.). The larger the residual, the greater the contribution of the cell to the extent of the resulting chi-square obtained value [5,6]. When the absolute value of the standardized residuals was greater than |1.96| in a cell, it was assumed that it contributed significantly to the test statistic [5,6].

For comparison of the two independent samples, that were not normal distributed (tested using Kolmogorov–Smirnov), a Mann–Whitney U-test was employed [7]. All statistical analyses were performed using SPSS 27 statistical analysis software. The level of significance was set at a = 0.05.

3. Results

Only seven farmers did adopt PV agriculture, accounting for 2.4%, which is far less than the proportion of people who did not adopt PV agriculture (97.6%). However, most respondents were knowledgeable of AVs, accounting for 79.8%. Out of a total of 287 respondents who participated in the survey, 133 (46.3%) farmers declared willingness to adopt AVs, whereas 154 (53.7%) were unwilling to adopt AVs. Table 1 reveals no significant gender difference in farmers' willingness to adopt AVs ($\chi^2 = 0.182$, df = 1, $p = 0.721$). On the contrary, the adoption of AVs was significantly influenced by the educational level of the respondents ($\chi^2 = 68.633$, df = 4, $p < 0.001$). When the educational level of respondents was higher, significantly more respondents than expected adopted AVs (stand. res. = +3.7 and +1.9), and significantly less respondents than expected did not adopt AVs (stand. res. = −3.4). Farmers' educational attainment is an explanatory variable that was found to have a positive influence on the adoption of eco-friendly approaches in agricultural lands [8,9]. Moreover, significantly more singles than expected were willing to adopt AVs (stand. res. = +1.8) ($\chi^2 = 13.367$, df = 3, $p = 0.004$). Farmers' knowledge concerning agro-energy had a positive influence on their acceptance of AV adoption ($\chi^2 = 32.631$, df = 1, $p < 0.001$). So, significantly more knowledgeable respondents on agro-energy were willing to adopt AV than expected (stand. res. = +1.9). Membership in agricultural associations was found as a strong driver in AV adoption ($\chi^2 = 18.160$, df = 1, $p < 0.001$) as well. When they were members of agricultural associations, significantly more respondents than expected were positive to adopt AV installation in their farms (stand. res. = +2.6). On the contrary, significantly less respondents than expected were negative to adopt AV (stand. res. = −2.4).

An additional demographic characteristic that can influence farmers' decision to adopt AVs on their agricultural land may be related to their age. A significant difference in the mean age of the respondents exists between those who adopt and those who do not adopt AVs (Mann–Whitney test = 15,902, $p < 0.001$). Particularly, the mean age of those who were willing to adopt AV (40.5 ± 10.1 years) is significantly lower than those who did not adopt (49.4 ± 12.3 years). The present finding agrees with previous outcomes, suggesting that younger ages are more willing to undertake the risk of participation in innovative agricultural practices [10,11].

Table 1. Demographic variables (%) for the adoption of agro-voltaics in the region of Western Macedonia *.

Variable	Adoption/Non-Adoption of AV		Statistic (χ^2)	d.f.	p-Value
	% Yes	% No			
Gender			0.182	1	0.721
Males	46.8	53.2			
Females	40.0	60.0			
Education			68.633	4	<0.001
Primary school	0.0 (−2.7)	100.0 (2.5)			
Middle school	11.9 (−3.3)	88.1 (3.0)			
High school	38.5	61.5			
University degree	70.3 (3.7)	29.7 (−3.4)			
Post-graduate	88.9 (1.9)	11.1			
Marital status			13.367	3	0.004
Single	61.5 (1.8)	38.5			
Married	42.2	57.8			
Divorced–Widowed	0.0	100.0			
N/A	60.0	40.0			
Knowledge about agro-energy			32.631	1	<0.001 *
Yes	55.0 (1.9)	45.0			
No	12.1 (−3.8)	87.9 (3.6)			
Participation in farmers' associations			18.160	1	<0.001 *
Yes	63.5 (2.6)	36.5 (−2.4)			
No	36.6 (−1.9)	63.4			

* Numbers within parentheses are standardized residuals. The larger the residual (>|1.96|), the greater the contribution of the cell to the magnitude of the resulting chi-square obtained value.

The main reason for adopting AVs, as reported by farmers, is income growth and stabilization (41%). The coverage of energy needs has been rated as the second most important factor by those who declared themselves as willing to adopt AVs (29%). Turning to another business activity has been reported as a motivation for the adoption of AVs (12%) as well, followed by farmers' environmental protection motivation (6%) and some other reasons that gathered very low percentages.

4. Discussion and Conclusions

AV agriculture is a promising choice for achieving green energy and crop production [12]. Based on empirical analysis, it was found that among the 287 surveyed farmers, their willingness to adopt AVs was relatively high; 46.3% of farmers were willing to adopt AVs, indicating, however, that most Greek farmers maintain the traditional view about the dominant food-producing role of agriculture. The analysis revealed that both education level and age are significant determinants of their intention to adopt AVs. In addition, knowledge about agro-energy is positively correlated with adoption willingness of the farmers. Our results point, unsurprisingly, to the fact that economic aspects dominate their decision. So, an overall understanding of farmers' views and attitudes can contribute to the optimal coexistence of crops and solar panels, with better results for farmers and the environment.

Author Contributions: Conceptualization, E.P. and D.L.; methodology, E.P. and D.L.; validation, E.P. and D.L; formal analysis, E.P. and D.L; investigation, E.P.; writing—original draft preparation, D.L.; writing—review and editing, E.P. and D.L. All authors have read and agreed to the published version of the manuscript.

Funding: This research received no external funding.

Institutional Review Board Statement: The study was conducted in accordance with the Declaration of Helsinki. Ethical review and approval were waived for this study since the study was conducted in accordance with the Declaration of Helsinki and the EU General Data Protection Regulation.

Informed Consent Statement: Informed consent was obtained from all subjects involved in the study.

Data Availability Statement: Data will be available upon request to the first author.

Conflicts of Interest: The authors declare no conflicts of interest. The funders had no role in the design of the study; in the collection, analyses, or interpretation of the data; in the writing of the manuscript; or in the decision to publish the results.

References

1. Dupraz, C.; Marrou, H.; Talbot, G.; Dufour, L.; Nogier, A.; Ferard, Y. Combining solar photovoltaic panels and food crops for optimising land use: Towards new agrivoltaic schemes. *Renew. Energy* **2011**, *36*, 2725–2732. [CrossRef]
2. Mamun, M.A.A.; Dargusch, P.; Wadley, D.; Zulkarnain, N.A.; Aziz, A.A. A review of research on agrivoltaic systems. *Renew. Sustain. Energy Rev.* **2022**, *161*, 112351. [CrossRef]
3. Frantal, B.; Prousek, A. It's not right, but we do it. Exploring why and how Czech farmers become renewable energy producers. *Biomass Bioenergy* **2016**, *87*, 26–34. [CrossRef]
4. Li, B.; Ding, J.; Wang, J.; Zhang, B.; Zhang, L. Key factors affecting the adoption willingness, behavior, and willingness-behavior consistency of farmers regarding photovoltaic agriculture in China. *Energy Pol.* **2021**, *149*, 112101. [CrossRef]
5. Agresti, A. *Categorical Data Analysis*, 2nd ed.; John Wiley: Hoboken, NJ, USA, 2002.
6. Agresti, A. *An Introduction to Categorical Data Analysis*; John Wiley: Hoboken, NJ, USA, 2007.
7. Nachar, N. The Mann Whitney U: A test for assessing whether two independent samples come from the same distribution. *Tutor. Quant. Methods Psychol.* **2008**, *4*, 13–20. [CrossRef]
8. Lazaridou, D.; Michailidis, A.; Trigkas, M. Socio-economic factors influencing farmers' intention to undertake environmental responsibility. *Environ. Sci. Pollut. Res.* **2019**, *26*, 14732–14741. [CrossRef] [PubMed]
9. McGurk, E.; Hynes, S.; Thorne, F. Participation in agri-environmental schemes: A contingent valuation study of farmers in Ireland. *J. Environ. Manag.* **2020**, *262*, 110243. [CrossRef]
10. Lazaridou, D.; Michailidis, A.; Mattas, K. Evaluating the willingness to pay for using recycled water for irrigation. *Sustainability* **2019**, *11*, 5220. [CrossRef]
11. Liontakis, A.; Sintori, A.; Tzouramani, I. The Role of the Start-Up Aid for Young Farmers in the Adoption of Innovative Agricultural Activities: The Case of Aloe Vera. *Agriculture* **2021**, *11*, 349. [CrossRef]
12. Chen, J.; Liu, Y.; Wang, L. Research on coupling coordination development for photovoltaic agriculture system in China. *Sustain. Times* **2019**, *11*, 1065. [CrossRef]

Disclaimer/Publisher's Note: The statements, opinions and data contained in all publications are solely those of the individual author(s) and contributor(s) and not of MDPI and/or the editor(s). MDPI and/or the editor(s) disclaim responsibility for any injury to people or property resulting from any ideas, methods, instructions or products referred to in the content.

Proceeding Paper

Lameness Identification System in Cattle Breeding Units †

Dimitrios Kateris [1,*], Anastasios Mitsopoulos [2], Charalampos Petkoglou [2] and Dionysis Bochtis [1]

1. Institute for Bio-Economy and Agri-Technology (iBO), Centre for Research and Technology-Hellas (CERTH), 6th km Charilaou-Thermi Rd., 57001 Thessaloniki, Greece; d.bochtis@certh.gr
2. Ergoplanning Ltd., Karatasou 7, 54626 Thessaloniki, Greece; tmitsop@ergoplanning.gr (A.M.); x.petkoglou@ergoplanning.gr (C.P.)
* Correspondence: d.kateris@certh.gr
† Presented at the 17th International Conference of the Hellenic Association of Agricultural Economists, Thessaloniki, Greece, 2–3 November 2023.

Abstract: Lameness is one of the most significant problems in cattle breeding. It is a major factor that causes discomfort and significantly reduces the welfare of affected animals. Lameness can result in a decrease in milk production or, if not detected early enough, may require the animal to be culled, leading to severe direct and indirect economic consequences for the business. The delayed recognition of lameness is often due to the methods used for detection, which mainly rely on the observation of animal mobility by the breeder. These methods almost exclude the early detection of the problem. This work aims to establish a new detection system that will be able to identify on time, reliably, and at an early stage the lameness symptoms based on the movement parameters of the animals.

Keywords: lameness; animal health; accelerator sensors; machine learning; cattle

1. Introduction

In the last decades, animal production in Greece has experienced rapid growth due to increased demand for animal products. This demand has led to a shift from traditional to intensive breeding systems, to intensify production. Cattle, being one of the most significant livestock animals, contribute significantly to the country's economic development. However, because of these changes, unknown health problems have emerged, or existing ones have worsened, mainly due to improper management of intensively farmed animals. Lameness, as one of the most significant issues concerning animal health, leads to a reduction in milk and meat production and an increase in healthcare costs for animals [1]. Therefore, it is crucial to detect lameness in a timely and reliable manner [2] to reduce costs and ensure animal welfare. Animals suffering from this disease exhibit various symptoms, such as difficulty walking [3], increased lying time compared to healthy animals [4], and reduced grazing behavior [5].

Until today, many studies have dealt with the accurate prediction of lameness. However, existing methods are unclear and unreliable [6], especially those attempting to approach lameness through computational analysis. The methods aiming to predict the condition vary from study to study. Some research endeavors seek to address the issue using optical technologies. Song et al. [7] attempted to diagnose lameness using high-resolution images and videos, while Viazzi et al. [8] tried to identify the condition using 2D and 3D cameras. Another way to determine the problem was the use of force sensors, with the help of which attempts were made to record and recognize animals displaying lameness [9].

In this direction, this work aims to present the architecture of a new detection system that will be able to identify lameness symptoms in different infection stages in cattle using machine learning algorithms capable of identifying lameness at an early stage, thus avoiding the economic consequences of production losses for livestock units.

2. Methodology

In our days, lameness detection in cattle is carried out through visual assessment of the mobility index. This happens using a simple five-level numerical scale that rates the animal's posture as "1" for normal stance and "5" for permanent elevation of the affected limb during walking. To assess the index, an "expert" observes the gait of each animal and based on the deviation from the normal movement pattern, assigns it the corresponding numerical score. However, this scale is subjective and lacks reliability, especially when the observation is made by cattle breeders lacking specialized knowledge. As a result, lameness issues are often underestimated and not on time detected for immediate treatment. This can be avoided by using a reliable lameness symptom recognition system in cattle.

The proposed system provides a solution to this problem. The system combines effectively and successfully several commercial wearable sensors (triaxial accelerometers) placed on the feet of animals for recording their kinesiological parameters (Figure 1). All this information feeds the proposed system to provide the user with an early diagnosis. The proposed system uses machine learning algorithms properly adapted to diagnose lameness at an early stage, even if the symptoms are not visually evident. This ability is due to the system's ability to highlight differences in the kinesiological signatures of diseased animals and compare them to those of healthy ones.

(a) (b)

Figure 1. (a) Locations of the four wearable sensors on the cattle legs and (b) the wearable sensors (Blue Trident, Vicon).

The Proposed System Architecture

This section presents the architecture of the proposed system, for detecting anomalies in a cattle's gait, which involves using four sensor devices placed on the cattle legs to capture motion data. The approach of the proposed system encompasses two key phases: the training phase and the detection phase. In the training phase, a machine learning model of the typical gait of cattle using a machine learning algorithm.

The next phase is the "detection phase". In this phase, the trained model is used to classify the cow's health condition as a healthy cow, a cow with lameness in the early stage, or a cow with lameness in an advanced stage by comparing its current gait with the model generated during the training phase. Both phases include the following computational steps:

- Data Acquisition: Sensor signals are collected from the sensor devices at the animal legs and stored in memory;
- Preprocessing: The collected data is filtered to remove any noise;
- Feature Extraction: Information characterizing the cow's gait is extracted from each stride. This information is imprinted in a set of features, which provide descriptive information about the cow's gait;
- Training and Testing data: The collected data are shared in two different groups, 70% training, and 30% testing, to use them for the training and the validation of the model. During the training phase, these features are used to train a machine-learning model;

- Classification: In the detection phase, this pre-trained machine learning model assigns a label of "normal" or "abnormal" to each set of features, which we call "Classification".

All these phases of the proposed system are presented in the following Figure (Figure 2).

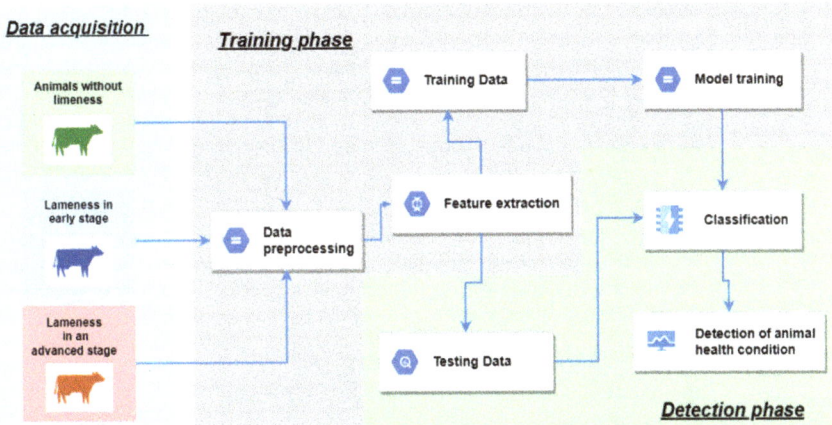

Figure 2. Overview of the proposed system architecture.

3. Results and Discussion

With the timely and rapid diagnosis of lameness, it becomes possible to avoid the discomfort of diseased animals and consequently reduce the daily milk production loss and premature culling. In addition to preserving the animal capital, given the high percentage of affected animals, between 5 and 7%, and a 20% reduction in milk production capacity, an increase of 1–1.4% in overall lifetime production is expected. The actual benefit is even greater and depends on how the situation is managed. If pharmacological treatment affecting milk production is administered, losses for a period are 100%, and if there is premature culling, the losses are even greater as a significant period of productive animal life is lost. Moreover, there are significant reductions in expenses associated with veterinary care for affected animals and a decrease in the transmission of lameness cases among livestock animals.

All of these can be achieved using the proposed system for early diagnosis of lameness. The breeders will have a reliable tool in their hands, which will enable them to improve the efficiency of the herd by avoiding the premature slaughter of animals with lameness. It will also significantly enhance the milk production capacity of the animals through rapid diagnosis of disease symptoms in their early stages. Optimization of estrus detection and reproductive management of the herd, as well as identification of nutritional and health disorders in the animals, will be achieved. The operational costs of the unit will be reduced due to a decrease in corresponding expenses for medication and veterinary services. These outcomes will result in the upgrading of local products, the improvement and expansion of business operations, and the strengthening of the local workforce.

Finally, it should be mentioned that the proposed system has been tested with data from public lameness data libraries and the results were highly encouraging. In the next stage, the proposed system will be tested in four different cattle breeding units to confirm the effectiveness of the system in real conditions.

4. Conclusions

In conclusion, the proposed system is being implemented for the first time in cattle breeding units, as it is a product of innovative laboratory research. Given that lameness detection is currently mainly based on subjective visual estimation by both breeders and veterinarians, the proposed system provides a reliable and affordable solution for timely lameness diagnosis. It has been proven that the use of a 5-point scoring scale does not provide significant reliability in diagnosing lameness, while other modern methods rely primarily on the use of visual means, such as images and videos, to diagnose lame animals. The drawback of these methods lies in their difficulty in continuously monitoring the animal in its natural environment. Therefore, the need to establish a new method for lameness detection with continuous recording of the animal's kinematic characteristics in its natural habitat, using new machine learning algorithms, provides a clear and accurate solution to the problem of early lameness diagnosis.

Author Contributions: Conceptualization, D.K.; methodology, D.K. and D.B.; resources, D.B.; writing—original draft preparation, D.K.; writing—review and editing, D.B., A.M. and C.P. All authors have read and agreed to the published version of the manuscript.

Funding: This research was funded by the Action «Rural Development Programme of Greece 2014–2020» under the call Measure 16 "Co-operation", Sub-Measure 16.1–16.2» that is co-funded by the European Regional Development Fund and Region of Central Macedonia, the project «CLARITY—Lameness identification system at milk- producing companies» (Project code: M16ΣΥΝ2-00034).

Institutional Review Board Statement: Not applicable.

Informed Consent Statement: Not applicable.

Data Availability Statement: Data is available on demand.

Conflicts of Interest: The authors Anastasios Mitsopoulos and Charalampos Petkoglou are employed by Ergoplanning Ltd. The remaining authors declare that the research was conducted in the absence of any commercial or financial relationships that could be construed as potential conflicts of interest.

References

1. Bruijnis, M.R.; Hogeveen, H.; Stassen, E.N. Assessing economic consequences of foot disorders in dairy cattle using a dynamic stochastic simulation model. *J. Dairy Sci.* **2010**, *93*, 2419–2432. [CrossRef] [PubMed]
2. Booth, C.J.; Warnick, L.D.; Grohn, Y.T.; Maizon, D.O.; Guard, C.L.; Janssen, D. Effect of lameness on culling in dairy cows. *J. Dairy Sci.* **2004**, *87*, 4115–4122. [CrossRef]
3. Walker, S.L.; Smith, R.F.; Routly, J.E.; Jones, D.N.; Morris, M.J.; Dobson, H. Lameness, Activity Time-Budgets, and Estrus Expression in Dairy Cattle. *J. Dairy Sci.* **2008**, *91*, 4552–4559. [CrossRef]
4. Ito, K.; von Keyserlingk M a, G.; Leblanc, S.J.; Weary, D.M. Lying behavior as an indicator of lameness in dairy cows. *J. Dairy Sci.* **2010**, *93*, 3553–3560. [CrossRef] [PubMed]
5. Miguel-Pacheco, G.G.; Kaler, J.; Remnant, J.; Cheyne, L.; Abbott, C.; French, A.P.; Pridmore, T.P.; Huxley, J.N. Behavioural changes in dairy cows with lameness in an automatic milking system. *Appl. Anim. Behav. Sci.* **2014**, *150*, 1–8. [CrossRef]
6. Schlageter-Tello, A.; Bokkers EA, M.; Koerkamp PW, G.G.; Van Hertem, T.; Viazzi, S.; Romanini CE, B.; Halachmi, I.; Bahr, C.; Berckmans, D.; Lokhorst, K. Manual and automatic locomotion scoring systems in dairy cows: A review. *Prev. Vet. Med.* **2014**, *116*, 12–25. [CrossRef] [PubMed]
7. Song, X.; Leroy, T.; Vranken, E.; Maertens, W.; Sonck, B.; Berckmans, D. Automatic detection of lameness in dairy cattle-Vision-based trackway analysis in cow's locomotion. *Comput. Electron. Agric.* **2008**, *64*, 39–44. [CrossRef]
8. Viazzi, S.; Bahr, C.; Van Hertem, T.; Schlageter-Tello, A.; Romanini CE, B.; Halachmi, I.; Lokhorst, C.; Berckmans, D. Comparison of a three-dimensional and two-dimensional camera system for automated measurement of back posture in dairy cows. *Comput. Electron. Agric.* **2014**, *100*, 139–147. [CrossRef]
9. Pastell, M.; Kujala, M.; Aisla, A.M.; Hautala, M.; Poikalainen, V.; Praks, J.; Veermäe, I.; Ahokas, J. Detecting cow's lameness using force sensors. *Comput. Electron. Agric.* **2008**, *64*, 34–38. [CrossRef]

Disclaimer/Publisher's Note: The statements, opinions and data contained in all publications are solely those of the individual author(s) and contributor(s) and not of MDPI and/or the editor(s). MDPI and/or the editor(s) disclaim responsibility for any injury to people or property resulting from any ideas, methods, instructions or products referred to in the content.

Proceeding Paper

Estimating Farmers' Creditworthiness under a Changing Climate †

Gregory Mygdakos [1], Panagiotis Tournavitis [2] and Emanuel Lekakis [1,*]

[1] AgroApps PC, Koritsas 34, 55133 Thessaloniki, Greece; mygdakos@agroapps.gr
[2] Cooperative Bank of Karditsa LLC, Kolokotroni 1, 43100 Karditsa, Greece; ptournavitis@bankofkarditsa.com
* Correspondence: mlekakis@agroapps.gr; Tel.: +30-6978897754
† Presented at the 17th International Conference of the Hellenic Association of Agricultural Economists, Thessaloniki, Greece, 2–3 November 2023.

Abstract: CreditScore combines the predictive power of crop growth models with future climatic scenarios, satellite images, and market data to form a comprehensive profile for each farmer-borrower, based on the future yields of their crops, with the ultimate goal of assessing long-term risks affecting yields which are related to climate change. The objective of this study is to present the tools and datasets that are employed operationally by CreditScore for future yield and profitability assessments. A modeling approach built on a fusion of satellite-derived vegetation indices, agro-meteorological indicators, and crop phenology is tested and evaluated in terms of data intensiveness for the prediction of wheat and cotton yields. AquaCrop, a process-based model, provided high to moderate accuracies by fully relying on freely available datasets as sources of input data. The findings introduce a promising framework that can support the financial institutions in evaluating potential customers' agribusinesses prior to and throughout the lending process.

Keywords: CreditScore; yield estimation; financial institutions; bank lending; loan

1. Introduction

In agriculture, where large down payments are required in conjunction with a lump sum payment at harvest, credit and access to credit for agricultural supplies and equipment are crucial to the sustainability and performance of farming enterprises. The volatility and uncertainty of agricultural income caused by climate change and the anarchic market situation places producers in the "High Risk Borrowers" category [1,2]. It is estimated that at European level, only one-sixth of farmers currently have access to credit, while simultaneously, young farmers face significant difficulties in accessing bank lending [3,4]. Therefore, with the majority of farmers having neither the guarantee nor the credit history to secure credit approval [2], there is a need to find alternative methods of assessing their creditworthiness.

CreditScore aspires to fill this gap, providing the right means for banking institutions to assess the real credit risk of potential borrowers and to exploit this potentially profitable sector. CreditScore strengthens the position of banks, which previously did not have access to information about farmers, by enabling them to assess their future solvency. In this way, every farmer, smallholder, or young farmer will be able to access financial products, which until now, banks have been unable to offer or they were offered at very high costs, making them unattractive. The objective of this study is to present the tools and datasets that are employed by CreditScore for future yield assessments. A process-based crop growth model was evaluated in estimating the yield of wheat and cotton.

2. Materials and Methods

CreditScore supports financial institutions in evaluating potential customers prior to and during the lending process. During evaluation, information concerning a farmer's financial details (past income, active mortgage payments, etc.), the requested capital, and complementary data, with regards to their holdings and cultivation plan, are taken into account for scoring their creditworthiness. The complementary data concern the number and location of parcels and whether these are the farmer's own capital or rent, the future crop plan per parcel, and possible subsidies. When registering a parcel in CreditScore, input data are retrieved automatically and are spatially aligned to allow for the estimation of yields. These are meteorological- (future climate based on the RCP scenarios) and soil (SoilGrids)-based. CreditScore employs the crop growth model AquaCrop to assess future yields.

To evaluate the performance of AquaCrop, wheat and cotton yield data were provided by the Cooperative Bank of Karditsa LLC, obtained from 15 farmers-borrowers. The data were fully anonymized and included the growing seasons (2019/20 and 2020/21 for wheat, 2021 for cotton), sowing dates, and yields for 87 wheat and 68 cotton parcels in the Thessaly region. The size of the parcels ranged from 0.1 to 13.4 ha. Gridded meteorological data for the growing seasons were derived from the ERA5-Land and ERA5 re-analysis, (Copernicus CDS). Daily weather data included T_{min}, T_{max}, ET_0, and precipitation. Soil data were retrieved from the SoilGrids database, up to a soil profile of 2 m, for each parcel. Sentinel-2 satellite images were acquired during the growing seasons. NDVI, GreenWDRVI, LAI, and Canopy Cover (CCRS) were assessed for pixels falling within each parcel. Using the representative CCRS curves, the CGC and CDC (canopy growth and decline coefficient) were calibrated for cotton and wheat. These parameters are provided in Table 1. A range of the average per parcel NDVI time series during the growing seasons is displayed in Figure 1. The simulations were executed for every parcel, and the yields obtained were compared to the actual yields provided by the farmers. The model efficiency (ME), the coefficient of determination (R^2), the root-mean-square error (RMSE), the normalized RMSE (nRMSE), the bias, and the Willmott's index of agreement (d) were selected as performance evaluation metrics.

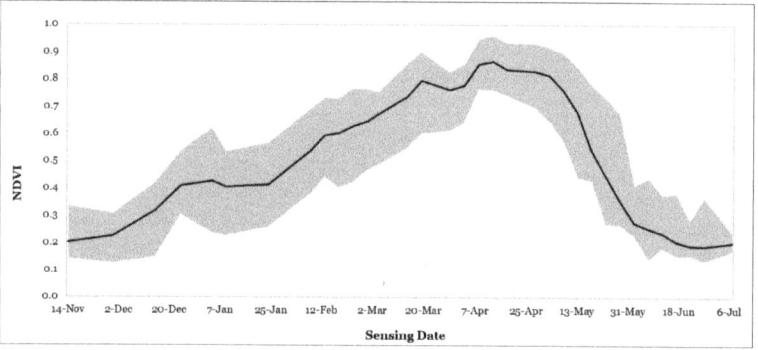

Figure 1. Reflected phenology, expressed as the NDVI, for the 2019/20 and 2020/21 seasons in wheat parcels.

Table 1. Calibrated parameters of AquaCrop.

Parameter	Unit	Wheat	Cotton
Soil surface covered by a seedling at 90% recover	cm^2/plant	1.50	6
Number of plants per hectare	Hm^{-2}	2,500,000	120,000
Maximum canopy cover (CC$_x$)	%	90	90
Calendar days: from sowing to emergence	d	13	14
Calendar days: from sowing to maximum root depth	d	93	98
Calendar days: from sowing to start senescence	d	178	144
Calendar days: from sowing to maturity	d	221	174
Calendar days: from sowing to flowering	d	150	64
Length of the flowering stage (days)	d	20	52

Table 1. Cont.

Parameter	Unit	Wheat	Cotton
Maximum effective rooting depth	m	0.3	1
Reference harvest index (HI$_o$)	%	42	35
Water productivity (WP)	gm^{-2}	17	15
Canopy growth coefficient (CGC)	Fraction d^{-1}	0.069	0.085
Canopy decline coefficient (CDC)	Fraction d^{-1}	0.0605	0.0605
Irrigation		Rainfed	Schedule

3. Results and Discussion

AquaCrop was evaluated in order to investigate whether the model has the potential to provide safe yield results through a simplified approach, with respect to data intensiveness and data sources. For cotton, the model was implemented under an irrigation scheduling mode. This is the only way to simulate future yields for irrigated crops, and the results were evaluated under this mode. The calculated statistical metrics summarized in Table 2 show that a very good agreement was obtained by AquaCrop regarding the simulation of the wheat and cotton yields. These average to high prediction accuracies are similar to those found in the literature using detailed weather station data or soils' physical properties derived from laboratory-analyzed samples as inputs [5,6].

Table 2. Statistical evaluation of AquaCrop results.

Statistical Metric	Units	Results for Wheat	Results for Cotton
Average estimates	kgha^{-1}	3430	4950
Average measured	kgha^{-1}	3450	4970
std estimates	kgha^{-1}	1128	87
std measured	kgha^{-1}	1233	541
Value range estimates	kgha^{-1}	750–6220	4820–5030
Value range measured	kgha^{-1}	1250–6700	4050–6370
ME	–	0.8	−0.1
RMSE	kgha^{-1}	616	566
nRMSE	%	17.9	11.4
bias	kgha^{-1}	1.3	−18.8
d	–	0.930	0.842
R^2	–	0.75	0.24

Many previous works supported that the use of calibration techniques based on remote sensing improves yield assessments from crop growth models [7,8]. The most logical pathway for a systematic calibration of AquaCrop is first and foremost to ensure a sound prediction of canopy cover. The key user-input parameters for this purpose are the coefficients defining the canopy development. In this study, the canopy cover data were derived from satellite images over the parcels and were used to safely simulate the future growth and yield of wheat and cotton crops for a particular area.

4. Conclusions

Apart from the identified significant contribution of remote sensing, the findings of this work also prove that gridded datasets on soil and environmental conditions can operationally be employed for yield prediction applications. The calibrated AquaCrop results, although obtained with a medium data input load and from publicly available datasets, were comparable with those reported in the literature for more detailed field experiments and treatments. It is challenging for simulation models to find relevance in real-world agriculture; however, the current work suggests that the combined simplicity and accuracy of AquaCrop make the model an indispensable tool within decision support systems. The investigated models and datasets are called upon to reduce the asymmetry of the available information and to cultivate trust and transparency between financial institutions and borrowers, in order to pave the way for credit to farmers. Particularly through CreditScore, financial institutions are able to:

1. Have at their disposal a long-term yield forecast, based on the near-future climate, that takes into account the effects of climate change on future yields, using crop growth models;
2. Assess borrowers' creditworthiness (for single- and multi-year loans), thus contributing to the formulation of personalized banking products and the regulation of contract terms.

Author Contributions: Conceptualization, G.M.; methodology, E.L.; validation, E.L.; investigation, G.M. and E.L.; data curation, P.T.; writing—original draft preparation, G.M., E.L. and P.T.; writing—review and editing, G.M., E.L. and P.T.; project administration, G.M.; All authors have read and agreed to the published version of the manuscript.

Funding: This research has been co-financed by the European Regional Development Fund of the European Union and Greek national funds through the Operational Program Competitiveness, Entrepreneurship and Innovation, under the call RESEARCH–CREATE–INNOVATE (project code: T2EDK-03246).

Institutional Review Board Statement: Not applicable.

Informed Consent Statement: Not applicable.

Data Availability Statement: No data is or will be shared.

Conflicts of Interest: Authors Gregory Mygdakos and Emanuel Lekakis were employed by company AgroApps PC. The remaining authors declare that the research was conducted in the absence of any commercial or financial relationships that could be construed as a potential conflict of interest.

References

1. EC (European Commission). *Modernizing and Simplifying the CAP-Economic Challenges Facing EU Agriculture*; Directorate-General for Agriculture and Rural Development: Brussels, Belgium, 2017; p. 15.
2. EIB (European Investment Bank). *Joint Initiative for Improving Access to Funding for European Union Young Farmers*; European Commission Directorate-General Agriculture and Rural Development, European Investment Bank Advisory Services fi-compass: Brussels, Belgium, 2019.
3. EC (European Commission). *Young Farmers in the EU-Structural and Economic Characteristics*; EU Agricultural and Farm Economic Briefs No 15; Directorate-General for Agriculture and Rural Development: Brussels, Belgium, 2017.
4. EIB (European Investment Bank). *Survey on Financial Needs and Access to Finance of EU Agricultural Enterprises*; European Commission Directorate-General Agriculture and Rural Development, European Investment Bank Advisory Services fi-compass: Brussels, Belgium, 2019.
5. Kale Celik, S.; Madenoglu, S.; Sonmez, B. Evaluating AquaCrop Model for Winter Wheat under Various Irrigation Conditions in Turkey. *J. Agric. Sci.* **2018**, *24*, 205–217.
6. Kheir, A.M.S.; Alkharabsheh, H.M.; Seleiman, M.F.; Al-Saif, A.M.; Ammar, K.A.; Attia, A.; Zoghdan, M.G.; Shabana, M.M.A.; Aboelsoud, H.; Schillaci, C. Calibration and Validation of AQUACROP and APSIM Models to Optimize Wheat Yield and Water Saving in Arid Regions. *Land* **2021**, *10*, 1375. [CrossRef]

7. Trombetta, A.; Iacobellis, V.; Tarantino, E.; Gentile, F. Calibration of the AquaCrop model for winter wheat using MODIS LAI images. *Agric. Water Manag.* **2015**, *164*, 304–316. [CrossRef]
8. Kasampalis, D.A.; Alexandridis, T.K.; Deva, C.; Challinor, A.; Moshou, D.; Zalidis, G. Contribution of Remote Sensing on Crop Models: A Review. *J. Imaging* **2018**, *4*, 52. [CrossRef]

Disclaimer/Publisher's Note: The statements, opinions and data contained in all publications are solely those of the individual author(s) and contributor(s) and not of MDPI and/or the editor(s). MDPI and/or the editor(s) disclaim responsibility for any injury to people or property resulting from any ideas, methods, instructions or products referred to in the content.

Proceeding Paper

Sectoral R&D Activities and Knowledge Production Functions: A Study Using International Data [†]

Penelope Gouta * and Christos T. Papadas

Department of Agricultural Economics and Rural Development, School of Applied Economics and Social Sciences, Agricultural University of Athens, 11855 Athens, Greece; cpap@aua.gr
* Correspondence: goutapenelope@gmail.com; Tel.: +30-6934595933
† Presented at the 17th International Conference of the Hellenic Association of Agricultural Economists, Thessaloniki, Greece, 2–3 November 2023.

Abstract: This study explores the relationship between the stock of knowledge generated by sectoral groups of the economy, calculated using Research and Development (R&D) expenditure, and knowledge output as measured by official patent applications. Using sector-specific R&D expenditure data published by the OECD, we calculate the total domestic R&D spending across the manufacturing, non-manufacturing, government, and educational sectors. Constructing a consistent 15-year panel dataset for the 17 most significant countries in R&D, we employ econometric subsampling, various estimators, and consider different rates of knowledge depreciation. Our findings reveal that the stock of knowledge in the private manufacturing, government, and educational sectors has a robust positive effect on patent generation.

Keywords: knowledge diffusion; R&D; patents; panel data

Citation: Gouta, P.; Papadas, C.T. Sectoral R&D Activities and Knowledge Production Functions: A Study Using International Data. *Proceedings* **2024**, *94*, 15. https://doi.org/10.3390/proceedings2024094015

Academic Editor: Eleni Theodoropoulou

Published: 23 January 2024

Copyright: © 2024 by the authors. Licensee MDPI, Basel, Switzerland. This article is an open access article distributed under the terms and conditions of the Creative Commons Attribution (CC BY) license (https://creativecommons.org/licenses/by/4.0/).

1. Introduction

This study uses OECD statistical datasets on research and development (R&D) expenditure and input–output tables. It classifies research data by economic sector across countries and harmonizes the sectors of the R&D expenditure dataset with those of the OECD input–output tables. Subsequently, based on the primary sources of R&D expenditure, it calculates and allocates these expenditures to four distinct groups of sectors within the domestic economy. This research aims to estimate the influence of the knowledge stock generated by each of these four domestic aggregate sectors through their R&D expenditures on domestic knowledge production, as measured by patents.

Our study intends to make a significant contribution to the relevant literature, where studies in knowledge production and diffusion considering sectoral inputs, e.g., sectoral R&D activity, are rather limited. Previous studies have predominantly focused on sampled firms within specific geographic areas, limiting the comprehensive and broadly applicable assessment of the impact of sectoral expenditure. In contrast, our study involves the total sectoral data from multiple countries over a time period. Additionally, the existing literature often attributes patent data to specific firms and sectors, overlooking the collaborative efforts among different firms that have led to patent production. The issue extends beyond sectoral studies, where patents stemming from collaborative efforts across regions are assigned to multiple entities, leading to an overestimation of knowledge production as the same patent is credited to different firms or sectors, meaning that they are counted repeatedly in observations. Our data and approach ascribe fractional values to patent applications based on the nationality of the contributing researchers. Distributing each patent application to the calculated fractional values according to the contributing researchers of each country enables the produced knowledge to be measured more accurately at the country level. Sectoral R&D expenditures, which are our independent variables, are used to measure the "stock of knowledge". For this reason, they are calculated accumulatively for each sector, in each country, for every year, over the examined period. This straightforward accumulation assumes no "knowledge depreciation". However, all calculations and accumulation take place by assuming also 5%, 10%, 15%, and 20% knowledge depreciation rates, as suggested by findings in the literature [1]. Depending on each scenario, the sectoral R&D expenditures of each year are transformed by the appropriate factors before they are added to yield the stock of knowledge in a certain year. This process is applied over all panel data observations. Econometric analysis takes place for all depreciation scenarios. Our exact dependent variable observation, affected by the stock of knowledge accumulated by each sector annually and per country throughout the examined period, is measured by the number of research patent applications originating from triadic patent families. We consider the priority day for each unique country–year pair and the residence of each researcher (inventor), enabling the assignment of fractional, non-integer patent numbers to different countries.

2. Materials and Methods

Following the aforementioned processing and calculations of our data, we constructed a panel data set, encompassing a span of 15 years and comprising the 17 preeminent countries engaged in R&D activities. This comprehensive dataset was crafted to encompass four distinct categories of domestic R&D expenditure stocks. For the purposes of our econometric analysis, we employed a Knowledge Production Function (KPF) adhering to the Cobb–Douglas framework, as originally articulated by Griliches in 1979. Drawing upon a review of the existing literature and the various econometric models utilized in the domain of R&D, the KPF framework was modified to align with the contemporary literature [2–5].

While the general framework of production functions traditionally construes inputs as being solely attributable to the agent under examination, without accounting for external sources (e.g., foreign countries, other firms, etc.), the landscape of R&D economics uniquely accommodates the inclusion of inputs from external entities, particularly in the context of capturing the influence of knowledge spillovers. While the R&D literature does not customarily bifurcate R&D expenditures by industry, a recurrent issue that emerges is that of multicollinearity; this is a phenomenon that is well documented in the pertinent literature [1,6,7] and is driven by linear correlations between independent variables. Scholarly discourse does not extensively engage with this predicament. Conversely, it is a common practice to introduce "foreign" R&D expenditure multiple times into the model, each instance weighted differently. This practice is employed to discern the impact of knowledge spillovers through distinct conduits. We posit that the utilization of varying weighting factors serves to mitigate the multicollinearity challenge. In our model, the inclusion of additional effects beyond our designated variables is achieved through the integration of pseudo-variables that are representative of spatial and temporal dimensions. Moreover, the calculation of such weighting factors proves impractical when applied to aggregated industry and country-level data sets, such as those characteristic of the OECD, owing to the unavailability of requisite statistical data.

3. Results and Discussion

Subsequent to the refinement of our Knowledge Production Function (KPF) framework, a series of tests was conducted to ascertain the selection of an appropriate econometric model. Both the Hausman test and the Breusch–Pagan Lagrange Multiplier test [8] were employed, concurring in their determination that the most suitable model was the fixed-effects model [9,10]. Next, the examination extended to encompass the identification of groupwise heteroscedasticity and autocorrelation phenomena. To assess the presence of these characteristics, we applied a modified Wald test and the Wooldridge test, respectively [11]. The outcomes of these tests yielded confirmation of groupwise heteroscedasticity and autocorrelation. In response, various estimators were considered in accordance with the established literature guidelines [12–14]. Additionally, the Newey–West estimator was deployed for estimation purposes, yielding outcomes consistent with those obtained through the previously mentioned estimators [15,16].

Nevertheless, it is crucial to acknowledge a limitation in the aforementioned estimators. While they demonstrate robustness in estimating specific parameters, they fail to adequately account for correlation between clusters. These estimators solely assume the existence of residual correlation within clusters. In light of this limitation, the Generalized Least Squares (GLS) estimator emerged as an alternative choice. Although the GLS estimator is traditionally applied in scenarios where the sample size (N) is less than the number of time periods (T), it remains an acceptable recourse in cases where the disparity between N and T is minimal, as is the case here ($N = 17$, $T = 15$). It should be noted that this situation, where N is approximately equal to 15, straddles a nuanced territory, and researchers adopt varying approaches to address it [17]. Additionally, we incorporated the Panel-Corrected Standard Error (PCSE) estimator, as posited by Beck et al. (1995). This estimator's appeal lies in its capacity to exhibit standard error estimates that are resilient to heteroskedastic disturbances, while simultaneously accounting for inter-cluster correlation and AR (1)-type autocorrelation. Lastly, we employed the Driscoll–Kraay methodology. This approach is lauded for its ability to adjust standard errors, ensuring the consistency of the estimator's covariance matrix irrespective of the stratified dimension represented by N. Furthermore, it effectively addresses the challenges of heteroscedasticity and autocorrelation within the analytical framework.

4. Conclusions

In our analysis, we applied our approach across different depreciation rates, ranging from 0% to 20%. A consistent observation emerged: the outcomes remained consistent across all depreciation rates, with no significant divergence, even with the utilization of various estimators. Our empirical analysis sheds light on the substantial and affirmative impact of Research and Development (R&D) expenditures on patent production. Specifically, our findings reveal that R&D spending within the private manufacturing sector, the educational sector, and the government sector have a significant impact on patent output. The estimated coefficients, reflecting the magnitudes of these effects, underscore the role of R&D spending in private manufacturing, surpassing the influence of R&D expenditure in other economic sectors. This outcome aligns with expectations, given the sectors' substantial allocation of resources and patent production. Additionally, our analysis underscores the role of R&D expenditure within the educational sector in driving patent production, surpassing the influence of the government sector; this is supported by the concentration of researchers and extensive research activities within educational institutions. Conversely, our results indicate that the impact of R&D expenditure on patent production in the private non-manufacturing sector is negligible and statistically insignificant.

Notably, while our focus centers on the sectors of private manufacturing, education, and government, it is important to highlight the indirect contributions of sectors closely aligned with agriculture, such as food processing and related manufacturing sectors. These sectors actively engage in R&D activities, investing in innovation and generating patents. The resulting knowledge diffusion from these sectors into agriculture underscores the interconnected nature of R&D efforts across the broader economic landscape, enhancing the agricultural sector's capacity for innovation and growth. Our findings emphasize the crucial role of strategic R&D investments in driving patent production, especially in the private manufacturing sector, and advocate for policies fostering educational–private sector collaboration that recognize the educational sector's significant influence on knowledge spillovers and innovation, while also highlighting the need to address sectoral R&D disparities for a balanced and sustainable distribution of innovation. Incorporating spatial and temporal dimensions with dummy variables in our analytical model yielded statistically significant results, confirmed by F-tests. Future research endeavors should construct indicators that aim to assess the disparities in the allocation of R&D expenditure among various sectors and countries. Analyzing sector interactions within countries unveils R&D

concentration levels, enhances the comprehension of patent geographical distribution, and identifies nations' knowledge production specialization.

Author Contributions: Methodology, C.T.P. and P.G.; data curation, P.G.; writing—original draft preparation, P.G.; writing—review and editing, C.T.P.; supervision, C.T.P. All authors have read and agreed to the published version of the manuscript.

Funding: This research received no external funding.

Institutional Review Board Statement: Not applicable.

Informed Consent Statement: Not applicable.

Data Availability Statement: Publicly available datasets were analyzed in this study. This data can be found here: https://data-explorer.oecd.org/.

Conflicts of Interest: The authors declare no conflicts of interest.

References

1. Griliches, Z. Issues in Assessing the Contribution of Research and Development to Productivity Growth. *Bell J. Econ.* **1979**, *10*, 92. [CrossRef]
2. Jaffe, A. Real Effects of Academic Research. *Am. Econ. Rev.* **1989**, *79*, 957–970. Available online: http://ideas.repec.org/a/aea/aecrev/v79y1989i5p957-70.html (accessed on 26 November 2019).
3. Bottazzi, L.; Peri, G. Innovation and spillovers in regions: Evidence from European patent data. *Eur. Econ. Rev.* **2003**, *47*, 687–710. [CrossRef]
4. Griffith, R.; Lee, S.; Van Reenen, J. Is distance dying at last? Falling home bias in fixed-effects models of patent citations. *Quant. Econom.* **2011**, *2*, 211–249. [CrossRef]
5. Drivas, K.; Economidou, C.; Karkalakos, S.; Tsionas, E.G. Mobility of knowledge and local innovation activity. *Eur. Econ. Rev.* **2016**, *85*, 39–61. [CrossRef]
6. Griliches, Z. The Search for R & D Spillovers. *Scand. J. Econ.* **1992**, *94*, 29–47. [CrossRef]
7. Hausman, J.; Hall, B.H.; Griliches, Z. Econometric Models for Count Data with an Application to the Patents-R & D Relationship. *Econometrica* **1984**, *52*, 909–938. Available online: http://www.jstor.org/stable/1911191 (accessed on 22 March 2016).
8. Baltagi, B.H.; Li, Q. Testing AR(1) against MA(1) disturbances in an error component model. *J. Econom.* **1995**, *68*, 133–151. [CrossRef]
9. Baum, C.F. Residual Diagnostics for Cross-section Time Series Regression Models. *Stata J. Promot. Commun. Stat. Stata* **2001**, *1*, 101–104. [CrossRef]
10. Drukker, D.M. Testing for serial correlation in linear panel-data models. *Stata J.* **2003**, *3*, 168–177. [CrossRef]
11. Wooldridge, J.M. *Econometric Analysis of Cross Section and Panel Data*; MIT Press: Cambridge, MA, USA, 2010.
12. Alvarez, B.J.; Arellano, M. The Time Series and Cross-Section Asymptotics of Dynamic Panel Data Estimators. *Econometrica* **2003**, *71*, 1121–1159. [CrossRef]
13. Stock, J.H.; Watson, M.W. Heteroskedasticity-robust standard errors for fixed effects panel data regression. *Econometrica* **2008**, *76*, 155–174. [CrossRef]
14. Lu, C.; Wooldridge, J.M. A GMM estimator asymptotically more efficient than OLS and WLS in the presence of heteroskedasticity of unknown form. *Appl. Econ. Lett.* **2020**, *27*, 997–1001. [CrossRef]
15. Newey, W.; West, K. A simple, positive semi-definite, heteroscedasticity and autocorrelation consistent covariance matrix. *Econometrica* **1986**, *55*, 703–708. [CrossRef]
16. Hoechle, D. Robust standard errors for panel regressions with cross-sectional dependence. *Stata J.* **2007**, *7*, 281–312. [CrossRef]
17. Reed, W.R.; Ye, H. Which panel data estimator should I use? *Appl. Econ.* **2011**, *43*, 985–1000. [CrossRef]

Disclaimer/Publisher's Note: The statements, opinions and data contained in all publications are solely those of the individual author(s) and contributor(s) and not of MDPI and/or the editor(s). MDPI and/or the editor(s) disclaim responsibility for any injury to people or property resulting from any ideas, methods, instructions or products referred to in the content.

Proceeding Paper

Exploring the Financial Viability of Greenhouse Tomato Growers under Climate Change-Induced Multiple Stress [†]

Giorgos N. Diakoulakis [1,*], Konstantinos Tsiboukas [1] and Dimitrios Savvas [2]

[1] Laboratory of Agribusiness Management, Department of Agricultural Economics and Rural Development, Agricultural University of Athens, Iera Odos 75, 11855 Athens, Greece; tsiboukas@aua.gr

[2] Laboratory of Vegetable Production, Department of Crop Production, Agricultural University of Athens, Iera Odos 75, 11855 Athens, Greece; dsavvas@aua.gr

* Correspondence: di_gi@aua.gr

[†] Presented at the 17th International Conference of the Hellenic Association of Agricultural Economists, Thessaloniki, Greece, 2–3 November 2023.

Abstract: In this study, we implement a linear programming farm model to explore the impact of climate change-induced multiple stress on the financial viability of greenhouse tomato growers. The main results are that new technologies and innovations can compensate growers for any profit loss associated with climate change. However, if the cost of adaptation is high enough, then its financial benefits are constrained by how efficient these innovations are in terms of productivity. We did not observe significant differences in input use between 'innovative' and 'conventional' production, and the yield under the adoption of new technologies was higher compared to 'conventional' production.

Keywords: linear programming; farm model; greenhouse tomato; climate change

1. Introduction

The relationship between climate change and agriculture has a long tradition in the scholarly literature, e.g., [1–3]. Additionally, during the last couple of years, the results of climate change, like high temperature and drought, have significantly affected the financial viability of producers [4,5]. To this end, many scholars call for the adoption of new technologies and innovations both as a mean towards environmental improvements but also as a mean towards producers' (or growers') financial stability [6].

Furthermore, mathematical programming farm models have been excessively used to understand farmers' (or growers) production choices, e.g., [7,8]. Among them, linear programming farm models (thereafter, LP-FM) have been used to analyze production plans in the agricultural sector, e.g., [9].

In this study, we are interested in the impact of climate change-induced multiple stress, namely increased heat, draught, and salinity. Particularly, we utilize a simple LP-FM to explore two vital questions. First, how climate change-induced multiple stress will affect the financial viability of Mediterranean greenhouse tomato growers. Second, how the adoption of new technologies and innovations can compensate growers for any profit losses due to climate change-induced multiple stress.

2. Materials and Methods

Our methodology can be divided into the following steps. First, we interviewed 22 greenhouse tomato growers (both in-person and online), where approximately 72.72% of the responders were located in Crete, whereas the remaining ones were located in the region of Peloponnese. The rationale of using Crete as the case study is because Crete, followed by Peloponnese, is the leading region in greenhouse vegetable production in Greece [10].

The second step was to design an LP-FM. Specifically, we assumed a representative greenhouse beef tomato grower whose objective was to choose their annual deci-hectare amount of fertilizers, chemical substances for pest management, the number of plants, and water consumption, such that her annual per deci-hectare gross margin was to be maximized, subject to both technical and financial constraints. The choice of these inputs (decision variables) was selected based on the answers given by the interviewed growers. Also, the upper and lower limits of the constraints were determined by the answers given by the growers.

The third step was to estimate the production coefficients. To do so, an approximated linear production function was used. The result of this estimation was used afterwards to the LP-FM to determine the optimal input use under the 'current situation' (or business-as-usual scenario). These values serve as a comparison between the current situation and our hypothetical scenarios.

The final step was to implement three hypothetical scenarios on the impact of climate change-induced multiple stress on both the production and financial efficiency of a 'conventional' production system: a low, a moderate, and a high impact scenario. In each of these three scenarios, further assumptions were made on the production and financial efficiency of a production system that utilizes new technologies and innovations that exhibit higher tolerance to climate change compared to 'conventional' one.

3. Results

The main results of our analysis can be summarized as follows. First, the adoption of new technologies and innovations can compensate greenhouse tomato growers, even in cases where the production efficiency of these technologies and innovations is close to the 'conventional' one.

Second, cost considerations might be important, especially when the production efficiency of these new technologies and innovations is close to the 'conventional' one.

Third, we did not find any significant difference in input use between 'conventional' production and production that utilizes new technologies and innovations. However, if the grower is constrained to produce a certain level of yield, then the adoption of new technologies and innovations that are more tolerant to climate change is likely to entail environmental improvements in terms of less input use, as well.

Finally, the yield been the produced crops in the latter cases exceeds that under the former one in almost every simulation. This result highlights potential social benefits because the adoption of new technologies and innovations can 'secure' a potential food supply under severe climate change conditions.

4. Conclusions

In this article, we tried to explore whether the adoption of new technologies and innovations can compensate greenhouse tomato growers for their profit losses due to climate change-induced multiple stress. The answer is yes, but the cost of adaptation should also be considered. Importantly, our analysis highlights that the adoption of new technologies and innovations can cover any excess demands for tomato. Thus, it might be down to policymakers to incentivize the transition to sustainable agriculture, especially if 'securing' food supply is their primal objective.

However, some limitations should be spelled out. First, our sample size is small, which may reduce the robustness of our estimated coefficients. Secondly, we gathered information by performing in-person interviews and by email. In most cases, growers did not keep a detail logbook regarding their production activities and the costs associated with them. Thus, our data are likely to exhibit some level of noise. The implication of these two limitations is that we exhibit high p-values, meaning that the estimated coefficients should be interpreted with caution. Finally, we focused our analysis on the identification of only four inputs. However, factors like labor, energy, and electricity consumption could be important as well. Thus, an extension of this study is left as an area for future research.

Author Contributions: Conceptualization, G.N.D. and K.T.; methodology, G.N.D. and K.T.; software, G.N.D.; formal analysis, G.N.D.; resources, D.S. and K.T.; data curation, G.N.D. and D.S.; writing—original draft preparation, G.N.D.; writing—review and editing, G.N.D., K.T. and D.S.; supervision, K.T. and D.S.; project administration, D.S.; funding acquisition, D.S. All authors have read and agreed to the published version of the manuscript.

Funding: This research was supported by PRIMA 2018-11 within the project 'VEGADAPT: Adapting Mediterranean vegetable crops to climate change-induced multiple stress', a Research and Innovation Action funded by the Greek General Secretariat for Research and Innovation (GSRI) and supported by the European Union.

Institutional Review Board Statement: Not applicable.

Informed Consent Statement: Informed consent was obtained from all subjects involved in the study.

Data Availability Statement: The raw data can be found at https://doi.org/10.5281/zenodo.8325730 (accessed on 7 September 2023). Also, we used a slightly modified version of the GAMS code that can be found at https://doi.org/10.5281/zenodo.7024627 (accessed on 26 August 2022).

Acknowledgments: We would like to express our sincere thanks to Dimitrios Kremmydas for his invaluable insights and guidance during the design of the linear programming farm model.

Conflicts of Interest: The authors declare no conflicts of interest. The funders had no role in the design of the study; in the collection, analyses, or interpretation of the data; in the writing of the manuscript; or in the decision to publish the results.

References

1. Aydinalp, C.; Cresser, M.S. The effects of global climate change on agriculture. *Am.-Eurasian J. Agric. Environ. Sci.* **2008**, *3*, 672–676.
2. Kurukulasuriya, P.; Rosenthal, S. Climate change and agriculture: A review of impacts and adaptations. In *Climate Change Series Paper No 91*; World Bank: Washington, DC, USA, 2003.
3. Mendelsohn, R. The impact of climate change on agriculture in developing countries. *J. Nat. Resour. Policy Res.* **2009**, *1*, 5–19. [CrossRef]
4. Dell, M.; Jones, B.F.; Olken, B.A. Temperature shocks and economic growth: Evidence from the last half century. *Am. Econ. J. Macroecon.* **2012**, *4*, 66–95. [CrossRef]
5. Pandey, S.; Bhandari, H. Drought: Economic costs and research implications. In *Drought Frontiers in Rice: Crop Improvement for Increased Rainfed Production*; World Scientific Publishing: Singapore, 2009; pp. 3–17.
6. O'sullivan, C.A.; Bonnett, G.D.; McIntyre, C.L.; Hochman, Z.; Wasson, A.P. Strategies to improve the productivity, product diversity and profitability of urban agriculture. *Agric. Syst.* **2019**, *174*, 133–144. [CrossRef]
7. Kaiser, H.M.; Messer, K.D. *Mathematical Programming for Agricultural, Environmental and Resource Economics*; John Wiley and Sons, Inc.: Hoboken, NJ, USA, 2011.
8. Norton, R.D.; Hazell, P.B. *Mathematical Programming for Economic Analysis in Agriculture*; Macmillan: New York, NY, USA, 1986.
9. Alotaibi, A.; Nadeem, F.A. Review of Applications of Linear Programming to Optimize Agricultural Solutions. *Int. J. Inf. Eng. Electron. Bus.* **2021**, *13*, 11–21. [CrossRef]
10. Savvas, D.; Ropokis, A.; Ntatsi, G.; Kittas, C. Current situation of greenhouse vegetable production in Greece. *Acta Hortic.* **2016**, *1142*, 443–448. [CrossRef]

Disclaimer/Publisher's Note: The statements, opinions and data contained in all publications are solely those of the individual author(s) and contributor(s) and not of MDPI and/or the editor(s). MDPI and/or the editor(s) disclaim responsibility for any injury to people or property resulting from any ideas, methods, instructions or products referred to in the content.

Proceeding Paper

In Search of the Agronomist as Trusted Advisor: A Farmer-Centric Case Study †

Eleni Pappa and Alex Koutsouris *

Department of Agricultural Economics & Rural Development, Agricultural University of Athens, 11855 Athens, Greece; elenip@aua.gr
* Correspondence: koutsouris@aua.gr
† Presented at the 17th International Conference of the Hellenic Association of Agricultural Economists, Thessaloniki, Greece, 2–3 November 2023.

Abstract: Given the interest in the new CAP in advisory services and the Agricultural Knowledge and Innovation System (AKIS), and the importance of trust development between farmers and advisors, in this piece of work we explore the issue of farmers' trust towards their sources of advice. The field research addressed professional farmers who were in contact with agronomist(s) in Ioannina. Overall, 51 farmers were interviewed using a snowball technique. The trust model was utilized to provide important insights about the antecedents of trust towards advisors on the part of farmers, focusing on three elements: ability, benevolence and integrity.

Keywords: key farmers; advisors; antecedents of trust; ability; benevolence; integrity

1. Introduction

The current advisory landscape in Greece is marked by the absence of a structured advisory system as well as a weak and fragmented AKIS [1]. Given the interest in the new CAP in advisory services and AKIS (Reg. (EU) 2021/2115), and the importance of trust development between farmers and advisors [2], we explored farmers' trust towards their sources of advice. We particularly aimed to identify the characteristics of the trusted agronomist with whom farmers would prefer to build advisory relationships. Trust between advisors and farmers has been underlined in the agricultural advisory literature to promote important outcomes such as advice seeking and usage [2] and knowledge exchange [3]. Relevant research in the field is deemed important [4,5]; nevertheless, it is scarce and limited in terms of analytical depth. The trust literature outside the 'agriculture/advisory' field(s) underlines that trust (willingness to depend on another party) is a highly complex and context-specific concept [6]. One of the most influential trust models is that of Mayer et al. [7]. The model argues about the importance of the antecedents of trust, focusing on three elements: ability, benevolence and integrity.

2. Materials and Methods

Our explorative research took place from January to March 2022 in the Prefecture of Ioannina (Epirus Region) which borders Albania and the Ionian sea. Out of its 8 municipalities, 5 municipalities were selected as they cover 86.8% of the cultivated land (with the other 3 being mountainous with much fewer farmers, mainly sheep and goat semi-nomadic breeders). The total cultivated land is 2654 ha., half of which (1345 ha.) is devoted to the cultivation of fodder crops; other important crops are vineyards and potatoes (in 1 out of the 5 municipalities each), corn and tree orchards. The field research addressed professional farmers who were in contact with agronomist(s)-as-advisors. For the research, an aide memoire was used comprising, among others, questions on the characteristics of the farmer and his/her farm, as well as the following open questions: (a) which are the characteristics that an (ideal) agronomist should have to trust him/her as your advisor? (b) what (and

Citation: Pappa, E.; Koutsouris, A. In Search of the Agronomist as Trusted Advisor: A Farmer-Centric Case Study. *Proceedings* **2024**, *94*, 17. https://doi.org/10.3390/proceedings2024094017

Academic Editor: Eleni Theodoropoulou

Published: 23 January 2024

Copyright: © 2024 by the authors. Licensee MDPI, Basel, Switzerland. This article is an open access article distributed under the terms and conditions of the Creative Commons Attribution (CC BY) license (https://creativecommons.org/licenses/by/4.0/).

how) would you like the trusted agronomist-advisor to (be able to) do for/with you? and (c) which would be the characteristics and actions of an agronomist whom you would not trust becoming your advisor? Overall, 51 farmers were interviewed following a snowball technique for each of the 5 municipalities. The interviews were recorded and transcribed to produce computer-generated documents using Google Docs. The research material comprised 570 pages analyzed per topic (exploratory analysis) [8,9]; some of the topics were based on the literature review, while others emerged from the primary material.

3. Results

Ability. The knowledge of the advisor is a given since all agronomists are university graduates. Nevertheless, trusted advisors should have in-depth knowledge and be continuously seeking to update it; their knowledge has to 'surpass' that of an experienced cultivator. Additionally, a trusted advisor should be an expert on farmers' specific crops. Advisors are judged by farmers according to their 'local level' and long-term experience; theoretical knowledge has to be integrated with practice. Specialization and experience are achieved, for the farmers, through farm visits or/and being cultivators themselves.

Ability is 'mediated' through the specific actions and behaviors of the agronomists that signal their ability, such as communication mode, roles and working methods. Of paramount importance is communication. Advisors' analytical and substantiated advice and answers, along with the ability to respond to complex questions, demonstrate their competence. Additionally, the advisors should be confident about their advice.

Concerning their role and working methods, agronomists should actively collect farmers' and regional data and base their advice on this analysis (including local experimentation/trials). The provision of concrete and stepwise advice based on data shows technical capacity and knowledge. The presence of the agronomist on the farm is the major 'criterion' which makes a good advisor. During field visits, agronomists can better demonstrate their knowledge to the farmer and thus farmers can better understand their competence. Finally, the effectiveness of the plan and/or suggested solutions is indicative of their ability.

Benevolence. With regard to 'benevolence', farmers underlined that a trusted advisor is one who strives to ensure farmers' interests. In turn, the advisor should not be entangled in any kind of interests (i.e., private companies) which might work against them. The above clearly favor the existence of a public extension service. Additionally, a trusted advisor must show his/her interest in farmers, i.e., to undertake concrete actions which benefit them. Finally, the advisor should have empathy and respect for farmers' efforts.

Benevolence is also 'mediated' through specific actions and behaviors. Advisors must initiate the establishment of communication, thus manifesting their interest in farmers. Additionally, the advisor should put questions to farmers and visit their farm to see the results of the implementation of the provided advice. Farmers appreciate an advisor's endeavors to provide tailor-made advice and devote substantial time to discuss with each farmer. The above underlines the advisor's interest in helping them. Farmers insist that advisors should be on the farm, signaling their interest to them.

Integrity. With regard to 'Integrity', farmers underlined that the advisor must treat them fairly, servicing all farmers on an equal footing and irrespective of their farm size, locality, etc. Moreover, advisors must be honest, transparent and accountable in their interventions. Additionally, integrity refers to reliable and predictable behavior in conjunction to the farmer's needs and values.

Integrity is 'mediated' through specific actions and behaviors. Trusted advisors are honest and transparent through the provision of information to farmers about the expected results of their interventions. Advisors should also be frank about a farmer's mistakes and dare to intervene when they notice them. Finally, advisors should transfer their knowledge and explain their recommendations to the farmer. For farmers, advisors' consistency and predictable behavior is mainly shown through farm visits. Farm visits must be frequent, especially in the critical moments of the cultivation period. In the same vein, the advisor must be available and promptly provide assistance, especially in case of crises.

4. Discussion and Conclusions

Our research in Ioannina Prefecture confirms, in the first place, previous findings in the extension/advisory literature, albeit in a more systematic way. Moreover, our research revealed mediators of trust, i.e., go-between 'variables' explaining the process through which two 'variables' are related. This piece of work in progress will allow for the better understanding of the degree to which farmers trust various types of agronomists (public, private, company representatives) and other actors-as-advisors, and thus of their (current and potential) role(s) in farmers' micro-AKIS [5]. It may also assist in the design of effective Innovation Support Services (ISS) in Greece. Furthermore, it may inform the Higher Educational Institutes (agronomic universities) curriculum in terms of several issues concerning future advisors' skills and approaches. It may be of interest to find out similarities or dissimilarities with farmers' views elsewhere and under different AKIS. Finally, this work aims to trigger more nuanced research on the topic of trust in advisory relationships. A farmer-centric contribution to the theory of trust is needed vis à vis the revival of the interest (in both research and policy orientations) for (plural but also inclusive and impartial) advisory services.

Author Contributions: Conceptualization E.P., methodology E.P., formal analysis E.P., Data curation A.K., writing—original draft preparation E.P. and A.K., writing—review and editing E.P and A.K., funding acquisition A.K. All authors have read and agreed to the published version of the manuscript.

Funding: This research received no external funding.

Institutional Review Board Statement: Not applicable.

Informed Consent Statement: Informed consent was obtained from all subjects involved in the study.

Data Availability Statement: The data presented in this study are available on request from the corresponding author. The data are not publicly available due to privacy restrictions.

Conflicts of Interest: The authors declare no conflicts of interest.

References

1. Koutsouris, A.; Zarokosta, E. Supporting bottom-up innovative initiatives throughout the Spiral of Innovations: Lessons from rural Greece. *J. Rural Stud.* **2020**, *73*, 176–185. [CrossRef]
2. Hilkens, A.; Reid, J.; Klerkx, L.; Gray, D. Money talk: How relations between farmers and advisors around financial management are shaped. *J. Rural Stud.* **2018**, *63*, 83–95. [CrossRef]
3. Hammersley, C.; Richardson, N.; Meredith, D.; Carroll, P.; McNamara, J.G. Supporting farmer wellbeing: Exploring a potential role for advisors. *J. Agric. Educ. Ext.* **2023**, *29*, 511–538. [CrossRef]
4. Klerkx, L. Advisory services and transformation, plurality and disruption of agriculture and food systems: Towards a new research agenda for agricultural education and extension studies. *J. Agric. Educ. Ext* **2020**, *26*, 131–140. [CrossRef]
5. Sutherland, L.A.; Labarthe, P. Introducing 'microAKIS': A farmer-centric approach to understanding the contribution of advice to agricultural innovation. *J. Agric. Educ. Ext.* **2022**, *28*, 525–547. [CrossRef]
6. Ezezika, O.C.; Oh, J. What is trust?: Perspectives from farmers and other experts in the field of agriculture in Africa. *Agric. Food Secur.* **2012**, *1*, S1. [CrossRef]
7. Mayer, R.; Davis, J.; Schoorman, F.D. An Integrative Model of Organizational Trust. *Acad. Manag. Rev.* **1995**, *20*, 709–734. [CrossRef]
8. Sarantakos, S. *Social Research*, 3rd ed.; Palgrave Macmillan: London, UK, 2005.
9. Gibbs, G.R. Thematic coding and categorizing. *Anal. Qual. Data* **2007**, *703*, 38–56.

Disclaimer/Publisher's Note: The statements, opinions and data contained in all publications are solely those of the individual author(s) and contributor(s) and not of MDPI and/or the editor(s). MDPI and/or the editor(s) disclaim responsibility for any injury to people or property resulting from any ideas, methods, instructions or products referred to in the content.

Proceeding Paper

Factors Influencing Consumer Receptivity to Sustainable Packaging: A Probit Regression Analysis [†]

Georgia S. Papoutsi [1,*] and Elena Kourtesi [2]

[1] Agricultural Economics Research Institute, Hellenic Agricultural Organization-DIMITRA, 11528 Athens, Greece
[2] MBA Food & Agribusiness Programme, Agricultural University of Athens, 11855 Athens, Greece; elenakourtesi@hotmail.com
* Correspondence: gpapoutsi@elgo.gr
[†] Presented at the 17th International Conference of the Hellenic Association of Agricultural Economists, Thessaloniki, Greece, 2–3 November 2023.

Abstract: The objective of this study is to investigate whether specific socioeconomic and attitudinal factors impact consumer receptivity to sustainable food packaging, with a particular focus on edible cups. A total of 1028 respondents completed an online questionnaire, and the data were analyzed using descriptive analysis and binary probit regression. The results reveal that demographic factors, such as household size and household economic position, have a positive influence on consumers' intention to consume edible packages. Additionally, attitudinal factors were found to be significant, with food technology neophobia negatively affecting consumers' willingness to try edible cups, while beliefs about the development of the sustainable packaging industry positively influence intention.

Keywords: sustainable packaging; edible cups; consumption intention; food technology neophobia scale

1. Introduction

The increasing concern for environmental sustainability and the need to reduce plastic waste have prompted researchers and food industry professionals to explore sustainable food packaging options. Sustainable food packaging aims to minimize the negative impact on the environment throughout the packaging's life cycle, including its sourcing, manufacturing, use, and disposal. As the demand for eco-friendly packaging grows, it becomes crucial to understand the factors that influence consumers' receptivity towards sustainable packaging. This paper aims to investigate the various demographic and attitudinal factors that play a role in shaping consumers' intention to adopt and consume environmental-friendly packaging options, such as edible cups, using a binary probit regression analysis.

Previous studies show that consumers have positive attitudes towards biodegradable materials [1] and they highlight positive attitudes towards sustainable packaging alternatives, including edible packaging [2]. Moreover, the literature indicates a positive willingness to pay for more environmentally friendly food containers made from biodegradable materials and sustainable packaging, in general [3,4]. Furthermore, consumer preferences towards eco-friendly and edible packaging may vary based on individual characteristics such as gender, age, education, and household size [5] as well as personal norms, attitudes, and environmental concerns [6].

The findings of this study highlight the importance of considering both demographic and attitudinal factors in understanding consumers' intention to adopt and consume edible cups as a sustainable packaging option, emphasizing the importance of targeted strategies. In the next section, we present our data collection and questionnaire design. We then present our results, and conclude with a discussion of our findings in the last section.

2. Materials and Methods

Survey data were collected electronically using an appropriately structured questionnaire. The questionnaire was created on the Google Forms platform, due to the recent COVID-19 pandemic, and was sent to a random sample of consumers during November 2020. The questionnaire was successfully completed by a total of 1028 consumers.

We created a questionnaire where at the beginning subjects were asked whether they knew what edible food packaging is. After this, all subjects, regardless of their response to the previous question, received information related to sustainable and edible packaging, and they were asked whether they are willing to consume (yes or no) an edible coffee cup produced from natural grain products. Besides the standard demographic information (age, gender, education, household size, and income level), the questionnaire also assessed the respondents' fear of novel food technology that was evaluated by applying the Food Technology Neophobia (FTN) scale [7].

We applied a binary probit model to reveal the demographic and attitudinal characteristics of those who are more likely to consume edible cups. The dependent variable for the probit analyses was the binary choice responses to the edible cup consumption question, where a positive answer was taken as 1 and a negative as 0.

3. Results

Beginning with the descriptive analysis, we initially assess the profile of our sample by considering a variety of observable characteristics of the subjects. The sample predominantly consists of younger participants, with a majority being female. Furthermore, a significant portion of the sample is pursuing a university degree. The reported household economic position falls within the average range. Lastly, many of the participants are familiar with the concept of edible packaging and hold the belief that the sustainable packaging industry will experience growth in the near future.

As for the FTN scale, the measured scores ranged from 14 to 91, with an average score of 45.27. The higher the score on this scale, the more likely it is the person to be afraid to consume foods produced by novel food technologies. Figure 1 provides additional insight, showing that the scores are clustered around and below the median. This suggests that a significant portion of the sample can be categorized as food technology neophiliac, indicating a higher tendency towards openness and acceptance of new food technologies.

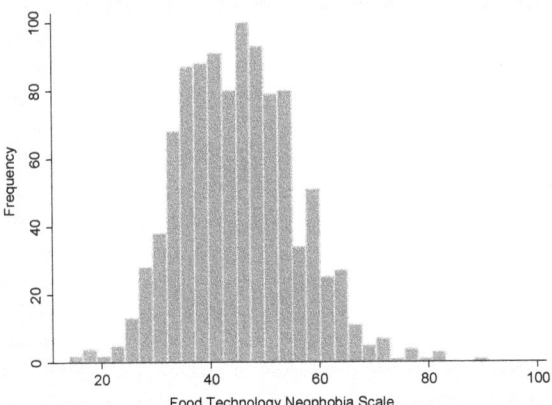

Figure 1. Distribution of WTP responses.

Table 1 presents the results from the estimated probit regression analysis. The likelihood ratio chi-square value of 56.55, with a p-value of zero, indicates that our model is statistically significant, suggesting that it fits significantly better than a model with no predictors. Regarding demographic characteristics, age, gender, and education do not have systematic effects on consumers' intention to taste edible cups made from natural grain products. However, household size has a positive and statistically significant effect, suggesting that consumers belonging to larger families are more likely to have the intention to consume edible cups. Additionally, a good household economic position increases the probability of consuming edible cups compared to a very bad economic position.

Regarding attitudinal factors, both food technology neophobia and beliefs about the development of sustainable packaging industry play significant roles in shaping consumers' behavior and their acceptance of innovative and sustainable packaging options, like edible cups. Specifically, for each one-unit increase in the Food Technology Neophobia scale, the z-score decreases by 0.01, indicating that consumers who are more afraid of new technologies in food production are less likely to try an edible cup. On the other hand, consumers who believe in the growth and advancement of the sustainable packaging industry in the near future are more likely to have the intention to consume edible cups.

Table 1. Binary probit model estimates.

	Coefficients	Standard Errors
Constant	0.27	(0.74)
Age	−0.01	(0.01)
Gender	−0.1	(0.14)
Household size	0.17 **	(0.07)
Household's economic position		
Bad	0.35	(0.37)
Average	0.45	(0.33)
Good	0.71 **	(0.35)
Very good	0.65	(0.45)
Education level		
University student	0.45	(0.41)
University graduate	0.32	(0.42)
Master/Ph.D. student	0.68	(0.43)
Master/Ph.D. graduate	0.25	(0.44)
Knows about edible food packaging	−0.11	(0.14)
FTN scale	−0.01 **	(0.01)
Development of sustainable packaging industry in the near future	0.8 ***	(0.15)
N	1028	
Chi-square	56.55	
Prob > chi2	0.00	

Notes: ** < 0.05, *** < 0.01.

4. Discussion and Conclusions

This study sheds light on the complex interplay between demographic characteristics and attitudinal factors in shaping consumers' intention to consume a sustainable packaging option. The findings have important implications for businesses, policymakers, and marketers seeking to promote sustainable packaging practices, enabling them to identify consumer segments willing to consume edible cups. By recognizing the significance of individual characteristics and addressing consumers' concerns related to food technology neophobia, businesses can better target their marketing efforts and design strategies to increase the adoption of edible cups.

While this study contributes to the understanding of consumer behavior regarding edible cups, it is important to note some limitations. The sample used is not representative, and the design could be further extended to include other sustainable packaging options. Future studies could consider factors such as cultural influences, pricing strategies, and sensory experiences to gain a more comprehensive understanding of consumers' acceptance and adoption of edible cups. Despite these limitations, our findings can be considered as a first perspective on consumer receptivity for edible cups in Greece, which may foster future studies in this field.

Author Contributions: Conceptualization, G.S.P. and E.K.; methodology, G.S.P. and E.K.; software, G.S.P. and E.K.; formal analysis, G.S.P. and E.K.; data curation, E.K.; writing—original draft preparation, G.S.P.; writing—review and editing, G.S.P. and E.K.; supervision, G.S.P. All authors have read and agreed to the published version of the manuscript.

Funding: This research received no external funding.

Institutional Review Board Statement: This study did not require ethical approval.

Informed Consent Statement: All subjects gave their informed consent for inclusion before they participated in the study.

Data Availability Statement: The data that support the findings of this study and the Greek questionnaire are available upon request from the authors.

Conflicts of Interest: The authors declare no conflicts of interest.

References

1. Orset, C.; Barret, N.; Lemaire, A. How consumers of plastic water bottles are responding to environmental policies? *Waste Manag.* **2017**, *61*, 13–27. [CrossRef] [PubMed]
2. Sonti, S. Consumer Perception and Application of Edible Coatings on Fresh-Cut Fruits and Vegetables. Master's Theses, Louisiana State University and Agricultural & Mechanical College, Baton Rouge, LA, USA, 2003.
3. Barnes, M.; Chan-Halbrendt, C.; Zhang, Q.; Abejon, N. Consumer Preference and Willingness to Pay for Non-Plastic Food Containers in Honolulu, USA. *J. Environ. Prot.* **2011**, *2*, 1264–1273. [CrossRef]
4. Boz, Z.; Korhonen, V.; Koelsch Sand, C. Consumer Considerations for the Implementation of Sustainable Packaging: A Review. *Sustainability* **2020**, *12*, 2192. [CrossRef]
5. Laroche, M.; Bergeron, J.; Barbaro-Forleo, G. Targeting Consumers Who Are Willing to Pay More for Environmentally Friendly Products. *J. Consum. Mark.* **2001**, *18*, 503–520. [CrossRef]
6. Prakash, G.; Pathak, P. Intention to buy eco-friendly packaged products among young consumers of India: A study on developing nation. *J. Clean. Prod.* **2017**, *141*, 385–393. [CrossRef]
7. Cox, D.; Evans, G. Construction and validation of a psychometric scale to measure consumers' fears of novel food technologies: The food technology neophobia scale. *Food Qual. Prefer.* **2008**, *19*, 704–710. [CrossRef]

Disclaimer/Publisher's Note: The statements, opinions and data contained in all publications are solely those of the individual author(s) and contributor(s) and not of MDPI and/or the editor(s). MDPI and/or the editor(s) disclaim responsibility for any injury to people or property resulting from any ideas, methods, instructions or products referred to in the content.

Proceeding Paper

The Implementation of Innovation and the Attitudes of Farmers towards Advisory Services: The Case of Western Macedonia, Greece [†]

Georgios Kountios [1,*], Ourania Notta [1], Nikolaos Kratimenos [1] and Ioannis Chatzis [2]

[1] Department of Agricultural Economics and Entrepreneurship, School of Agriculture, International Hellenic University, 57400 Sindos, Greece; ournotta@ihu.gr (O.N.); nikos.kratimenos@gmail.com (N.K.)
[2] Payment and Control Agency for Guidance and Guarantee Community Aid, Domokou 5, 10445 Athens, Greece; ioannis.chatzis@opekepe.gr
* Correspondence: gkountios@ihu.gr
[†] Presented at the 17th International Conference of the Hellenic Association of Agricultural Economists, Thessaloniki, Greece, 2–3 November 2023.

Abstract: This paper investigates the implementation of innovation and the advisory needs of farmers in agricultural holdings in Western Macedonia. The research carried out was divided into two parts. Initially, we investigated how the programs under analysis related to innovation in the agricultural sector, specifically with respect to the improvement plans that were implemented in the region. Specifically, parameters such as age, gender, place of residence, and the types and number of investments were examined. The analysis revealed farmers' characteristics and the types of their investments, in addition to their degrees of adoption of innovation and their attitudes towards innovation. We also tried to clarify the reasons leading farmers toward innovation and the importance of consulting services in their orientation towards innovative ideas.

Keywords: agricultural knowledge and innovation systems; productive investments for the modernization of agricultural holdings; Western Macedonia; advisory services; sustainability

1. Introduction

Innovation is considered the heart of value creation for small businesses [1] and a key strategy for improving productivity and promoting the sustainable use of resources, and it is also a resilient tool for rural development (OECD, 2006–2013). Ref. [2] developed agricultural innovation indicators to measure levels of innovation in the agricultural sector. The overall research findings showed that the effort to innovate varies among agricultural systems. Factors such as farm size and intensity as well as access to credit and agricultural training appear to promote innovation. Profitable agricultural businesses are generally more innovative [3]. Conversely, increasing age and working outside the agricultural sector seem to hinder innovation [4]. Internal variables such as flexibility and adaptation have the greatest influence on an agricultural enterprise's propensity to innovate. Training levels, the exchange of knowledge with research centers, and firm location have an impact on the propensity to innovate [5], while technological progress and resource sufficiency improve the quality of life of producers [6].

One the other hand, the agricultural sector faces challenges posed by climate change, soil erosion, and biodiversity loss, as well as changing consumer preferences for food and concerns about how it is produced [7]. Sustainability in agriculture has a positive impact on opening markets and ensures that farmers can set affordable prices for the products they sell and receive fair profits [8,9]. Productive investments for the modernization of agricultural holdings are an important tool for supporting the agricultural sector with investments that contribute to increasing competitiveness, the use of renewable energy

sources (RES), and environmental protection. Western Macedonia constitutes an important strength of Greece in terms of the production of agricultural products, even under difficult conditions, due to its soil and climatic characteristics.

2. Methods

After reviewing the Greek and foreign literature, a questionnaire was created and used as a tool to collect primary data on producers' views on the investment plans concerning their joining of the Regional Unit of Grevena. The investments that they proceeded to make were funded by the AGRICULTURAL DEVELOPMENT PROGRAM (2014–2022) and more specific for Sub-Measure 4.1.1 (concerning the implementation of investments that contribute to the competitiveness of a farm) and Sub-measure 4.1.3 (concerning the implementation of investments that contribute to the use of Renewable Energy Sources (RES) and environmental protection).

Some of the sections that the questionnaire included concerned demographic characteristics such as gender, age, marital status, and education level and factors that lead to innovation related to consulting services. Additionally, there was a section regarding farmers' attitudes toward innovation as well as a specific section regarding the degree of adoption of innovative technologies. Special reference was made to the barriers and reasons preventing a farmer from adopting technological innovation. At the end of the interview, the interviewees were asked about the main factor that prevented them from participating in funding programs.

The starting point of factor analysis is to explore and identify the underlying dimensions that lie behind the original variables in a set. In the case of our research, we studied 29 variables. Specifically, the innovation attitude, learning orientation, and market orientation factors consist of 6 subfactors, while the consulting services factor consists of 11.

The factor analysis was performed using the IBM SPSS statistics data editor program.

The variables were grouped into 5 factors, and they concern 67.7% (cumulative %) of the sample. According to the data quality indicators of SPSS, we noted that the sample can be considered adequate, as the value of the KMO index was equal to 0.826 > 0.50, and the evaluation of the correlations between the variables allowed for factorial analysis, as Bartlett's Test of Sphericity had a value of $p = 0.00 < 0.05$

The extraction of the factors was conducted using principal component analysis, with the aim of studying all existing variation in order to extract the largest percentage of variation from the fewest possible factors. To extract the factors, the corresponding research was based on the Guttman–Kaiser criterion, according to which selectable factors are those with an eigenvalue greater than 1. The rotation of the factors is necessary as it allows for the easier interpretation of the factors that emerge from the analysis.

The first factor, 'Consulting services', explains 36.45% of the total variance. It is characterized by high correlation coefficients: (a) when a technical product does not perform well, I analyze the reasons for the failure; (b) I receive technical advice from my suppliers; and (c) I consult private geotechnical offices in the area so that I can implement the best practices.

The second factor, 'Innovation Attitude', which explains 13.9% of the total variance, shows a high correlation with the following parameters: (a) innovations improve the results of my farm, (b) innovation is worthwhile, (c) adopting innovation is a useful decision, (d) I am motivated to innovate, and (e) I am informed by private geotechnical offices about research and innovation programs.

The third factor is called market orientation and explains 6.57% of the total variance. The parameters that are highly correlated are as follows: (1) my pursuit of producing cheaper products gives me advantages over other farms; (2) I look for new customers every year; (3) customer satisfaction is the main goal of my farm; (4) I am interested in my customers' preferences for product quality and follow them during the production process; (5) customers direct me with respect to which varieties to grow; and (6) my pursuit of producing high-quality products gives me advantages over other holdings.

The fourth factor, 'Learning Orientation', explains 6% of the total variance, and the parameters for which it shows a high correlation are as follows: (1) I usually consult universities and research centers, (2) I like to read magazines or other media from reputable bodies about new crops or methods that I could introduce in my business, and (3) I take part in research and innovation projects supported by public or private bodies.

Finally, the fifth factor, 'Consulting services from the public', explains 4.78% of the total variance and shows a high correlation with the following parameters: (a) I am informed by public geotechnical services about research and innovation programs and (b) I consult public geotechnical services in this region to apply best practices.

3. Results and Discussion

The farmers that participated in a funding program (sub-measure 4.1.1 monopolized the interest of farmers, as only 5% participated in sub-measure 4.1.3) were mainly people between the ages of 35 and 60 (68%) and mostly men (67%).

Regarding the types of investments that were supported, the farmers were oriented toward the acquisition of mechanical equipment (74.5%), meaning that they had either not yet reached the level required to be able to use modern innovative technologies or practices such as precision agriculture, the use of GPS, drones, weather stations, etc., or that they might not have trusted such innovations or even known of their existence.

Regarding the attitudes of the farmers towards innovation, the results showed that most of them considered innovation useful, finding that it improves the economic outcome of a farm and simultaneously admiring people who proceed to adopt an innovation.

The research also showed that most of the sample was accustomed to being informed by magazines and other media (websites and social media) of reputable organizations while enjoying participating in exhibitions and seminars to discover new ideas and share experiences with colleagues.

The processing of the farmers' answers regarding consulting services showed that most of them had a positive opinion of public extension services and the Common Agricultural Policy (CAP). These factors allow agricultural holdings to innovate. In the contrary, farmers do not consult universities and research centers, nor do they participate in research and innovation projects. It also seems that the farmers in the sample were informed about research and innovation programs as well as agricultural practices that they could adopt mainly from the private sector and not from public geotechnical services.

Finally, it is worth mentioning that, according to the sample, the main barriers to innovation adoption are purchase costs, lack of financing, lack of credit, and lack of equity.

In closing this section, it is worth mentioning that in the factor analysis, we studied 29 variables, which were finally grouped into five factors that concern 67% of the sample. The factors resulting from the processing of the data are shown below:

1. Consulting services;
2. Innovation attitude;
3. Market orientation;
4. Orientation toward learning;
5. Consulting services from the state;

4. Conclusions

There is no doubt that innovation is considered the heart of value creation for small businesses and a key strategy for improving productivity and promoting the sustainable use of resources. Farmers need valid access to knowledge and information as well as training and education. This is included in the framework of the European Union strategies and should be facilitated by policies wherever advisory services play a very important role.

Investments in agricultural holdings are important tools for supporting the agricultural sector, contributing to increasing competitiveness, the use of RES, and environmental protection.

Western Macedonia is an important strength of Greece in terms of the production of agricultural products, even under difficult conditions, due to its soil and climatic characteristics. Innovation is perhaps the answer to these difficulties, and for this reason, it was chosen as an area of research.

In conclusion, farmers of Western Macedonia turned to traditional investments and showed no specific interest in new innovative technologies. Nevertheless, the majority had a very positive opinion and were in favor of innovation.

Perhaps this means that they have not yet reached the level required to be able to use modern innovative technologies or practices such as precision agriculture, GPS, drones, weather stations, etc. On the other hand, it may mean that they do not trust such innovations or even know of their existence. The only thing that is certain is that their attitude towards innovation should be investigated further.

Author Contributions: Conceptualization and methodology, N.K. and O.N.; investigation, data curation, writing—original draft preparation N.K.; writing—review and editing, G.K.; visualization, I.C.; supervision, O.N. All authors have read and agreed to the published version of the manuscript.

Funding: This research received no external funding.

Informed Consent Statement: Informed consent was obtained from all subjects involved in the study.

Data Availability Statement: Data are contained within the article.

Conflicts of Interest: Author Ioannis Chatzis was employed by the Payment and Control Agency for Guidance and Guarantee Community Aid. The remaining authors declare that the research was conducted in the absence of any commercial or financial relationships that could be construed as a potential conflict of interest.

References

1. Lussak, A.; Abdurachman, E.; Gautama, I.; Setiowati, R. The influence of financial performance and innovation of services and products on the survival of small businesses in food and beverage in the Jakarta city with mediation of operational improvement. *Manag. Sci. Lett.* **2020**, *10*, 463–468. [CrossRef]
2. Renwick, A.; Läpple, D.; O'Malley, A.; Thorne, F. Innovation in the Irish agrifood sector. University College Dublin. 2014. Available online: http://www.ucd.ie/t4cms/BOI_Innovation_report.pdf (accessed on 10 September 2023).
3. Alam, K.; Adeyinka, A.A. Does innovation stimulate performance? The case of small and medium enterprises in regional Australia. *Aust. Econ. Pap.* **2021**, *60*, 496–519.
4. Läpple, D.; Thorne, F. The role of innovation in farm economic sustainability: Generalised propensity score evidence from Irish dairy farms. *J. Agric. Econ.* **2019**, *70*, 178–197. [CrossRef]
5. Castillo-Valero, J.S.; Villanueva, E.C.; García-Cortijo, M.C. Regional reputation as the price premium: Estimation of a hedonic model for the wines of Castile-La Mancha. *Rev. De La Fac. De Cienc. Agrar. UNCuyo* **2018**, *50*, 293–310.
6. Jurjević, Ž; Bogićević, I.; Đokić, D.; Matkovski, B. Information technology as a factor of sustainable development of Serbian agriculture. *Strateg. Manag.* **2019**, *24*, 41–46. [CrossRef]
7. Knickel, K.; Ashkenazy, A.; Chebach, T.C.; Parrot, N. Agricultural modernization and sustainable agriculture: Contradictions and complementarities. *Int. J. Agric. Sustain.* **2017**, *15*, 575–592. [CrossRef]
8. Kountios, G. The role of agricultural consultants and precision agriculture in the adoption of good agricultural practices and sustainable water management. *Int. J. Sustain. Agric. Manag. Inform.* **2022**, *8*, 144–155. [CrossRef]
9. Kovách, I.; Megyesi, B.G.; Bai, A.; Balogh, P. Sustainability and Agricultural Regeneration in Hungarian Agriculture. *Sustainability* **2022**, *14*, 969. [CrossRef]

Disclaimer/Publisher's Note: The statements, opinions and data contained in all publications are solely those of the individual author(s) and contributor(s) and not of MDPI and/or the editor(s). MDPI and/or the editor(s) disclaim responsibility for any injury to people or property resulting from any ideas, methods, instructions or products referred to in the content.

Proceeding Paper

Online Sales Promotion of Geographical Indication Products: The Case of Evia PDO Dried Figs [†]

Argyrios Georgilas [1] and Zacharoula Andreopoulou [2,*]

[1] Department of Forestry and Natural Environment, Aristotle University of Thessaloniki, 54124 Thessaloniki, Greece; ageorgila@for.auth.gr

[2] Department of Forestry and Natural Environment, Lab. of Forest Informatics, Aristotle University of Thessaloniki, P.O. Box 247, 54124 Thessaloniki, Greece

* Correspondence: randreop@for.auth.gr; Tel.: +30-2310992714

[†] Presented at the 17th International Conference of the Hellenic Association of Agricultural Economists, Thessaloniki, Greece, 2–3 November 2023.

Abstract: Through a literature review and secondary research on the internet, the aim of this paper is to investigate the dynamics of the internet presence of the producers, packers, traders, and online sellers of PDO dried figs in the Kymi and Taxiarchi regions of the Evia Regional Unit, Greece. With the use of big data, an attempt is made to identify internet users' preferences concerning the dried figs. Suggestions for improved internet presence that will match demand with supply can be subsidised from EU regional development funds and contribute to the increase in internet sales of PDO Evia dried figs.

Keywords: geographical indication products; online sales; PDO; dried figs; Evia

Citation: Georgilas, A.; Andreopoulou, Z. Online Sales Promotion of Geographical Indication Products: The Case of Evia PDO Dried Figs. *Proceedings* **2024**, *94*, 20. https://doi.org/10.3390/proceedings2024094020

Academic Editor: Eleni Theodoropoulou

Published: 23 January 2024

Copyright: © 2024 by the authors. Licensee MDPI, Basel, Switzerland. This article is an open access article distributed under the terms and conditions of the Creative Commons Attribution (CC BY) license (https://creativecommons.org/licenses/by/4.0/).

1. Introduction

Since 1992, the initiative providing the legislative framework for the certification of geographical indication products in the European Union has been launched and is continually improving. By the beginning of 2021, 3306 products with a geographical indication had been registered in the European Union, with most of them relating to countries in the European south. GI-certified products have more than doubled the added value compared to the same products without GI certification, which contributes to the improvement in the development of rural areas of the European Union [1].

The internet plays an important role in sales promotion. The online marketing of products and services is an essential tool for rural development. Various methods have been developed to evaluate the online presence of businesses, considering various criteria [2]. The integration of information technologies and online marketing practices by Greek agribusinesses shows continuous growth [3].

The recognition of the importance of the production, promotion, and protection of products with a geographical indication is among the priorities of the European Union for the period of 2021–2027, as mentioned in the proposal (EC, 2022/0089 COD) of the European Commission [4]. Protecting GI products at the online sales level has low success rates and requires additional efforts and initiatives (Figure 1).

An attempt is made through this research to evaluate and optimize the integration of the e-commerce of producers, packers, traders, and online sellers of the PDO dried figs produced in the regional unity of Evia, at the areas of Kymi and Taxiarchis. To match supply with demand and promote online sales, trends were explored in consumer searches on the internet and digital social networks for GI products.

	Completely agree	Somewhat agree	Neutral	Somewhat disagree	Completely disagree
Production stage	60%	36%	4%	0%	0%
Processor stage	52%	48%	0%	0%	0%
Wholesale stage	22%	48%	26%	4%	0%
Retailer stage	33%	50%	13%	4%	0%
Online sales	23%	32%	23%	18%	5%

Figure 1. Effectiveness of controls on GIs/TSGs. Source: Evaluation support study on GI/TSG [5].

2. Materials and Methods

The online sales promotion of PDO Evia figs was investigated through a secondary research approach to evaluate the online presence of the 10 companies approved by ELGO DIMITRA (as of September 2022) as beneficiaries of the PDO indication for Evia dried figs production, packaging, and marketing. Using keywords for PDO Evia dried figs in English and Greek, 18 additional online sellers were spotted on the first page of each search at Google Search engine and were added to the evaluation [6].

The detailed assessment of the online presence of businesses was carried out by checking the existence (grade 1) or not (grade 0) of 30 specific characteristics ($X1$ to $X30$) [7,8]. The 30 characteristics are: $X1$ = More than two languages; $X2$ = Information about the products and services; $X3$ = Contact information; $X4$ = Information about the area; $X5$ = Digital Map; $X6$ = Audiovisual material; $X7$ = Live Web Camera; $X8$ = Search Engine; $X9$ = Site Map; $X10$ = News feedback; $X11$ = Online Survey; $X12$ = Online Contact Form; $X13$ = Weather Forecast; $X14$ = Traffic meter app; $X15$ = Frequently asked questions; $X16$ = Useful Links; $X17$ = General Information; $X18$ = Available files for download; $X19$ = Date/Time app; $X20$ = Events calendar; $X21$ = Calendar of Holidays and nominal holidays; $X22$ = Share buttons; $X23$ = Social Media Profile; $X24$ = Forum; $X25$ = Relative to products information; $X26$ = Third party ads; $X27$ = Newsletter; $X28$ = RSS feed; $X29$ = Member registration; $X30$ = Customization and security.

In the second stage of the research, the Google Trends tool was used to explore user search statistics on Google Search Engine and YouTube for the keywords "Figs", "Dried Figs", and "PDO". Google Trends is used in scientific surveys for big data research [9]. The generated results do not include searches performed within websites.

3. Results

The producers, packers, traders, and online sellers of Evia PDO dried figs that were spotted in Google Search and were concluded at the assessment were are follows: Agricultural Cooperative of Taxiarchis; Deli Carpous; Oxilipro; Cuma; Sykakymis; Kumilio; Wikifarmer producer; Farmasarli; Askada; Food we love; Elliniko.ch; Xmesazontes producer; Go healthy farmacy; Think bio store; Nomeefoods; Gr-ocery; Fromthenatureshop; Brinkys organic groceries; Amalikerasmata; Foodtrails; Terrapura; Elenianna; Thefoodmarket; Ubuy.ci; Mediterranean gourmet; Grecian purveyor; Lazada.

A summary of the evaluation for each one of the 30 characteristics examined is shown below (Figure 2). The numbers vary between 0 (absent in all sites) to 28 (present in all sites).

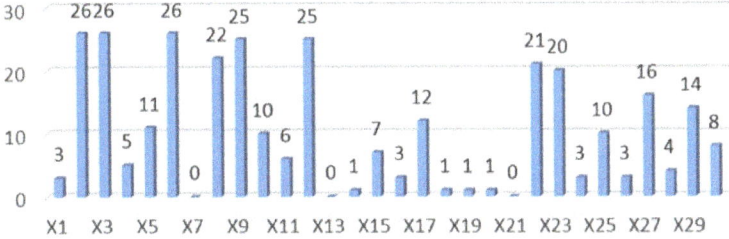

Figure 2. The total of 30 assessment features identified across the 28 sites [1].

Provision of information about the products offered (X2), the availability of photographic and audio-visual material (X6), the ease of navigation (X9), the provision of contact information (X3), and the availability of online contact forms (X12) were the dominant features identified in 89–93% of the websites (Figure 2).

Google Trend results for the keyword "figs" revealed a peak in the searches of the users between August and September, shortly after the harvesting period of the figs (Figure 3).

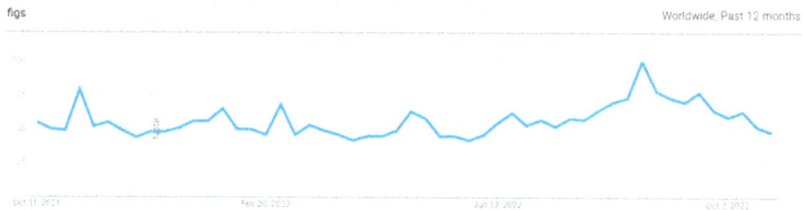

Figure 3. Search for the term "figs" worldwide, November 2021–October 2022 [1].

Top searches in Google Search and YouTube are related to dried figs' health benefits and dried figs recipes (Figure 4).

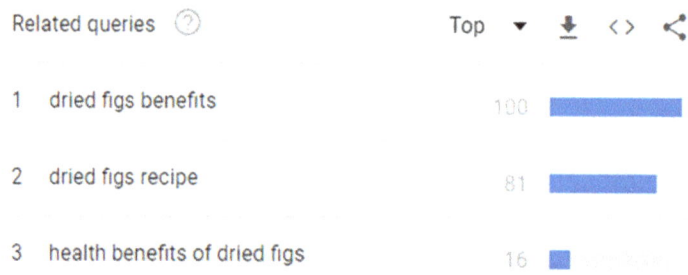

Figure 4. Searches related to dried figs between November 2021–October 2022 [1].

4. Discussion

Only 3 out of 28 sites use more than two languages (X1), a feature that can increase site traffic and boost online sales worldwide. Useful links (X16) can increase site traffic and improve search engine optimization. Available files for download (X18), such as recipes, nutrition value information, and dietary benefits of figs, can boost sales by matching demand with supply. Company profiles on social media and "share" buttons on websites (X22 and 23) have become essential features, taking into consideration the continuous increase in social media usage. Relative to product information (X25) is a feature than can be utilised to add nutrition value information and recipes for Evia dried figs.

GI products, such as the PDO dried figs of the regional unity of Evia, are associated with specific geographical areas of the European Union. An increase in the demand for Evia PDO dried figs on the internet can increase not only sales and, therefore, production but can also create more opportunities for rural development. The use of live web cameras (X7) at the figs' cultivation sites (none found in research), the provision of information about the area (X4), and the online presence of an up-to-date events calendar (X20) can contribute to increasing tourism demand for the area. Increased tourist numbers at the area can also boost PDO dried figs sales.

The European Union, recognising the importance of digital transformation for local and regional development, has located distinct packages of funds for its less-developed regions, such as the regional unity of Evia. The digital literacy of the dried figs producers and the improvement in their internet presence can be subsidised by EU funds.

Author Contributions: Conceptualization, A.G. and Z.A.; methodology, A.G. and Z.A.; software, A.G. and Z.A.; validation, A.G. and Z.A.; formal analysis, A.G. and Z.A.; investigation, A.G. and Z.A.; resources, A.G. and Z.A.; data curation, A.G. and Z.A.; writing—original draft preparation, A.G. and Z.A.; writing—review and editing, A.G. and Z.A.; visualization, A.G. and Z.A.; supervision, A.G. and Z.A.; project administration, A.G. and Z.A.; funding acquisition, A.G. and Z.A. All authors have read and agreed to the published version of the manuscript.

Funding: This research received no external funding.

Data Availability Statement: All data from this research are available after communication with the corresponding author.

Conflicts of Interest: The authors declare no conflicts of interest.

References

1. Georgilas, A. *The Online Promotion of PDO Products as a Regional Development Factor: The Case of Evia Figs*; Aristotle University: Thessaloniki, Greece, 2023. [CrossRef]
2. Andreopoulou, Z.; Tsekouropoulos, G.; Koliouska, C.; Koutroumanidis, T. Internet marketing for sustainable development and rural tourism. *Int. J. Bus. Inf. Syst.* **2014**, *16*, 446–461. [CrossRef]
3. Tsekouropoulos, G.; Andreopoulou, Z.; Koliouska, C.; Vatis, S.E. The role of internet in the promotion of agri-food enterprises: E-marketing, management and organizational functions. In Proceedings of the International Conference on Contemporary Marketing Issues, Thessaloniki, Greece, 13–15 June 2012.
4. Rur-Lex. Available online: https://eur-lex.europa.eu/legal-content/EN/TXT/PDF/?uri=CELEX:52022PC0134R(01) (accessed on 14 September 2023).
5. European Commission. Available online: https://ec.europa.eu/transparency/documents-register/detail?ref=SWD(2021)428&lang=en (accessed on 14 September 2023).
6. Leverage Marketing. Available online: https://www.theleverageway.com/blog/how-far-down-the-search-engine-results-page-will-most-people-go/ (accessed on 14 September 2023).
7. Andreopoulou, Z.; Arabatzis, G.; Manos, B.; Sofios, S. Promotion of rural regional development through the WWW. *Int. J. Appl. Syst. Stud.* **2007**, *1*, 290–304. [CrossRef]
8. Andreopoulou, Z.; Koliouska, C.; Zopounidis, C. *Multicriteria and Clustering: Classification Techniques in Agrifood and Environment*; Springer International Publishing: Berlin/Heidelberg, Germany, 2017.
9. Jun, S.P.; Yoo, H.S.; Choi, S. Ten years of research change using Google Trends: From the perspective of big data utilizations and applications. *Technol. Forecast. Soc. Chang.* **2018**, *130*, 69–87. [CrossRef]

Disclaimer/Publisher's Note: The statements, opinions and data contained in all publications are solely those of the individual author(s) and contributor(s) and not of MDPI and/or the editor(s). MDPI and/or the editor(s) disclaim responsibility for any injury to people or property resulting from any ideas, methods, instructions or products referred to in the content.

 *proceedings*

Proceeding Paper

Identifying Veterinary Students' Attitudes on Entrepreneurial Intentions: A Two-Step Cluster Analysis †

Athanasios Batzios [1,*], Vagis Samathrakis [2], Alexandros Theodoridis [3], Georgia Koutouzidou [4] and Alexandros Kakouris [5]

1. School of Agriculture, Aristotle University of Thessaloniki, 54124 Thessaloniki, Greece
2. Department of Accounting and Information Systems, International Hellenic University, 57400 Thessaloniki, Greece; samavagi@gmail.com
3. School of Veterinary Medicine, Aristotle University of Thessaloniki, 54124 Thessaloniki, Greece; alextheod@vet.auth.gr
4. Department of Agriculture, University of Western Macedonia, 53100 Florina, Greece; koutouzidoug@gmail.com
5. Department of Management Science and Technology, University of Peloponnese, 22100 Tripolis, Greece; akakouris@uop.gr

* Correspondence: thanos.batzios@gmail.com; Tel.: +30-6972352326
† Presented at the 17th International Conference of the Hellenic Association of Agricultural Economists, Thessaloniki, Greece, 2–3 November 2023.

Abstract: In this paper, the attitudes of veterinary students concerning the "factors driving their entrepreneurial intentions" and the "effects of family and wider environment on starting a business" were analyzed using Two-Step Cluster Analysis. A survey was conducted on 105 veterinary students who were asked to indicate their "agreement" on certain individual issues. The analysis of the data collected resulted in two students' profiles with respect to the factors driving their entrepreneurial intentions ("The cautious students" and "The reluctant students"), and in three students' profiles with respect to the influence of family and the wider environment on starting a business ("The conscious students", "The cautious and conservative students" and "The well informed and decisive students"). The study's findings could contribute to reinforcing the actions of educational institutions towards targeted training of students on entrepreneurship/market issues.

Keywords: veterinary students' attitudes; Two-Step Cluster Analysis; entrepreneurial intentions

1. Introduction

Entrepreneurship is a dominant element of economic growth, promoting business innovation and technology adoption, creating new jobs and supporting the development of managerial talents [1–4]. Particularly during economic recession, enhancing entrepreneurship is an important tool of response [5] inversely related to unemployment [6,7], enabling young people to create their own employment opportunities and develop business ideas [8]. On the other hand, the intensification of knowledge and the increasing importance of lifelong learning are shaping a particularly complicated employment framework.

In this context, the choice of professional career by veterinary graduates is an important process that occurs mainly in the pre-degree stage, associated with a variety of work opportunities. In Greece, the veterinary profession is expanding and the prospects for employment are rather favorable due to the increasing demand for animal health and medical care services [9]. Simultaneously, a shift of veterinarians towards the private sector has been recently recorded [10], launching new market conditions and considerably influencing the students' entrepreneurial intentions.

Therefore, higher veterinary education institutions, in addition to the professional knowledge provided, should be integrated into the curricula of concrete education/training

on the entrepreneurial mindset and its role in establishing veterinary business ventures [4]. Entrepreneurship is not simply about setting up a business; it is primarily an entrepreneurial mindset, in terms of the ability to identify opportunities, evaluate them, and take action on those opportunities [11,12]. An entrepreneurial mindset is a combination of skills and features that can be used to create new sustainable business, while not being afraid of taking risks when needed [13]. Thus, entrepreneurial education/training would prepare veterinary students with skills and knowledge to be potential entrepreneurs [14,15], enhancing their entrepreneur intentions.

This study aims to investigate the entrepreneurial intentions of veterinary students and to categorize students into homogeneous groups, with the belief that this could contribute to developing targeted educational activities that would effectively guide students, enhancing their career intentions.

2. Materials and Methods

A survey was conducted on 105 veterinary students who declared their agreement on individual issues related to the "factors driving their entrepreneurial intentions" and the "effects of family and wider environment on starting a business". Four levels of agreement were used and the respective scores were attributed.

The collected data were statistically analyzed with Two-Step Cluster Analysis (TSCA) to identify possible students' profiles with respect to the variables of interest. TSCA is an exploratory multivariate method designed to identify natural groups of similar records within a dataset [16,17]. The method uses an algorithm that handles both categorical and/or continuous variables, and automatically determines the optimal number of clusters, based on values of either the Schwarz's Bayesian Criterion (BIC) or Akaike Information Criterion (AIC) [18–21]. The log-likelihood distance measure is used for categorical variables [22]. The clustering criterion (e.g., the BIC) is calculated for each potential model solution and the changes in BIC and in distance measure are assessed to determine clusters [17]. A model's "goodness" is assessed by the Silhouette coefficient of cohesion and separation, with values >0.2 being acceptable [23–25].

3. Results and Discussion

Regarding the sample structure, 62.9% of the respondents were female, indicating an increased female preference for the veterinary profession, which has also been reported by Henry and Jackson [26] in the UK. In total, 81.9% of the students were ≤24 years old and 48.6% had work experience, while 23.8% and 17.1% of the students' fathers and mothers, respectively, were entrepreneurs/freelancers, approaching the national rate of 22% [27].

The TSCA on the "Factors driving respondents' entrepreneurial intentions" resulted in two clusters (48.6% and 51.4% of respondents), with ratios of sizes and silhouette measures being satisfactory (Figure 1). The students' attitude on the issue "I have skills to start a sustainable business" is the most important (predictor importance: PI = 1.00) for cluster formation, followed by issues "I can manage the process of setting up a business" (PI = 0.59), "I know the practical details necessary to start a business" (PI = 0.55), and "It is easy to start/run a business" (PI = 0.48). Work experience and gender affect the students' attitudes, contrary to their family's residence and father's profession.

The model outcomes revealed that the first cluster "The cautious students" consists of students who to some extent agree that they have the skills to start a business (slightly: 76.5%; fairly: 23.5%), they could manage setting up a business (100.0%), they know the practical details to start a business (88.2%) and that they would find it easy to start/run a business (slightly: 60.8%, fairly: 21.6%; highly: 3.9%) (Figure 2).

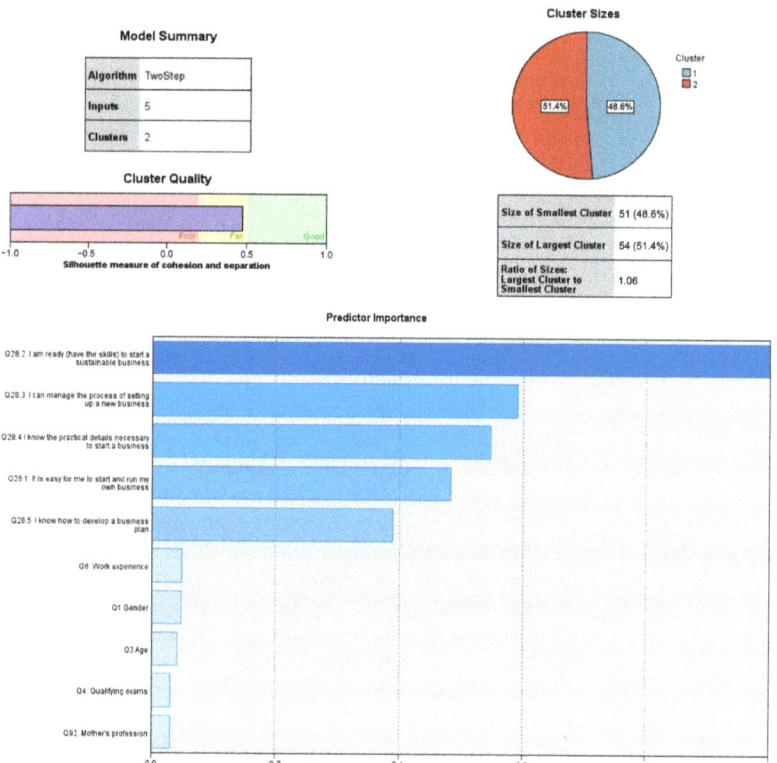

Figure 1. TSCA model summary and predictor importance of the "factors driving respondents' entrepreneurial intentions".

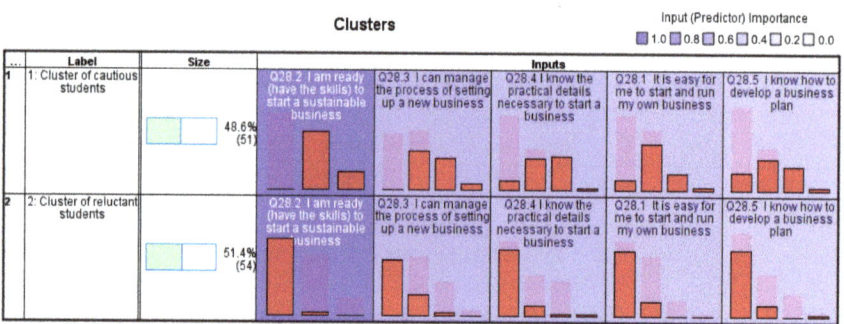

Figure 2. Input importance and cell absolute distributions of TSCA of the "Factors driving the respondents' entrepreneurial intentions".

Women dominate the cluster (54.9%), the most frequent students' age is ≤24 years old and 56.9% of the students have work experience. The second cluster "The reluctant students" includes students who do not believe that they have the skills to start a business (96.3%), could manage the process of setting up a business (70.4%), know the practical details to start a business (83.3%), would find it easy to start/run a business (81.5%), or even that they could develop a business plan (83.3%). Women dominate the cluster (70.4%), the most frequent age is ≤24 years old (81.5%), 59.3% of the students have no work experience and 48.1% have a mother working in the public sector.

Regarding the "Effects of family and wider environment on starting a business", the TSCA identified three clusters (37.1%, 27.6% and 35.2% of respondents), with satisfactory ratios of sizes and silhouette measures showing a good fit of cluster quality (Figure 3). The students' attitude on the issue "Acceptance by parents of any decision to start a business" is the most important (PI = 1.00) for cluster formation, followed by their attitudes on issues such as "Acceptance by colleagues..." (PI = 0.80), and "Acceptance by friends..." (PI = 0.44). Their mother's profession and gender affect the students' attitudes, contrary to work experience.

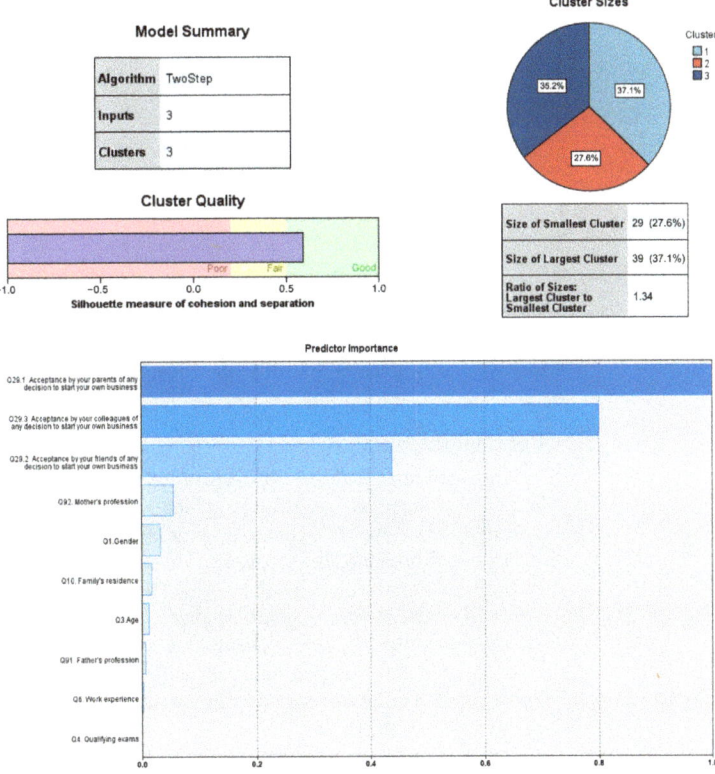

Figure 3. Model summary and predictor importance of the TSCA of the "Effects of family & wider environment on starting a new business".

In particular, the 1st cluster "The conscious students" consists of students who believe that starting a business would be highly accepted by their family (82.1%) and friends (53.8%), and fairly (59.0%) by their colleagues (Figure 4). Women dominate the cluster (69.2%), the most frequent age is ≤24 years old (89.7%), and 51.3% have work experience. The 2nd cluster "The cautious and conservative students" consists of students expecting moderate support from their family (100.0%), friends (75.9%) and colleagues (58.6%). Women dominate the cluster (69.0%), 51.7% have no work experience and 44.8% have fathers and 62.1% mothers working in the public sector. The 3rd cluster "The well informed and decisive students" is composed of students expecting a high level of support from their families, friends and colleagues to start a business. Women account for 51.4% of the cluster, 54.1% have no work experience, and 29.7% have an entrepreneur/freelance professional father.

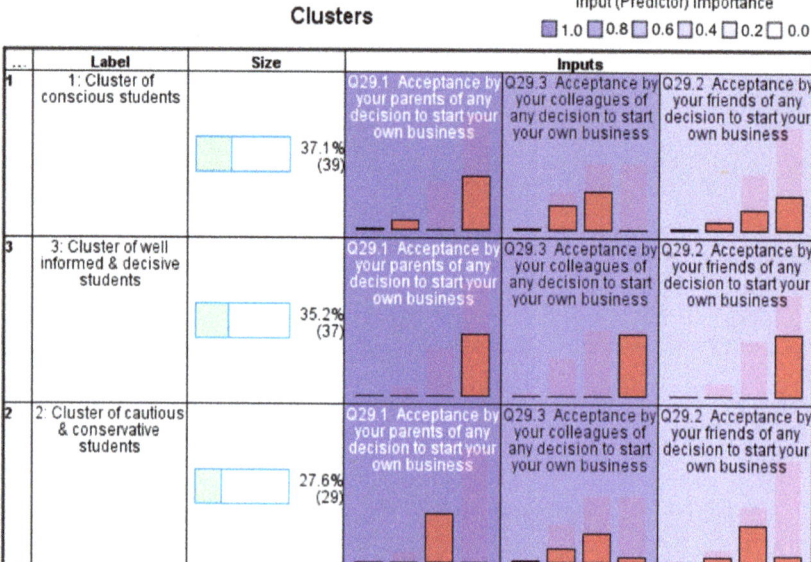

Figure 4. Input importance and cell absolute distributions of TSCA of the "Effects of family & wider environment on starting a new business".

4. Conclusions

The Two-Step Cluster Analysis resulted in two clusters, "The cautious students" and "The reluctant students" regarding the factors driving students' entrepreneurial intentions, and in three clusters, "The conscious students", "The cautious & conservative students" and "The well informed & decisive students" regarding the influence of family and the wider environment on starting a business. The segmentation of students highlights that the majority of students appear skeptical about the concept of creating a sustainable business, expressing uncertainty, perhaps due to the complex business framework in Greece. A high rate of students declared ignorance about the procedures/practical details of setting up a business, while family and the wider environment play a crucial role, affecting the potential decision to start a business.

The study's findings confirm the need for education/training of veterinary students on business and marketing issues. Based on the described profiles, a response to these aspects could be the implementation of targeted training seminars and workshops on entrepreneurship and marketing issues. The integration of entrepreneurship education with lectures on selected marketing and entrepreneurship issues, the organization of open career-days with the participation of students and veterinary professionals, and visits to relative business entities could be beneficial for the students, helping them to effectively utilize the experience gained through their planned internships in livestock farms and veterinary clinics.

Author Contributions: Conceptualization, A.B., V.S., A.T., G.K. and A.K.; methodology, A.B., V.S., A.T., G.K. and A.K.; validation, A.B., V.S. and A.T.; formal analysis, A.B., V.S. and A.T.; investigation, A.B., V.S. and A.T.; resources, A.B. and V.S.; data curation, A.B., V.S., A.T., G.K. and A.K.; writing—original draft preparation, A.B.; writing—review and editing, A.B., V.S., A.T., G.K. and A.K.; Visualization, A.B.; supervision, A.T. All authors have read and agreed to the published version of the manuscript.

Funding: This research received no external funding.

Institutional Review Board Statement: Not applicable.

Informed Consent Statement: Not applicable.

Data Availability Statement: The data presented in this study will be available upon request from the corresponding author.

Conflicts of Interest: The authors declare no conflicts of interest.

References

1. Azhar, A.; Javaid, A.; Rehman, M.; Hyder, A. Entrepreneurial Intentions among Business Students in Pakistan. *J. Bus. Syst. Gov. Ethics* **2014**, *5*, 13–21. [CrossRef]
2. Kowang, T.O.; Apandi, S.Z.B.A.; Ong, C.H.; Fei, G.C.; Saadon, M.S.I.; Othman, M.R. Undergraduates' entrepreneurial intention: Holistic determinants matter. *Int. J. Eval. Res. Educ. (IJERE)* **2021**, *10*, 57–64. [CrossRef]
3. Feakes, A.; Lindsay, N.; Palmer, E.; Petrovski, K. Business Intentions of Australian Veterinary Students—My Business or Yours? A Cluster Analysis. *Animals* **2023**, *13*, 1225. [CrossRef] [PubMed]
4. Zemlyak, S.; Naumenkov, A.; Khromenkova, G. Measuring the Entrepreneurial Mindset: The Motivations behind the Behavioral Intentions of Starting a Sustainable Business. *Sustainability* **2022**, *14*, 15997. [CrossRef]
5. Sarri, A.; Laspita, S.; Panopoulos, A. Drivers and barriers of entrepreneurial intentions in times of economic crisis: The gender dimension. *South-East. Eur. J. Econ.* **2018**, *16*, 147–170.
6. Ahmed, I.; Nawaz, M.M.; Ahmad, Z.; Shaukat, M.Z.; Usman, A.; Rehman, W.U.; Ahmed, N. Determinants of Students' Entrepreneurial Career Intentions: Evidence from Business Graduates. *Eur. J. Soc. Sci.* **2010**, *15*, 14–22.
7. Thurik, R.; Carree, M.A.; Audretsch, D.B. Does Self-Employment Reduce Unemployment? *J. Bus. Ventur.* **2008**, *23*, 673–686. [CrossRef]
8. Bagheri, A. University students' entrepreneurial intentions: Does education make a difference?: Perspectives on Trends, Policy and Educational Environment. In *Entrepreneurship Education and Research in the Middle East and North Africa (MENA)*; Faghih, N., Zali, M.R., Eds.; Springer International Publishing: Berlin/Heidelberg, Germany, 2018; pp. 131–154. [CrossRef]
9. Kirmizoglou, I. Marketing Research in a Private Veterinary Clinic to Determine the Quality of Services Provided and the Satisfaction Felt by Its Clients. Diploma Thesis, Master's Program in Business Administration. University of Macedonia, Thessaloniki, Greece, 2020. (In Greek, Abstract in English).
10. GEOTEE. Distribution of Sectors by Branch and Occupation 2022. Available online: https://www.geotee.gr/geo/stats2022/stats2022_pin_katan_klado_ergasia.htm (accessed on 4 April 2023).
11. Naumann, C. Entrepreneurial mindset: A synthetic literature review. *Entrep. Bus. Econ. Rev.* **2017**, *5*, 149–172. [CrossRef]
12. Pidduck, R.J.; Clark, D.R.; Lumpkin, G.T. Entrepreneurial mindset: Dispositional beliefs, opportunity beliefs, and entrepreneurial behavior. *J. Small Bus. Manag.* **2021**, *61*, 1–35. [CrossRef]
13. Hattenberg, D.Y.; Belousova, O.; Groen, A.J. Defining the entrepreneurial mindset and discussing its distinctiveness in entrepreneurship research. *Int. J. Entrep. Small Bus.* **2021**, *44*, 30–52. [CrossRef]
14. Liu, X.; Lin, C.; Zhao, G.; Zhao, D. Research on the effects of entrepreneurial education and entrepreneurial self-efficacy on college students' entrepreneurial intention. *Front. Psychol.* **2019**, *10*, 869. [CrossRef] [PubMed]
15. Khalid, B.; Chaveesuk, S.; Chaiyasoonthorn, W. MOOCs adoption in higher education: A management perspective. *Pol. J. Manag. Stud.* **2021**, *23*, 239–256. [CrossRef]
16. SPSS Inc. *SPSS Statistics Base 17.0 User's Guide*; Polar Engineering and Consulting: Chicago, IL, USA, 2008; ISBN 978-1-56827-400-3.
17. BPSA MALANG. Two Step Cluster Analysis. Brawijaya Professional Statistical Analysis. 2018. Available online: https://www.coursehero.com/file/35722421/ebook-038-tutorial-spss-two-step-cluster-analysispdf/ (accessed on 22 April 2023).
18. Punj, G.; Stewart, D.W. Cluster Analysis in Marketing Research: Review and Suggestions for Application. *J. Mark. Res.* **1983**, *20*, 134–148. [CrossRef]
19. Chiu, T.; Fang, D.; Chen, J.; Wang, Y.; Jeris, C. A robust and scalable clustering algorithm for mixed type attributes in large database environment. In Proceedings of the Seventh ACM SIGKDD International Conference on Knowledge Discovery and Data Mining—KDD '01, San Francisco, CA, USA, 26–29 August 2001; Association for Computing Machinery: New York, NY, USA, 2001; pp. 263–268, ISBN 978-1-58113-391-2.
20. Daniela, S. Applying TwoStep Cluster Analysis for Identifying Bank Customers' Profile. *Bul. Univ. Pet. Gaze Ploiesti* **2010**, *LXII*, 66–75.
21. Biomedical Statistics. Two-Step Cluster Analysis. Available online: https://biomedicalstatistics.info/en/cluster-analysis/two-step-cluster.html (accessed on 21 June 2023).
22. Chan, Y.H. Biostatistics 304. Cluster analysis. *Singapore Med. J.* **2005**, *46*, 153. [PubMed]
23. Papaoikonomou, A. Greek educational system and the formation of national identity: A Two-Step Cluster Analysis model on a teachers' sample. *İMGELEM* **2002**, *6*, 351–364. [CrossRef]
24. Han, J.; Kamber, M.; Pei, J. *Data Mining: Concepts and Techniques*, 3rd ed.; Morgan Kaufmann Publishers: Waltham, MA, USA, 2012; ISBN 978-0-12-381479-1.
25. Ahmed, S.A.S. Evaluating Students' Performance of Social Work Department Using K-means and Two-step Cluster: "A Case Study of Mogadishu University". PH.D. Curriculum. *Mogadishu Univ. J.* **2021**, *7*, 13–33.

26. Henry, C.; Jackson, E. Women's entrepreneurship and the future of the veterinary profession. *e-Organ. People* **2015**, *22*, 34–42.
27. Tsakanikas, A.; Stavrakaki, S.; Valavanioti, E.; Yiotopoulos, I. *Annual Entrepreneurship Report 2020–2021: Mild Impacts of Pandemic on Early-Stage Entrepreneurship*; In the context of the research program "Global Entrepreneurship Monitor" (GEM); Foundation for Economic and Industrial Research: Athens, Greece, 2022; ISSN 2653-889X. (In Greek)

Disclaimer/Publisher's Note: The statements, opinions and data contained in all publications are solely those of the individual author(s) and contributor(s) and not of MDPI and/or the editor(s). MDPI and/or the editor(s) disclaim responsibility for any injury to people or property resulting from any ideas, methods, instructions or products referred to in the content.

Proceeding Paper

Consumers' Trust and Preferences Regarding Local Plant Varieties and Indigenous Farm Animal Breeds in Western Macedonia, Greece [†]

Dimitrios Kyriazoglou [1], Vasiliki Makri [1], Martha Tampaki [1], Katerina Melfou [1], Athanasios Ragkos [2] and Ioannis A. Giantsis [1,*]

[1] Faculty of Agricultural Sciences, University of Western Macedonia, 53100 Florina, Greece; kyriazog@yahoo.gr (D.K.); makrivasil@bio.auth.gr (V.M.); tampakimartha@gmail.com (M.T.); kmelfou@uowm.gr (K.M.)

[2] Agricultural Economics Research Institute, Hellenic Agricultural Organization—DIMITRA, 11528 Athens, Greece; ragkos@elgo.gr

* Correspondence: igiantsis@uowm.gr

[†] Presented at the 17th International Conference of the Hellenic Association of Agricultural Economists, Thessaloniki, Greece, 2–3 November 2023.

Abstract: The value of rearing indigenous animal breeds and cultivating local plant varieties is extremely high in terms of regional economy and heritage preservation. The purpose of the present research was to investigate the preferences and opinions of consumers in Western Macedonia regarding local varieties and indigenous breeds. For this purpose, an appropriate questionnaire was designed and distributed to a sample of 80 consumers from Western Macedonia. The questions combined the demographic, psychographic and institutional characteristics of consumers. According to our findings, most participants recognize the importance of the conservation of indigenous animal breeds and local plant varieties as well as the products derived from them. Additionally, a large percentage showed a preference for these products for the purpose of supporting the local economy. Nevertheless, particularly for indigenous animal breeds, despite the recognition of their high value and need for conservation, only a small proportion of the participants could name some of the indigenous breeds correctly. Conversely, this was not observed concerning local plant varieties, of which participants were more aware. Thus, better promotion and overall better marketing could enhance the recognition of these resources, emphasizing their high value.

Keywords: indigenous animals; local plants; Western Macedonia; consumer preference

1. Introduction

Biodiversity is the measurement of genetic variation, species diversity and ecosystem diversity levels [1]. According to Sahney et al. [2], biodiversity is divided into four levels: (a) taxonomic diversity (usually measured at the level of species diversity), (b) ecological diversity (the level of ecosystem diversity), (c) morphological diversity (genetic diversity and molecular diversity) and (d) functional diversity (which is a measure of the number of functionally different species in a population). Greece includes 6600 species and subspecies of angiosperm plants and more than 23,000 species of terrestrial and freshwater animals. The breeding of indigenous animals and the cultivation of local plants are inextricably linked to the social and economic life of each region. Native products are very important for the diet as well as the economy of a region. The agricultural sector is an important economic activity for the region of Western Macedonia. The purpose of this study was to investigate the preference and opinion of consumers regarding local products derived from indigenous livestock breeds and local plant varieties in the region of Western Macedonia.

2. Materials and Methods

Research was carried out using questionnaires ($N = 80$). The questionnaires included sections about (a) demographic characteristics, (b) psychographic characteristics and (c) institutional characteristics of the consumers. The above sections were created with the aim to investigate the knowledge level concerning the breeding of indigenous animals and the cultivation of local plants. Collected data were categorized using Excel 2010.

3. Results and Discussion

The majority of participants were women, with the largest percentage belonging to the age group 26–44 and the population category of 5.000–20.000 inhabitants per city. In addition, the educational level with the largest percentage was that of higher education followed by high school diploma holders. Furthermore, the largest percentage corresponded to the 801–1500 income per month category. The majority of the participants (90%) recognized the value of local plant varieties and indigenous animal breeds. Buyers were triggered mainly by underlying factors when purchasing a new product. The main reasons for consumers' preference for local products were (a) contribution to the local economy, (b) reasons of loyalty to employment policies in the local market and (c) saving resources for future generations. In addition, the participants considered local products to be worth their money and wholesome. Also, the majority of participants believed that (a) people should buy local products rather than imported ones and (b) only products that cannot be produced in Greece should be imported. In general, however, the participants did not have basic knowledge regarding the status of indigenous animal breeds and animal-related husbandry, a fact that was not observed in local plant varieties. Menger and Hamm [3] report in their research that the participants possessed little awareness of issues concerning animal diversity and sustainability, the risk of animal breeds' extinction and conservation. However, in Germany, the biodiversity and conservation of indigenous animals is a specialized sensitive topic [4]. The results of the study by Menger and Hamm [3] showed that conservation plays a key role in the production of native products. According to our results, in agreement with the aforementioned studies from other European countries, the labeling and proper marking of local products would highly benefit the reliable recognition of local resources, of both animal and plant origin (Figure 1).

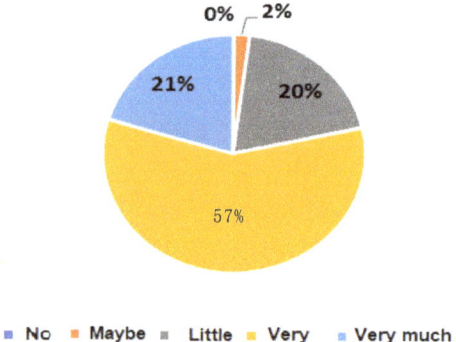

Figure 1. Consumers' agreement with labeling of local products.

4. Conclusions

The cultivation of local varieties of plants and rearing indigenous animal breeds may contribute significantly to the strengthening of local economy. In addition, it enhances the preservation of local heritage and the conservation of local genetic resources. The region of Western Macedonia is associated with the cultivation of local varieties of plants and the breeding of indigenous animals. Efforts should be conducted to achieve the better and more reliable recognition of these products on the local and global market. Advertising, public relations and direct sales are among the proposed methods for effective promotion. Marketing can be viewed as an organizational function and a set of processes for creating, delivering and communicating the nutritional value of local products to customers. The organization of events with the purpose of promoting and advertising local products to the wider public is another useful proposal.

Author Contributions: Conceptualization, I.A.G.; methodology, D.K.; software, V.M.; validation, I.A.G.; formal analysis, M.T.; investigation, M.T.; resources, M.T.; data curation, K.M. and A.R.; writing—original draft preparation, D.K. and V.M.; writing—review and editing, A.R., K.M. and I.A.G.; visualization, M.T.; supervision, I.A.G.; project administration, I.A.G.; funding acquisition, K.M. All authors have read and agreed to the published version of the manuscript.

Funding: This research received no external funding.

Institutional Review Board Statement: The study was conducted in accordance with the Declaration of Helsinki, and approved by the Institutional Review Board (or Ethics Committee) of University of Western Macedonia (protocol code 154/2023-01.02.2023).

Informed Consent Statement: Not applicable.

Data Availability Statement: All data of this research are available after communication with the corresponding author.

Conflicts of Interest: The authors declare no conflicts of interest.

References

1. United National Research Council (US) Committee on Noneconomic and Economic Value of Biodiversity. Perspectives on Biodiversity: Valuing Its Role in an Everchanging World. Washington (DC): National Academies Press (US); 2. What is Biodiversity? 1999. Available online: https://www.ncbi.nlm.nih.gov/books/NBK224405/ (accessed on 14 January 2024).
2. Sahney, S.; Benton, M.J.; Ferry, P. Links between global taxonomic diversity, ecological diversity and the expansion of vertebrates on land. *Biol. Lett.* **2010**, *6*, 544–547. [CrossRef] [PubMed]
3. Menger, A.; Hamm, U. Consumers knowledge and perceptions of endangered livestock breeds: How wording influences conservation efforts. *Ecol. Econ.* **2021**, *188*, 107117. [CrossRef]
4. Sponenberg, D.P.; Martin, A.; Couch, C.; Beranger, J. Conservation Strategies for Local Breed Biodiversity. *Diversity* **2019**, *11*, 177. [CrossRef]

Disclaimer/Publisher's Note: The statements, opinions and data contained in all publications are solely those of the individual author(s) and contributor(s) and not of MDPI and/or the editor(s). MDPI and/or the editor(s) disclaim responsibility for any injury to people or property resulting from any ideas, methods, instructions or products referred to in the content.

Proceeding Paper

Evaluation of the Certification Procedure of Farm Advisors in Greece [†]

Ekaterini Alexaki [1], Ioannis Dimitriadis [1], Efstratios Michalis [2,*], Christina-Eleni Giatra [2] and Athanasios Ragkos [2]

[1] Directorate of Development and Extension, General Directorate of Quality Assurance of Agricultural Products, Hellenic Agricultural Organization–DIMITRA, 111 45 Athens, Greece; kalexaki@elgo.gr (E.A.); idimitriadis@elgo.gr (I.D.)

[2] Agricultural Economics Research Institute, Hellenic Agricultural Organization-DIMITRA, 111 45 Athens, Greece; chrgiatra@gmail.com (C.-E.G.); ragkos@elgo.gr (A.R.)

* Correspondence: efstratiosmichalis@gmail.com

[†] Presented at the 17th International Conference of the Hellenic Association of Agricultural Economists, Thessaloniki, Greece, 2–3 November 2023.

Abstract: Farm Advisory constitutes one of the most important tools to support rural development in the European Union and is also an integral part of Agricultural Knowledge and Innovation System (AKIS). The purpose of this paper is to present the results of the evaluation of the two calls for certification of Farm Advisors in Greece, which were addressed to individuals. The evaluation was based on a questionnaire survey of candidates who participated to the online certification procedure. The analysis is based on descriptive statistics methods and shows that overall most respondents were satisfied with most Modules, although they suggest to provide better links between scientific evidence and practical applications. Although there are serious limitations that do not permit to draw generalized conclusions, the evaluation procedure pointed out specific domains that require improvements and, especially, that a more robust evaluation system is required.

Keywords: Agricultural Knowledge and Innovation System (AKIS); questionnaire survey; training material

Citation: Alexaki, E.; Dimitriadis, I.; Michalis, E.; Giatra, C.-E.; Ragkos, A. Evaluation of the Certification Procedure of Farm Advisors in Greece. *Proceedings* **2024**, *94*, 23. https://doi.org/10.3390/proceedings2024094023

Academic Editor: Eleni Theodoropoulou

Published: 23 January 2024

Copyright: © 2024 by the authors. Licensee MDPI, Basel, Switzerland. This article is an open access article distributed under the terms and conditions of the Creative Commons Attribution (CC BY) license (https://creativecommons.org/licenses/by/4.0/).

1. Introduction

Farm Advisory constitutes one of the most important tools to support rural development in the European Union and is also an integral part of Agricultural Knowledge and Innovation System (AKIS) [1]. Farm Advisors are qualified to give farmers sound advices on a variety of issues, including but not limited to land eligibility, conditionality, and scheme applications. They can also assist farmers in meeting their obligations and avoiding financial penalties under EU and national funded Schemes [2].

Under Article 15 of Reg (EU) 1305/2013, Greece programmed two out of the three possible options for the 2014–2020 period, i.e., advisory services provision (Sub-measure 2.1 of the Rural Development Programme (RDP) of Greece "2014–2020"), and training of Farm Advisors (FAs) (sub-measure 2.3) [3,4]. Advisory services measures (art. 15—Measure 02 of the RDP) and co-operation/innovation (art. 35 and 56—M16 of the RDP) were also put into place with a broader application field, while the budget allocated to Advisory Services Measure 2 was more than double compared to the 2007–2013 period [4,5].

Based on this framework and also on the FA legislation under Reg (EU) 1306/2013, the Greek Ministry of Rural Development and Food established a new framework in 2018 (Decision 163/13692/1 February 2018 of the Minister of Rural Development and Food), by means of which the Hellenic Agricultural Organization (ELGO)-DIMITRA was the designated Organization for training, certifying, and controlling FAs [6,7]. At its core was the introduction of the National Registers of certified FAs and advisory bodies, which were put under the responsibility of the ELGO-DIMITRA. With their certification and their registration in the Register of Agricultural Advisors of ELGO-DIMITRA, the national Farm Advisory Service (FAS) was effectively put into operation, thus fulfilling an institutional obligation for Greece and also introducing an important development driver for the support of Greek farmers. In addition to the existing legal framework, the new Law 5035/2023 states that one of the objectives of ELGO-DIMITRA is "... the advisory aid of famers", while the responsibilities of ELGO-DIMITRA include "... the design, organization and implementation of education, training and information activities" (Article 4) [8]. Under the new legal framework, the General Directorate of Strategic Advisory and Rural Development is also introduced.

FAs can be certified in up to ten (10) of the following thematic fields (modules), depending on their specialization:

- Module 1. Cross Compliance—good agricultural and environmental conditions.
- Module 2. Requirements for implementing Article 11 of the Water Framework Directive.
- Module 3. Requirements for implementing Article 55 of Regulation (EC) No. 1107/2009, in particular compliance with the general principles of integrated pest management as referred to in Article 14 of Directive 2009/128/EC.
- Module 4. Climate change mitigation and adaptation.
- Module 5. Organic farming.
- Module 6. Modernization of agricultural farms—improvement to sustainability—competitiveness.
- Module 7. Risk management in agriculture and animal husbandry.
- Module 8. Implementation of standards for workspace safety.
- Module 9. Management of rural environment—integrated management in agricultural production. Part 2: Requirements for the application in crop production (national standards AGRO2).
- Module 10. Advisory for young farmers: farm management; cooperation and market access; regulatory obligations; new technologies.

The certification is of indefinite duration in compliance with the obligations arising from relevant EU and national legislation.

The certification and registration of FAs in the Register of ELGO-DIMITRA is subject to the successful attendance of a training program (Decision No. 163/13692/01.02.2018 of the Minister of Rural Development and Food) [9]. Until 2023, ELGO-DIMITRA has published three Calls for the expression of interest for certification as FAs (two for individuals—2018 and 2021—and one for legal entities—2021) [10–12]. Candidates followed the program exclusively on an e-learning platform, through which they had access to the thematic fields they applied for and also to training material (an e-book) [13]. After the completion of each thematic field, trainees were evaluated with an online test, which included multiple-choice questions and true/false statements. Participants were graded on a scale of 0 to 100%, and a minimum score of 75% was required in order to successfully finalize the attendance of the training (with the possibility for a re-evaluation).

In both Calls, the success rates of participants were over 95% for all thematic fields, while participation was 87% in the first Call and 78% in the second. As a result, by the end of October 2022, 3980 individuals were registered as FAs, most of which were agronomists (63.8%), followed by agricultural technologists (various expertise) (21.8%), foresters (3.9%), and veterinarians (2.3%), while the remaining ones came from several other backgrounds. Additionally, 98 legal persons were registered.

The purpose of this paper is to present the results of the evaluation of the two Calls that were addressed to individuals. The evaluation is based on a questionnaire that was addressed to all participants.

2. Methods

During the posting week of each thematic field, an evaluation questionnaire was sent to participants along with the training material. The questionnaire included the following eight closed-ended questions (evaluation items): clear, complete content; structure and organization of content; modern—topical knowledge; links between scientific knowledge and practice; suitability for e-learning; met expected results and training needs; effectiveness of the training method; and general impression. Participants could answer using a five-point Likert (1–5) scale (1 = not at all; 2 = a little; 3 = quite a bit; 4 = a lot; 5 = very much). In addition, during the first call, an open field was available to respondents, where they were free to record their observations and comments about the certification process. In the second Call, it was obligatory for all participants to fill out the questionnaire and submit it after the examination test, while in the first one, this was optional. In both Calls, the evaluation process was fully anonymous. The analysis was based on descriptive statistical methods (means).

3. Results and Discussion

In total, only 867 responses were received in the first Call (for all sections), and 8358 responses (average 836/module) in the second Call (obligatory in both Calls; the same person may have answered more than once, but in each case in the context of a different section). This difference does not allow to make comparative assumptions and conclusions. However, some basic observations can be derived, which are useful for future Calls.

- The small number of responses on the first Call compared to the second implies that candidates were not highly motivated to share their comments about the process. It is also interesting to note that while 218 people participated in the evaluation of the first Module, in the following ones only 60–111 responses were received.
- The section of trainees who were satisfied with all eight evaluation items (rated 4 or 5) increased in size in the second Call (from slightly over 50% to more than 67%), while, similarly, negatively satisfied trainees (rated 1 or 2) were between 10 and 20% per Module in the first Call but less than 10% in the second.
- All average scores per Module were higher on the second Call compared to the first.
- In the first Call, Modules seven and eight received the highest scores among all Modules in four and two items, respectively. In the second, Modules five and eight received the highest scores in four and three items, respectively. On the other hand, Modules 9 (first call) and 6 (second Call) were ranked the lowest for all eight items.
- While participants recognized that the program was characterized by "modern and up-to-date knowledge", the lack of connection between scientific knowledge and practical application was identified as a key problem in both Calls.

4. Conclusions

The results of descriptive statistics capture some indicative trends in participants' opinions. However, there are serious limitations that do not permit drawing generalized conclusions. First, the sociodemographic characteristics of respondents were not recorded (e.g., age, gender, specialty, and employment status), which does not allow us to draw conclusions for different disciplines or professional backgrounds. Second, most qualitative observations and comments (in text form) that were submitted do not refer to specific parts of the evaluation and thus do not allow for the drawing of relevant conclusions. Given these limitations, the following actions could contribute to the improvement of the evaluation but also of the whole certification process:

1. A more robust evaluation procedure, with a redesigned questionnaire to include more questions and respondents' sociodemographic profile.
2. Improve the links between the training material and practical applications (interactive exercises, audiovisual demonstration material).
3. Regular update of the content of all Modules; revision of existing or addition of new ones.
4. A post-certification survey of registered FAs.
5. Development of a monitoring system for the action in order to record the professional activities of FAs.

Author Contributions: Conceptualization, A.R., E.A., and I.D.; methodology, A.R., E.A., I.D., and E.M.; formal analysis, A.R. and I.D.; writing—original draft preparation, A.R., C.-E.G., E.M., and E.A. All authors have read and agreed to the published version of the manuscript.

Funding: This research received no external funding.

Institutional Review Board Statement: National laws exempt this type of studies from requiring ethical review and approval.

Informed Consent Statement: Not applicable.

Data Availability Statement: Data are available from the authors.

Conflicts of Interest: The authors declare no conflicts of interest.

References

1. Prager, K.; Creaney, R.; Lorenzo-Arribas, A. Criteria for a system level evaluation of farm advisory services. *Land Use Policy* **2017**, *61*, 86–98. [CrossRef]
2. Hammersley, C.; Richardson, N.; Meredith, D.; Carroll, P.; McNamara, J.G. Supporting farmer wellbeing: Exploring a potential role for advisors. *J. Agric. Educ. Ext* **2022**, *29*, 511–538. [CrossRef]
3. Official Journal of the European Union Regulation (EU). No 1305/2013 of the European Parliament and of the Council of 17 December 2013 on Support for Rural Development by the European Agricultural Fund for Rural Development (EAFRD) and repealing Council Regulation (EC) No 1698/2005. Available online: https://eur-lex.europa.eu/LexUriServ/LexUriServ.do?uri=OJ:L:2013:347:0487:0548:en:PDF (accessed on 16 September 2023).
4. Ministry of Rural Development and Food. Brief Presentation of Rural Development Programme of Greece 2014–2020. Available online: http://agrotikianaptixi.gr/sites/default/files/%CE%A3%CE%A5%CE%9D%CE%9F%CE%A8%CE%97_%CE%A0%CE%91%CE%91_2014-2020.pdf (accessed on 16 September 2023).
5. Government Gazette of the Hellenic Republic Volume B 2124/14.10.2008. Available online: https://www.geotee.gr/lnkFiles/YA_328821_2008.pdf (accessed on 16 September 2023).
6. Official Journal of the European Union Regulation (EU). No 1306/2013 of the European Parliament and of the Council of 17 December 2013 on the Financing, Management and Monitoring of the Common Agricultural Policy and Repealing Council Regulations (EEC) No 352/78, (EC) No 165/94, (EC) No 2799/98, (EC) No 814/2000, (EC) No 1290/2005 and (EC) No 485/2008. Available online: https://eur-lex.europa.eu/LexUriServ/LexUriServ.do?uri=OJ:L:2013:347:0549:0607:EN:PDF (accessed on 16 September 2023).
7. Government Gazette of the Hellenic Republic Volume B 267/1.02.2018. Available online: https://www.geotee.gr/MainNewsDetail.aspx?CatID=1&RefID=21006&TabID=4 (accessed on 16 September 2023).
8. Government Gazette of the Hellenic Republic Volume A 76/28.03.2023. Available online: https://www.karagilanis.gr/files/nomos_5035_2023.pdf (accessed on 16 September 2023).
9. Government Gazette of the Hellenic Republic Volume B 267/01.02.2018. Available online: https://www.e-nomothesia.gr/kat-agrotike-anaptukse/georgia/upourgike-apophase-163-13692-2018.html (accessed on 16 September 2023).
10. ELGO–DEMETRA. Call for Expression of Interest for Natural Persons Who Would Like to Be Certified as Agricultural Advisers 07.11.2018. Available online: https://www.elgo.gr/images/georgikoi_symvouloi/prosklisi_fp.pdf (accessed on 16 September 2023).
11. ELGO–DEMETRA. Call for Expression of Interest for Natural Persons Who Would Like to Be Certified as Agricultural Advisers 26.07.2021. Available online: https://www.elgo.gr/images/ioanna/anakoinwseis/2%CE%97_%CE%A0%CE%A1%CE%9F%CE%A3%CE%9A%CE%9B%CE%97%CE%A3%CE%97_%CE%93%CE%A3.pdf (accessed on 16 September 2023).
12. ELGO–DEMETRA. Call for Expression of Interest for Egal Persons Who Would Like to Be Certified as Agricultural Advisory Providers 30.03.21. Available online: https://www.elgo.gr/images/events/Deltia_Typou_2021/%CE%A0%CE%A1%CE%9F%CE%A3%CE%9A%CE%9B%CE%97%CE%A3%CE%97_%CE%93%CE%A3.pdf (accessed on 16 September 2023).
13. ELGO–DEMETRA. Educational Materials for the Training of Farm Advisors. Available online: https://www.elgo.gr/index.php?option=com_content&view=article&id=2208:anartisi-ekpaideftikoy-ylikoy-epimorfosis-georgikon-symvoylon&catid=160&Itemid=1249 (accessed on 16 September 2023).

Disclaimer/Publisher's Note: The statements, opinions and data contained in all publications are solely those of the individual author(s) and contributor(s) and not of MDPI and/or the editor(s). MDPI and/or the editor(s) disclaim responsibility for any injury to people or property resulting from any ideas, methods, instructions or products referred to in the content.

Proceeding Paper

Assessing Economic and Social Security in Agricultural Cooperatives: Two Case Studies from Cooperatives in Northern and Southern Greece [†]

Stefania Tselempi *, Michalis Frantzis and Panagiota Sergaki

School of Agriculture, Forestry and Natural Environment, Aristotle University of Thessaloniki, 54124 Thessaloniki, Greece; frantzism@agro.auth.gr (M.F.); gsergaki@auth.gr (P.S.)
* Correspondence: stselempi@agro.auth.gr
[†] Presented at the 17th International Conference of the Hellenic Association of Agricultural Economists, Thessaloniki, Greece, 2–3 November 2023.

Abstract: Agricultural cooperatives have an important role in supporting agricultural development and improving the well-being of their members. They provide farmers with financial and social security, as well as fostering an environment that is supportive of collective actions. This study aims to assess the economic and social safety of female cooperative members by looking at their experiences and perceived improvements over time. It examines how gender dynamics, social capital, and cooperative engagement affect women's perceptions of economic and social security through field surveys and structured interviews. According to the preliminary findings, active engagement in cooperatives improves women's feelings of social security, belonging, and empowerment. They might not be as confident in their ability to make economic judgments due to societal prejudices, resource access restrictions, and cultural norms. This study emphasizes the potential of women to break down traditional gender norms and obstacles as well as the economic gains associated with cooperative activity. These findings provide empirical support and inform efforts to promote empowerment and gender equality in agricultural cooperatives.

Keywords: agricultural cooperatives; women; social security; economic security

Citation: Tselempi, S.; Frantzis, M.; Sergaki, P. Assessing Economic and Social Security in Agricultural Cooperatives: Two Case Studies from Cooperatives in Northern and Southern Greece. *Proceedings* **2024**, *94*, 24. https://doi.org/10.3390/proceedings2024094024

Academic Editor: Eleni Theodoropoulou

Published: 23 January 2024

Copyright: © 2024 by the authors. Licensee MDPI, Basel, Switzerland. This article is an open access article distributed under the terms and conditions of the Creative Commons Attribution (CC BY) license (https://creativecommons.org/licenses/by/4.0/).

1. Introduction

Agricultural cooperatives are collective actions and remarkable business models that have been an integral part of the agricultural sector, offering numerous benefits to their members. These cooperative groups play a crucial role in improving the well-being of their members while also enhancing the value and development of agriculture [1]. They serve as socioeconomic institutions that give farmers the chance to work together, overcome the challenges they face, and address problems as a group [2]. Agricultural cooperatives do this by establishing a positive environment that promotes cooperation, offers necessary tools and resources, and aids in the development and success of individual farmers [2–4].

An important aspect of a cooperative membership is the sense of economic and social security it offers to its members. Economic security includes aspects like financial stability, access to input and output markets, an improvement in bargaining power, and income protection [5]. Social security is the result of strong social relationships, mutual trust, and a sense of community within the cooperative, which creates a support system that may offer emotional and social assistance when necessary [6,7]. For evaluating the cooperatives' success in achieving their goals and promoting the well-being of their participants, it is essential to comprehend the level of economic and social security provided to the cooperative members.

This study focuses on evaluating the female members' social and economic safety in two agricultural cooperatives in Northern and Southern Greece. Our specific research objective is to determine whether women, as members of the agriculture cooperative, feel economically and socially secure as a result of their participation and to investigate any acknowledged improvements to their social well-being over time.

2. Theoretical Framework and Research Questions

A cooperative is a user-owned and user-controlled business from which benefits are derived and distributed on the basis of use [8]. Based on this principle of cooperatives, it is clear that collaboration and social ties between the members are important factors in the cooperative's success. In fact, while farmers have many economic reasons to establish a cooperative, a high amount of social capital among the potential members initiates their cooperative actions [6]. This results in members developing expectations regarding the advantages or outcomes that they believe they will receive from joining the cooperative [9]. Their expectation can regard both their financial performance and their social interactions, which drives the sense of belonging to a community.

In Greece, the family is a very important institution, emphasizing the value of mutual obligations within kinship and marriage bonds. These elements have a big impact on women's overall position in the agri-food industry. There is broad agreement that gender continues to influence who leads agricultural cooperatives and who represents farmers in political discussions. The greater involvement of women in farm decision making throughout time has resulted in a favorable change [10]. Their participation in cooperatives strengthens their position within their local society [11], and this is why we wanted this study to target a group of female members of agricultural cooperatives.

Through this research, we aimed to investigate how these women view the connection between their gender and their position in the workforce. More specifically, our research questions were as follows:

- To what extent do agricultural cooperatives act as platforms for women's economic and social empowerment?
- How does women's active involvement in agricultural cooperatives affect their overall sense of economic and social security?
- Can women overcome societal barriers and traditional gender norms through active engagement in agricultural cooperatives?
- Does the economic well-being of women benefit from their significant contributions to the agricultural sector through active engagement in agricultural cooperatives?

Thus, the main objective of this paper is to provide empirical evidence by answering the above questions.

3. Methodology and Sampling

Because of the lack of statistical or previous research on this topic in agricultural cooperatives in Greece, qualitative analysis was used. Two agricultural cooperatives in Greece—one in the North (Amyntaio) and one in the South (Santorini)—were chosen for their sizeable proportion of female members and cooperative coiling activities, allowing for a comparison of mainland and insular regions.

From each cooperative, a total of 20 questionnaires were gathered to evaluate the economic and social security drawn from 20 women among 36 Amyntaio AC members and 20 out of 68 women among Santorini AC members. Purposive sampling was used in the study to pick cooperatives and individual farmers, and both qualitative and statistical methods, such as the SPSS analysis, cross-tabulation, and graph analysis of member profiles, were used for data analysis. Between July and September 2022, semi-structured questionnaires were used to collect primary data focused on demographics and 5-point Likert scale inquiries. The variables included the size of the farm enterprise, the age of the farm, the degree of specialization in farm activities, profitability, gross income, subjective norms, challenges, and cooperative experiences in order to determine what influences members' sense of security.

4. Results

In both Amyntaio and Santorini cooperatives, female members, averaging around 41 years old with similar years of cooperative membership (around 12 years) but varying farm sizes (Amyntaio: 23.55 acres, Santorini: 14.9 acres) were included. Leadership roles were limited in both the cooperatives (Amyntaio: 80%, Santorini: 85% have not held leadership positions), but active participation in cooperative activities was high (Amyntaio: 75%, Santorini: 70%). Women viewed the two cooperatives positively, and they seemed to empower them both economically and socially. Notably, they both perceived their economic empowerment to a very big extent and their social empowerment as positive, with women in Santorini appearing more socially empowered (Figures 1 and 2).

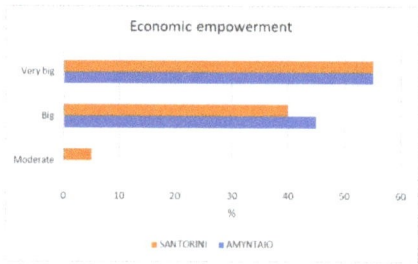

Figure 1. Economic empowerment.

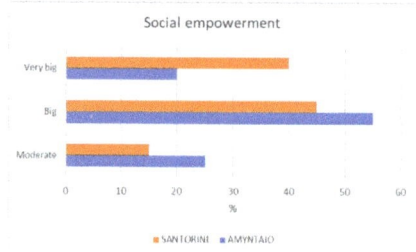

Figure 2. Social empowerment.

Also, women's active participation and engagement in the cooperatives' activities seemed to foster their economic growth and improve their social standing (Figures 3 and 4).

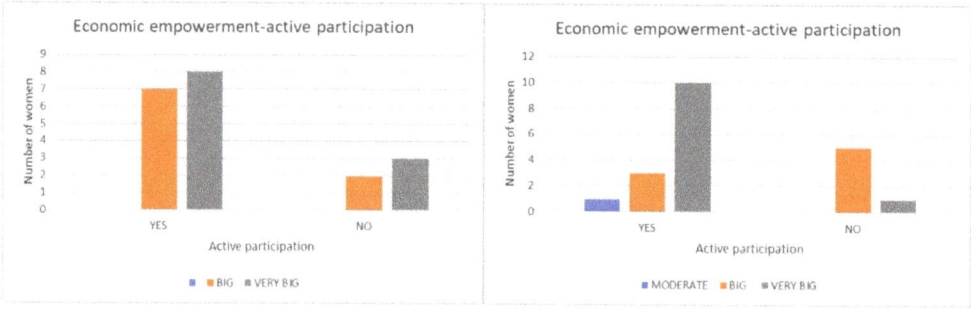

Figure 3. Economic empowerment—active participation in Amyntaio (**left**) and in Santorini (**right**).

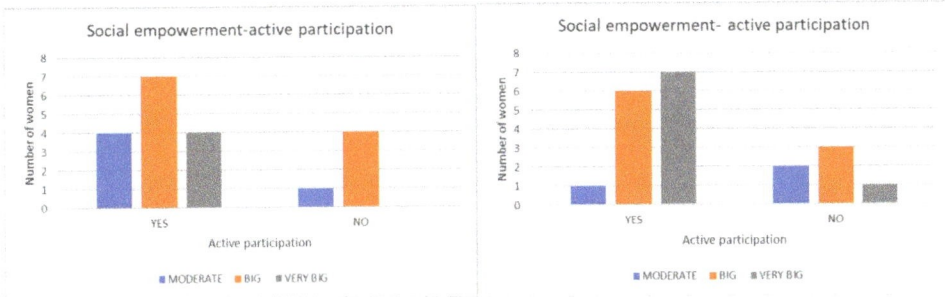

Figure 4. Social empowerment—active participation in Amyntaio (**left**) and in Santorini (**right**).

Moreover, a few social barriers were faced based on gender biases in Amyntaio (70%), but more were faced in Santorini (85%) within the context of the cooperative, which they believed had a changing perception over time (Figures 5 and 6).

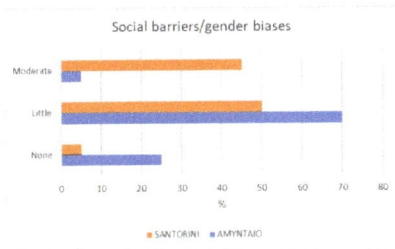

Figure 5. Social barriers/gender biases.

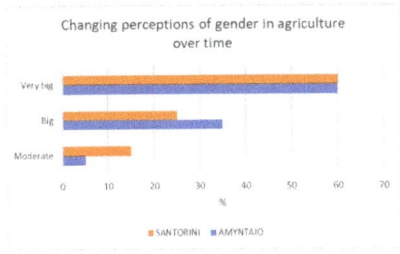

Figure 6. Changing perception of gender in agriculture over time.

Finally, they seemed to be overall satisfied financially by their engagement in the agriculture cooperatives as they received, to a large extent, economic benefits from them (Figures 7 and 8).

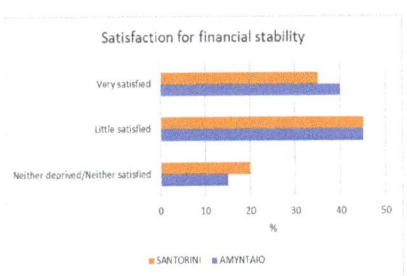

Figure 7. Satisfaction with financial stability.

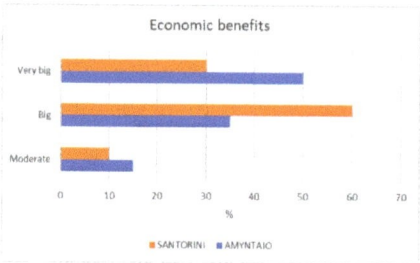

Figure 8. Economic benefits.

5. Conclusions

It is considered that cooperatives have been recognized as important platforms for women's economic and social empowerment. Women who actively participate in agriculture cooperatives report a heightened sense of both economic and social security (Figures 1 and 2). Through their active participation, women gain a stronger sense of belonging and empowerment within their communities, which contributes to their overall social security. Due to obstacles like restricted access to resources, cultural norms, and societal expectations, women's confidence and perceived agency in making economic decisions have been impacted. Biases against women may prevent them from participating fully in agriculture cooperatives and affect how secure they feel financially.

Author Contributions: Conceptualization, S.T., M.F. and P.S.; methodology, S.T. and M.F.; software, S.T. and M.F.; validation, S.T., M.F. and P.S.; formal analysis, S.T. and M.F.; investigation, S.T. and M.F.; data curation, S.T. and M.F.; writing—original draft preparation, S.T. and M.F.; writing—review and editing, S.T., M.F. and P.S.; visualization, S.T.; supervision, P.S.; project administration, P.S. All authors have read and agreed to the published version of the manuscript.

Funding: This research received no external funding.

Institutional Review Board Statement: Not applicable.

Informed Consent Statement: Informed consent was obtained from all subjects involved in the study.

Conflicts of Interest: The authors declare no conflict of interest.

References

1. Batzios, A.; Kontogeorgos, A.; Chatzitheodoridis, F.; Sergaki, P. What Makes Producers Participate in Marketing Cooperatives? The Northern Greece Case. *Sustainability* **2021**, *13*, 1676. [CrossRef]
2. Mojo, D.; Fischer, C.; Degefa, T. The determinants and economic impacts of membership in coffee farmer cooperatives: Recent evidence from rural Ethiopia. *J. Rural. Stud.* **2017**, *50*, 84–94. [CrossRef]
3. Ahmed, M.H.; Mesfin, H.M. The impact of agricultural cooperatives membership on the wellbeing of smallholder farmers: Empirical evidence from eastern Ethiopia. *Agric. Food Econ.* **2017**, *5*, 1–20. [CrossRef]
4. Wassie, S.B.; Kusakari, H.; Masahiro, S. Inclusiveness and effectiveness of agricultural cooperatives: Recent evidence from Ethiopia. *Int. J. Soc. Econ.* **2019**, *46*, 614–630. [CrossRef]
5. Sergaki, P.; Michailidis, A. Small-Scale Food Producers: Challenges and Implications for SDG2. In *Zero Hunger. Encyclopedia of the UN Sustainable Development Goals*; Leal Filho, W., Azul, A., Brandli, L., Özuyar, P., Wall, T., Eds.; Springer: Cham, Switzerland, 2020; pp. 787–799, ISBN 978-3-319-95674-9.
6. Deng, W.; Hendrikse, G.; Liang, Q. Internal social capital and the life cycle of agricultural cooperatives. *J. Evol. Econ.* **2021**, *31*, 301–323. [CrossRef]
7. Kustepeli, Y.; Gulcan, Y.; Yercan, M.; Yıldırım, B. The role of agricultural development cooperatives in establishing social capital. *Ann. Reg. Sci.* **2020**, *70*, 681–704. [CrossRef]
8. Dunn, J.R. Basic cooperative principles and their relationship to selected practices. *J. Agric. Coop.* **1988**, *3*, 83–93.
9. Hansen, M.H.; Morrow, J.; Batista, J.C. The impact of trust on cooperative membership retention, performance, and satisfaction: An exploratory study. *Int. Food Agribus. Manag. Rev.* **2002**, *5*, 41–59. [CrossRef]

10. Tsiaousi, A.; Partalidou, M. Female farmers in Greece: Looking beyond the statistics and into cultural–social characteristics. *Outlook Agric.* **2021**, *50*, 55–63. [CrossRef]
11. Sergaki, P.; Partalidou, M.; Iakovidou, O. Women's agricultural co-operatives in Greece: A comprehensive review and swot analysis. *J. Dev. Entrep.* **2015**, *20*, 1550002. [CrossRef]

Disclaimer/Publisher's Note: The statements, opinions and data contained in all publications are solely those of the individual author(s) and contributor(s) and not of MDPI and/or the editor(s). MDPI and/or the editor(s) disclaim responsibility for any injury to people or property resulting from any ideas, methods, instructions or products referred to in the content.

Proceeding Paper

The Annual Maintenance Costs of Draft Horses as a Part of Fixed Costs in Horse-Powered Agriculture: A Case Study from Požega, Croatia [†]

Ranko Gantner [1,*], Igor DelVechio [2], Zvonimir Steiner [1], Maja Gregić [1] and Vesna Gantner [1]

[1] Faculty of Agrobiotechnical Sciences Osijek, J. J. Strossmayer University of Osijek, V. Preloga 1, 31000 Osijek, Croatia; zsteiner@fazos.hr (Z.S.); mgregic@fazos.hr (M.G.); vgantner@fazos.hr (V.G.)
[2] Croatian Federation of Heavy Draft Horse Breeders Association, 44317 Popovača, Croatia; zujo.macak@gmail.com
* Correspondence: rgantner@fazos.hr; Tel.: +385-31-554-823
[†] Presented at the 17th International Conference of the Hellenic Association of Agricultural Economists, Thessaloniki, Greece, 2–3 November 2023.

Abstract: The aim of this research is to estimate the fixed costs of the maintenance of draft horses in a low-input farm. Research has revealed that in the investigated case, the fixed costs of maintenance of three draft mares were EUR 5115.39 annually, with human working hours having the greatest share of 73.6%. Income from sales of foals partially offsets the total fixed costs, thus virtually lowering the costs to the level of EUR 1215.39 annually. At the investigated farm (operating on 1.3 ha of arable field crops), the fixed costs per worked arable area were very high, amounting 934.92 EUR/ha, mainly because of little total arable area worked. The theoretical capacity of horse-powered farming with three mares historically was 15 ha, and at such an area, the fixed costs per hectare would fall to the acceptable level of 81 EUR/ha. However, the acceptance of horse-powered farming could face much hesitance, mainly because it is a labor-intensive way of farming, far from the attitudes of modern people. Personal inner transformation might help make this option more attractive.

Keywords: low-input farming; animal work; fixed costs; personal inner transformation

Citation: Gantner, R.; DelVechio, I.; Steiner, Z.; Gregić, M.; Gantner, V. The Annual Maintenance Costs of Draft Horses as a Part of Fixed Costs in Horse-Powered Agriculture: A Case Study from Požega, Croatia. *Proceedings* **2024**, *94*, 25. https://doi.org/10.3390/proceedings2024094025

Academic Editor: Eleni Theodoropoulou

Published: 24 January 2024

Copyright: © 2024 by the authors. Licensee MDPI, Basel, Switzerland. This article is an open access article distributed under the terms and conditions of the Creative Commons Attribution (CC BY) license (https://creativecommons.org/licenses/by/4.0/).

1. Introduction

In the light of need for lesser reliance on fossil energy resources, as well as for lesser environmental impact, horse-powered farming might offer a fairly sustainable option. Namely, draft horses are being fueled with fodder, which comprises organic compounds rich in energy captured from recent photosynthesis [1]. Moreover, this fodder is produced close to the place of consumption, most often at the same farm where it is utilized, thus avoiding distant transport. Technology for producing the fodder is quite simple and cheap, as are the horse-drawn implements used in traditional farming. Horse-powered agriculture also offers the benefits of lessening soil compaction, which has become a serious problem in arable farming [2], thus helping farmers to recover the soil capacity for water accumulation and improve the drought resistance and soil fertility [3]. Despite the obvious attractiveness of animal-powered farming, there are many economic issues unknown to modern decision makers in the farming sector. Among the issues certainly are the fixed costs, in this case related to the maintenance of draft horses on an annual basis. These costs include the feed costs, watering, housing (shelter and fencing), and care. The aim of this research is to reveal the fixed costs of draft horse maintenance in a small family farm near Požega, Croatia.

2. Materials and Methods

Data for conducting this research have been obtained by on-farm observations of horse feeding, care, and work, and from the records of the investigated family farm near Požega (in the hilly region of the central Slavonia, Croatia). The farm constantly (in the long run) keeps three mares of the Croatian Heavy Draft Horse Breed (average body weight of about 650 kg/head), and raises three foals each year. Foals are being sold each year after weaning (at the age of 7 months) as a source of income or, in some circumstances, are kept on the farm for the replacement of an old mare (rarely). Mares on the farm usually give a draft force for powering the field agrotechnical operations on the entire 1.3 ha of arable land (for soil preparation, seeding and the cultivation of oats, maize, and green-manure crops, like crimson clover and brassicas), for transport of hay and manure and, on some occasions, for pulling a carriage in wedding procession. Forages fed to horses come from the nearby abandoned grasslands and lucerne crop (hay for winter feeding and fresh green herbage for summer feeding), at virtually no cost, since the use of the meadow and lucerne was free, whilst the oats come from their own oats crop. The costs of mowing the meadow, hay gathering, baling and transport of bales amounted for total of 13 EUR/220 kg round bale in the year 2022, which equals 0.052 EUR/kg of hay. The meadow gives between 38 and 60 bales of hay annually, depending on the year. The cost of the oats is assumed to be equal to the average market price during period from the last harvest to the forthcoming harvest in July 2023, which was 0.30 EUR/kg. The cost of working hours of the farmer is assumed to be equal to the per-hour net Croatian average salary (1094 EUR/month, with 21 working days of 8 h per day [4]), which amounts for 6.51 EUR/hour. The average Eurodiesel fuel price was assumed to be 1.40 EUR/l.

3. Results

During the winter season (from mid-October to mid-March), the horses are kept in the winter coral, with a shelter near the farmer's home (a distance about 150 m) and, therefore, it took only one hour daily (Table 1) for the farmer to serve the fodder and water to the horses, and to take care of them (grooming, checking the fencing, checking their health). During the summer season (from mid-March to mid-October) the horses are kept in summer coral near the orchard, meadow, and small arable field of the farmer (the distance from farmer's home is about 2.5 km). During the summer period, the farmer needed two hours daily (Table 1) to get to the meadow or lucerne crop, to mow the fresh herbage, to load the herbage into a private van, to take and serve it to the horses, for watering, care for the horses, and for checks of the fencing.

Table 1. Daily consumption of fresh herbage, hay, and oats, working hours spent by the farmer, and the related monetary costs for the three mares (in winter) plus three foals (in summer).

	Winter Period			Summer Period		
Daily Consumption and Costs	Total	Per Mare	Total EUR	Total	Per Mare and Foal	Total EUR
Hay consumption (kg)	73	24	3.80			
Fresh green herbage (kg)				300	100	0.00
Oats consumption (kg)	6	2	1.80	6	2	1.80
Farmers working hours (h)	1	0.33	6.51	2	0.67	13.02
Diesel fuel for private van (l)				0.4	0.13	0.56
Total daily			12.11			15.38

The total costs during the summer period were much greater than during the winter period (Table 2), due to the longer period of summer feeding, and doubled the farmer's working time spent on serving the horses, mainly because of the time needed for everyday mowing and transporting the herbage to the summer coral.

Table 2. Annual consumption of fresh herbage, hay, and oats, working hours spent by the farmer, and the related monetary costs for the three mares (in winter) plus three foals (in summer).

Seasonal Consumption and Costs	Winter Period			Summer Period		
	Total	Per Mare	Total EUR	Total	Per Mare and Foal	Total EUR
Hay consumption (kg)	11,096	3699	576.99			
Fresh green herbage (kg)				63,900	21,300	0.00
Oats consumption (kg)	912	304	273.60	1278	426	383.00
Farmers working hours (h)	152	51	989.52	426	142	2773.00
Diesel fuel for private van (l)				85.2	28.4	119.28
Total seasonal			1840.11			3275.28
Total annual			5115.39			

The farmer's working hours resulted in the greatest share in the total annual costs of three draft mares' maintenance (Figure 1), followed by cost of oats, hay, and diesel fuel.

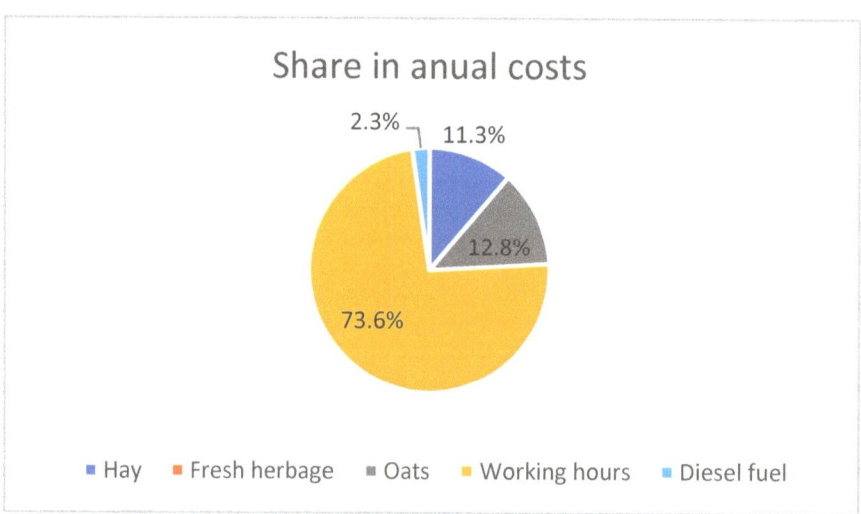

Figure 1. The share of various costs in the total annual cost of maintenance of the three draft mares.

The income from the foals depends on the sex of the foals. Namely, males are sold for 1000 EUR/head, whilst females for 1600 EUR/head. Under the assumption that the ratio of male: female is 1:1, then the average income per foal would be 1300 EUR/head, which totals EUR 3900 annually per three mares. The income from the sales of foals partially offsets the total annual costs of keeping the draft mares, so the rest of EUR 1215.39 should be charged to the use of the mares in field work, i.e., agrotechnical operations. On the investigated farm, these three draft mares give the draft power for cultivating 1.3 ha of arable land, thus giving an average fixed cost of 934.92 EUR/ha. This can be deemed as relatively high when compared to the average value of field arable crops, like maize, wheat, and oats (between 1000 and 3000 EUR/ha), but in the case of the studied farm, it is because the farm operates on a small area. Under the assumption that a well-trained pair of horses can give power to up to 15 ha of field crops, the fixed costs diminish with an increase in the farm size, down to 81 EUR/ha for a farm size of 15 ha (Figure 2).

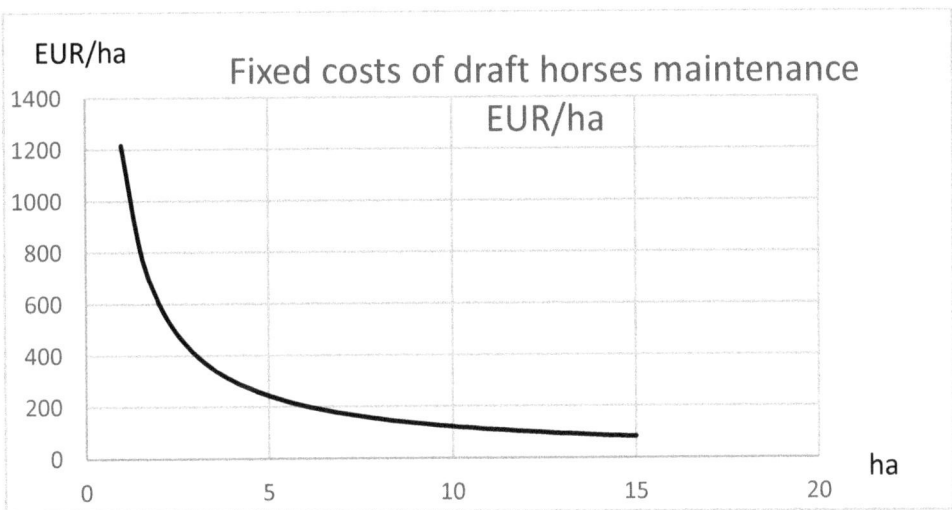

Figure 2. Projected fixed cost per ha, depending on the farm size.

4. Discussion

Despite the fact that fixed costs can be lowered to an acceptable level with an increase in the farm's arable area (up to the historical maximum of 15 ha per pair of draft horses), the acceptance of horse-powered farming could face much hesitance, mainly because it is a labor-intensive way of farming, far from the attitudes of modern people. Personal inner transformation [5] might help make this option more attractive.

Author Contributions: Conceptualization, R.G.; methodology, R.G.; validation, Z.S.; formal analysis, R.G.; investigation, R.G.; resources, I.D.; data curation, R.G.; writing—original draft preparation, R.G.; writing—review and editing, V.G.; visualization, M.G.; supervision, V.G.; project administration, V.G.; funding acquisition, V.G. All authors have read and agreed to the published version of the manuscript.

Funding: This research and dissemination were supported by the Fund for Bilateral Relations within the Financial Mechanism of the European Economic Area and Norwegian Financial Mechanism for the period 2014-2021 (Grant number: 04-UBS-U-0031/23-14).

Institutional Review Board Statement: This study did not require ethical approval.

Informed Consent Statement: Informed consent was obtained from all subjects involved in the study.

Data Availability Statement: Data are withheld due to personal privacy.

Conflicts of Interest: The authors declare no conflicts of interest. The funders had no role in the design of the study; in the collection, analyses, or interpretation of data; in the writing of the manuscript; or in the decision to publish the results.

References

1. Gantner, R.; Šumanovac, L.; Zimmer, D.; Ronta, M.; Glavaš, H.; Ivanović, M.; Jarić, D. Fodder as a biofuel: Cost effectiveness of powering horses in plowing operation. In Proceedings of the 6th International Scientific Symposium "Economy of Eastern Croatia—Vision and Growth, Osijek, Croatia, 25–27 May 2017; University of Osijek, Faculty of Economics: Osijek, Croatia, 2017; pp. 618–625.
2. Hamza, M.A.; Anderson, W.K. Soil compaction in cropping systems: A review of the nature, causes and possible solutions. *Soil Tillage Res.* **2005**, *82*, 121–145. [CrossRef]
3. Gantner, R.; Baban, M.; Glavaš, H.; Ivanović, M.; Schlechter, P.; Šumanovac, L. Indices of sustainability of horse traction in agriculture. In Proceedings of the 3rd International Scientific Symposium on Economy of Eastern Croatia—Vision and Growth, Osijek, Croatia, 22–24 May 2014; University J. J. Strossmaayer in Osijek, Faculty of Economy: Osijek, Croatia.

4. Croatian Bureau of Statistics. Prosječna Mjesečna Neto Plaća za Siječanj 2023. Iznosila Je 1094 Eura. Available online: https://dzs.gov.hr/vijesti/prosjecna-mjesecna-neto-placa-za-sijecanj-2023-iznosila-je-1-094-eura/1486 (accessed on 15 June 2023).
5. Woiwode, C.; Schäpke, N.; Bina, O.; Veciana, S.; Kunze, I.; Parodi, O.; Schweizer-Ries, P.; Wamsler, C. Inner transformation to sustainability as a deep leverage point: Fostering new avenues for change through dialogue and reflection. *Sustain. Sci.* **2021**, *16*, 841–858. [CrossRef]

Disclaimer/Publisher's Note: The statements, opinions and data contained in all publications are solely those of the individual author(s) and contributor(s) and not of MDPI and/or the editor(s). MDPI and/or the editor(s) disclaim responsibility for any injury to people or property resulting from any ideas, methods, instructions or products referred to in the content.

Proceeding Paper

The Effect of Farm Size on the Differences in Mastitis Prevalence and Its Consequences on Milk Production in Holstein Cows [†]

Vesna Gantner [1,*], Ivana Jožef [2], Ranko Gantner [1], Maja Gregić [1] and Zvonimir Steiner [1]

1. Department of Animal Production and Biotechnology, Faculty of Agrobiotechnical Sciences Osijek, University of J. J. Strossmayer Osijek, V. Preloga 1, 31000 Osijek, Croatia; rgantner@fazos.hr (R.G.); mgregic@fazos.hr (M.G.); zsteriner@fazos.hr (Z.S.)
2. Faculty of Agriculture, University in Novi Sad, 21000 Novi Sad, Serbia; vanjajozef@gmail.com
* Correspondence: vgantner@fazos.hr
† Presented at the 17th International Conference of the Hellenic Association of Agricultural Economists, Thessaloniki, Greece, 2–3 November 2023.

Abstract: Aiming to determine the prevalence of mastitis and its consequences on milk production, 3,953,637 test-day records of Holstein cows (period 01/2005 to 12/2022) were analyzed. The obtained analyses indicate differences in mastitis prevalence and the consequences on successive milk production depending on herd size. The lowest mastitis prevalence was observed on the largest farms (>500), while the most pronounced recovery potential was observed for farms with 200–500 cows. Higher mastitis prevalence and lower recovery potential observed at smaller farms indicate the necessity of education and knowledge transfer to those farms.

Keywords: mastitis prevalence; Holstein breed; somatic cell count; daily milk yield

Citation: Gantner, V.; Jožef, I.; Gantner, R.; Gregić, M.; Steiner, Z. The Effect of Farm Size on the Differences in Mastitis Prevalence and Its Consequences on Milk Production in Holstein Cows. *Proceedings* **2024**, *94*, 26. https://doi.org/10.3390/proceedings2024094026

Academic Editor: Eleni Theodoropoulou

Published: 24 January 2024

Copyright: © 2024 by the authors. Licensee MDPI, Basel, Switzerland. This article is an open access article distributed under the terms and conditions of the Creative Commons Attribution (CC BY) license (https://creativecommons.org/licenses/by/4.0/).

1. Introduction

One of the most expensive and frequent disorders on a dairy cattle farm is the inflammation of the udder or mastitis. The prevalence of mastitis could occur in a subclinical or clinical form, including a set of various changes in the animal's udder and the deteriorating overall health status of an animal. This could be caused by bacterial infections, mechanical injuries or irritation (inadequate milking), inadequate hygiene, etc. Furthermore, mastitis prevalence is correlated with the use of antibiotics and considerable financial loss due to a decline in the quality and quantity of milk [1]. In addition, mastitis prevalence has a negative effect on the environment by increasing GHG emissions from dairy farms [2]. The application of various mastitis detection methods and the prevention of mastitis prevalence represents an efficient way of enabling economically and environmentally satisfactory dairy farming. Study [3] stated that mastitis prevalence damages the udder tissue, resulting in a raised somatic cell count (SCC). Therefore, SCC, as an integral parameter in regular milk recording, could be used as an accurate indicator of mastitis prevalence without any additional cost [4,5].

Due to the increasing significance of the prevention of various disorders/diseases in dairy cattle, this research aimed to determine mastitis prevalence and its consequences on milk production in the Holstein population considering farm size.

2. Methods Section

After logical control, the analyzed data set consisted of 3,953,637 test-day records of Holsteins referring to the period 01/2005 to 12/2022. The daily somatic cell count (SCC) was used as a mastitis indicator (healthy animals (SCC < 200,000/mL) for cows at mastitis risk (SCC = 200,000–400,000/mL) and cows with mastitis (SCC > 400,000/mL)). The prevalence was defined as the percentage (%) of cows in each class from the total population (analysis was performed considering herd size). For the analysis of the mastitis consequence, only cows with determined mastitis (SCC > 400,000/mL) were considered. The daily milk yield on the day when mastitis was detected was taken as a reference value. The mastitis index was defined concerning the number of days after mastitis as follows: D-0 (detecting date), A-1 (within 35 days), A-2 (36–70 days), A-3 (71–105 days), and A-4 (>105 days). The effect of mastitis on successive milk production was tested using the MIXED procedure of SAS [6] with a statistical model that included the effects of lactation stage, parity, age at first calving, the milk recording season, and mastitis index. The statistical analysis was performed separately for each herd size class (<5, 5–10, 10–50, 50–200, 200–500, >500 cows).

3. Results and Discussion

The prevalence of healthy cows at mastitis risk and cows with mastitis concerning the farm size is presented in Figure 1. The highest percentage of healthy animals was observed for the largest farms with more than 500 cows in lactation at the amount of 73.6%, while the highest percentage of animals at risk and with mastitis was observed in farms with less than 5 cows (16.5%; 27.8%). Furthermore, there was a visible trend of an increase in the risk of mastitis and mastitis prevalence depending on the number of animals in lactation, i.e., with increasing farm size as the prevalence decreased. The observed trend could be explained by significantly better management strategies (higher level of investments in equipment, knowledge, feeding quality, etc.) and the higher genetic potential of animals in production on larger farms.

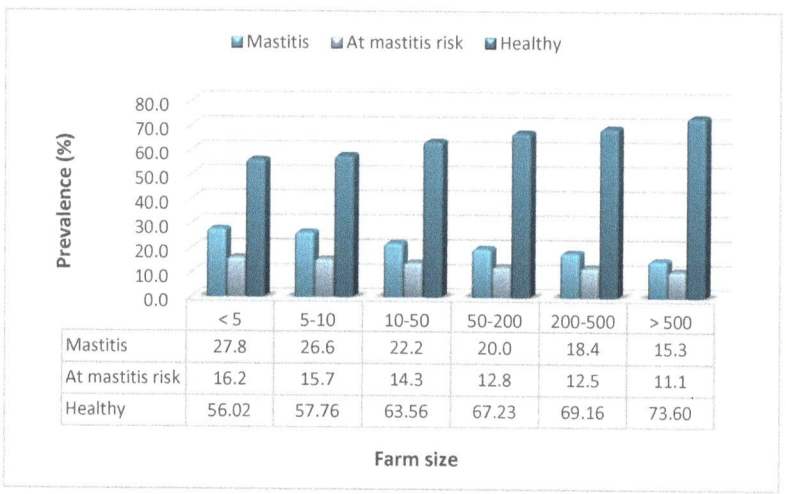

Figure 1. Prevalence of mastitis in Holstein cows depending on farm size.

Estimated differences in the quantity (kg) and value (euro) of milk at successive milk recordings after the detection of mastitis (SCC > 400,000/mL) depending on herd size are shown in Table 1. The highest difference was recorded at the first successive milk recording after mastitis detection (A-1 milk recording) regardless of the farm size, with the highest difference observed on farms with 200–500 cows (64.239 kg; 33.40 euro) and the lowest on farms with less than 5 cows. In the other analyzed periods between successive milk recordings, the differences varied but were mainly negative (indicating a decrease in milk production). The highest total estimated difference was recorded in herds with 200 to 500 cows (66.654 kg; 34.66 euro), while the lowest total estimated difference was recorded in herds with less than 5 cows (22.30 kg; 11.60 euro).

Table 1. Estimated differences in quantity (kg) and value (euro) of milk from Holstein cows with mastitis (SCC > 400,000/mL) depending on farm size.

Farm Size	A-1		A-2		A-3		A-4		Total Difference	
	kg	eur	kg	eur	kg	eur	kg	eur	kg	eur
<5	20.587	10.71	1.282	0.67	0.599	0.31	−0.164	−0.09	22.304	11.60
5–10	31.424	16.34	−2.443	−1.27	0.624	0.32	−1.843	−0.96	27.762	14.44
10–50	42.620	22.16	−4.812	−2.50	−3.818	−1.99	−0.857	−0.45	33.134	17.23
50–200	44.088	22.93	7.829	4.07	−4.646	−2.42	−8.511	−4.43	38.760	20.16
200–500	64.239	33.40	5.939	3.09	3.058	1.59	−6.582	−3.42	66.654	34.66
>500	56.042	29.14	−0.532	−0.28	−7.928	−4.12	4.931	2.56	52.513	27.31

Note: A-1, A-2, A-3, A-4—successive milk recordings.

A determined increase in daily production for successive milk recordings indicates the potential for recovery of animals after the prevalence of mastitis. From the point of view of the total difference in the analyzed period (four successive milk recordings after mastitis prevalence), the highest increase in the daily productivity was determined in herds with 200–500 cows in lactation, indicating the highest recovery potential of animals at those farms. Furthermore, it was observed that the recovery from mastitis risk varied regarding farm size, with the lowest observed in small farms with less than five cows in production.

Ref. [7] states that herd size affects the prevalence of any disorder/disease within the herd, including mastitis, and that a higher prevalence was found in smaller herds (30–99 cows). Similarly, Ref. [8] also reports a higher frequency of subclinical mastitis in small herds compared to medium and large herds and explains the same with less attention paid to cow management when the farm is small. Furthermore, Ref. [7] states that an increase in herd size is associated with increased milk production and productivity. Refs. [9,10] suggest that season, herd management, average production, somatic cell counts, and herd size could be related to mastitis prevalence rate in dairy herds.

4. Conclusions

The obtained analyses indicate differences in mastitis prevalence and consequences on successive milk production depending on herd size. The lowest mastitis prevalence was observed on the largest farms (>500), while the most pronounced recovery potential was observed at farms with 200–500 cows. Higher mastitis prevalence and lower recovery potential observed in smaller farms indicate the necessity of education and knowledge transfer to those farms.

Author Contributions: Conceptualization, V.G. and I.J.; methodology, V.G.; software, V.G.; validation, V.G., I.J., and R.G.; formal analysis, V.G.; investigation, V.G. and I.J.; resources, V.G. and R.G.; data curation, V.G., I.J. and M.G.; writing—original draft preparation, V.G.; writing—review and editing, R.G. and Z.S; visualization, V.G. and M.G.; supervision, V.G.; project administration, R.G. and Z.S.; funding acquisition, V.G. and R.G. All authors have read and agreed to the published version of the manuscript.

Funding: Research and dissemination were supported by the Fund for Bilateral Relations within the Financial Mechanism of the European Economic Area and Norwegian Financial Mechanism for the period 2014–2021 (Grant number: 04-UBS-U-0031/23-14).

Institutional Review Board Statement: This study did not require ethical approval.

Informed Consent Statement: Not applicable.

Data Availability Statement: The data used in this research are unavailable due to privacy restrictions.

Conflicts of Interest: The authors declare no conflicts of interest.

References

1. Hill, A.E.; Green, A.L.; Wagner, B.A.; Dargatz, D.A. Relationship between herd size and annual prevalence of and primary antimicrobial treatments for common diseases on dairy operations in the United States. *Prev. Vet. Med.* **2009**, *88*, 264–277. [CrossRef] [PubMed]
2. Özkan Gülzari, Ş.; Vosough Ahmadi, B.; Stott, A.W. Impact of subclinical mastitis on greenhouse gas emissions intensity and profitability of dairy cows in Norway. *Prev. Vet. Med.* **2018**, *150*, 19–29. [CrossRef] [PubMed]
3. Pyorala, S. Indicators of inflammation in the diagnosis of mastitis. *Vet. Res.* **2003**, *34*, 565–578. [CrossRef] [PubMed]
4. Alhussien, M.N.; Dang, A.K. Milk somatic cells, factors influencing their release, future prospects, and practical utility in dairy animals: An overview. *Vet. World* **2018**, *11*, 562–577. [CrossRef]
5. ICAR, International Committee for Animal Recording. ICAR Guidelines: Section 2—Guidelines for Dairy Cattle Milk Recording. ICAR. 2022. Available online: https://www.icar.org/Guidelines/02-Overview-Cattle-Milk-Recording.pdf (accessed on 13 January 2024).
6. SAS Institute Inc. *SAS User's Guide*; Version 9.4.; SAS Institute Inc.: Cary, NC, USA, 2019.
7. Fesseha, H.; Mathewos, M.; Aliye, S.; Wolde, A. Study on Prevalence of Bovine Mastitis and Associated Risk Factors in Dairy Farms of Modjo Town and Suburbs, Central Oromia, Ethiopia. *Vet. Med. Res. Rep.* **2021**, *12*, 271–283. [CrossRef] [PubMed]
8. Gantner, V.; Mijić, P.; Kuterovac, K.; Solić, D.; Gantner, R. Temperature-humidity index values and their significance on the daily production of dairy cattle. *Mljekarstvo* **2011**, *61*, 56–63.
9. Tomazi, T.; Ferreira, G.C.; Orsi, A.M.; Gonçalves, J.L.; Ospina, P.A.; Nydam, D.V.; Moroni, P.; dos Santos, M.V. Association of herd-level risk factors and incidence rate of clinical mastitis in 20 Brazilian dairy herds. *Prev. Vet. Med.* **2018**, *161*, 9–18. [CrossRef]
10. Weber, C.T.; Corrêa Schneider, C.L.; Busanello, M.; Bandeira Calgaro, J.L.; Fioresi, J.; Gehrke, C.R.; da Conceição, J.M.; Haygert-Velho, I.M.P. Season effects on the composition of milk produced by a Holstein herd managed under semi-confinement followed by compost bedded dairy barn management. *Semin. Cienc. Agrar.* **2020**, *41*, 1667–1678. [CrossRef]

Disclaimer/Publisher's Note: The statements, opinions and data contained in all publications are solely those of the individual author(s) and contributor(s) and not of MDPI and/or the editor(s). MDPI and/or the editor(s) disclaim responsibility for any injury to people or property resulting from any ideas, methods, instructions or products referred to in the content.

Proceeding Paper

Exploring the Impact of the Greening of the Agri-Food Sector on Economic Growth: An Empirical Approach in the BVAR Framework for the EU [†]

Eleni Zafeiriou [1,*], Garyfallos Arabatzis [2], Georgios Tsantopoulos [2], Spyros Galatsidas [2] and Stavros Tsiantikoudis [2]

1. Department of Agricultural Development, Democritus University of Thrace, 68200 Orestiada, Greece
2. Department of Forestry and Management of the Environment and Natural Resources, Democritus University of Thrace, 68200 Orestiada, Greece; garamp@fmenr.duth.gr (G.A.); tsantopo@fmenr.duth.gr (G.T.); sgalatsi@fmenr.duth.gr (S.G.); stsianti@fmenr.duth.gr (S.T.)
* Correspondence: ezafeir@agro.duth.gr; Tel.: +30-6932627501
† Presented at the 17th International Conference of the Hellenic Association of Agricultural Economists, Thessaloniki, Greece, 2–3 November 2023.

Abstract: The Greening in agro—food sector has become within the last decade a high priority issue given the 17 Sustainable targets set by OECD. More specifically, the Sustainable Development Goals (SDGs) by 2030, intend to promote using environmental resources in close correlation with measures to reduce non-environmental human pressure on the planet as well as in agro-food sector. The present work studies the greening of agro-food sector as synopsized in emissions per capita by agro-food sector for the EU and its relation to economic growth per capita with the assistance of a BVAR framework. Our findings do not validate success in greening of agro—food sector since the emissions reduction is not accompanied by economic growth a result that rejects the hypothesis of eco efficiency. Future research could involve the construction of an index that should incorporate more variables that will reflect more accurately the greening efforts in agro—food sector.

Keywords: ecoefficiency; farm to fork strategy; BVAR; impulse response; agro-food industry

1. Introduction

Modern lifestyles worldwide are constantly putting pressure on natural resources that are increasingly at risk of depletion. The vertical growth of the population and the continuous strengthening of industrial and agricultural production have for years created concerns about the ability of future societies to cover their basic needs. More specifically, food production must double by 2050 to meet the world's growing population's expected demand and that the global population will number approximately 9.8 billion by 2050 and 11.2 billion by 2100 [1,2].

Therefore, an organized and gradual shift towards green production processes that can ensure the sustainability of the future is an option. At the same time, those methods are identified that can adapt green entrepreneurship to the requirements of the necessary economic development. The agri-food sector is decisive for the survival of the people and through it the largest volume of food is produced. Therefore, applying green practices in this area as well can ensure sustainable economic growth.

Well-organized and resilient agro-food systems can ensure the survival of societies in the future [1]. Agenda 2030, namely ESG of the United Nations, with 17 complex and interrelated objectives, provides a useful tool for sustainability. The European Union makes a great effort to cope with this new reality and therefore, in this direction, governments have promoted policies for a green transition through which ecoefficiency may be a feasible result satisfying societal demand [2–6].

Having in mind all the above, the present work makes an effort to analyse the impact of utilizing green entrepreneurship (as synopsized in emissions per capita in tonnes generated by the agrifood system for EU as an entity) and its linkage to economic development (as reflected to GDP per capita generated by agriculture Forestry and Fishing).

2. Materials and Methods

The data of the present work are annual for the time period 1990–2020. As mentioned above, we selected the emissions per capita to be represented by the agri-food sector's intensity (as proxy for environmental degradation) and GDP per capita (to describe EU economic growth).

The data employed in our Model are illustrated in the next Figure 1.

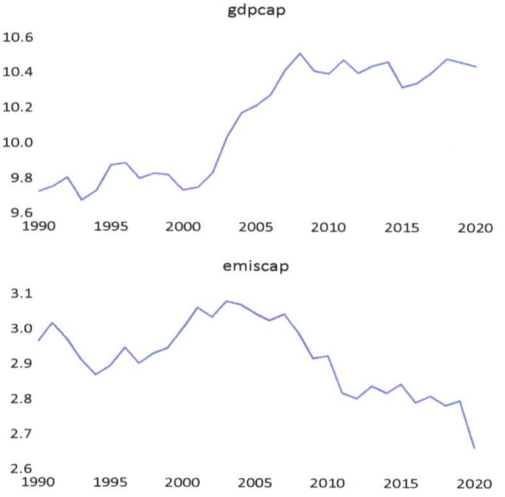

Figure 1. Evolution of the model variables employed (1990–2020).

The break unit root test is the first analysis employed for our data [7]. Then, we employed the BVAR methodology in order to detect the interlinkages among green energy and economic growth in agriculture [8–14]. The mathematical form of a BVAR model is the same though the parameters' estimation and interpretation do not coincide. Actually, the BVAR models, by incorporating prior information about model parameters, secure reliable results since the particular process stabilizes parameter estimation. BVAR model estimation is based on the Minnesota prior specification, while all the information is incorporated in the parameters' estimations. Based on the maximum likelihood function, we estimate the posteriors [15,16].

Based on the BVAR estimation model, we generate a tractable posterior density function that is similar to that of the prior. The prior selected is the Litterman/Minnesota algorithm for the target parameter. The next step in our BVAR analysis involves the specification of the prior covariance or the target parameter, having incorporated a set of hyperparameters [14–17].

The last step in our analysis involves impulse response function estimation (IRF) for each variable as well as forecast error variance decomposition analysis (FEVD). Impulse response analysis is a significant tool in econometric analysis, since it may well describe the evolution of the estimated VAR model's variables as a response to a shock in one or more variables. In other words, this step allows the analyst to trace the transmission of a single shock within the noisy system of equations and therefore we can make an assessment of the economic policy impacts on the model variables' evolution within a period that may be 10 or 20 years in the case the data employed are annual [6,7]. In a similar vein, variance decomposition or in other words 'forecast error variance decomposition is a specific tool that may adequately and precisely interpret the relations between variables described by the model estimated. This methodology will amplify impulse response analysis since it further quantifies the contribution rates of all variables to the impact on the dependent variable [18,19].

The model's evaluation was based on forecast accuracy performance for the classic VAR and BVAR specifications, respectively, with the assistance of the following indices, the root mean square error (RMSE) and the mean absolute error (MAE) [17]. The forecast accuracy measures were selected on the basis of sensitivity extending to the deviations from the true values.

3. Results

The break unit root test provided the results illustrated in Table 1.

Table 1. ADF break unit root results.

Variables	ADF Break Unit Root	Break Date
CEM	−3.33 (0.778)	1999
ΔCEM	−5.5 *** (0.00)	2001
GDP	−3.8 (0.48)	2002
ΔGDP	4.82 *** (0.0)	2003

*** Reject unit root test for 1% level of significance with critical values −4.94, −4.44, and −4.19 for 1, 5 and 10% levels of significance. CEM denotes carbon emissions per capita for the agri-food system for the EU; GDP is denoted as GDP per capita; ΔCEM ΔGDP denotes the first differences of the variables.

Based on the aforementioned findings for the EU, all the respective variables are found to be I(1) with the years 1999 and 2002 being identified as structural breaks. The Kyoto Protocol (1996–1999 signing period) as well as the different financial crises may well explain the breakpoints identified. Impulse response analysis was also employed to detect and identify the interlinkages among the variables employed, as illustrated in Figure 2.

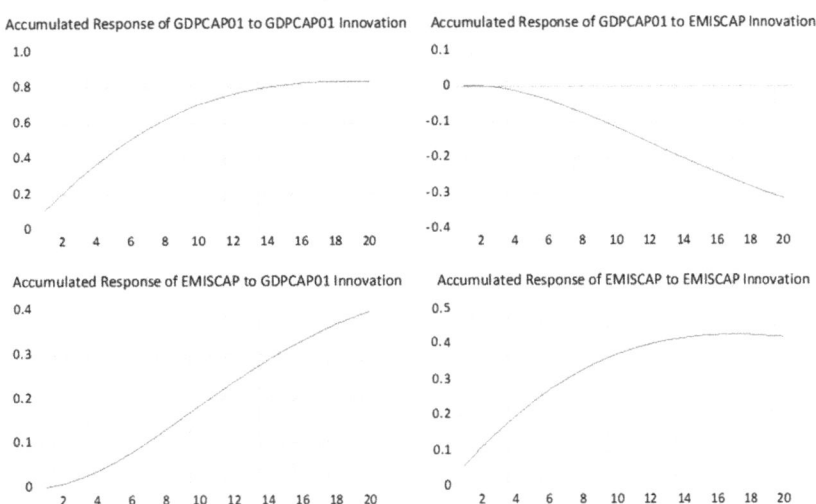

Figure 2. Impulse response analysis of the variables employed.

The figures constructed were based on the Bayesian methodology of Gibbs sampling while 1000 iterations were implemented to acquire the results [18]. GDP is increasing with a declining trend for a time period of twenty years while emissions are increasing with a declining trend in the first decade, though then the slope of the curve begins to change and increases. This means that the greening of the agri-food sector cannot provide steadily increasing growth and therefore that more steps need to be taken for ecoefficiency to become an achievable objective in EU in Figure 3.

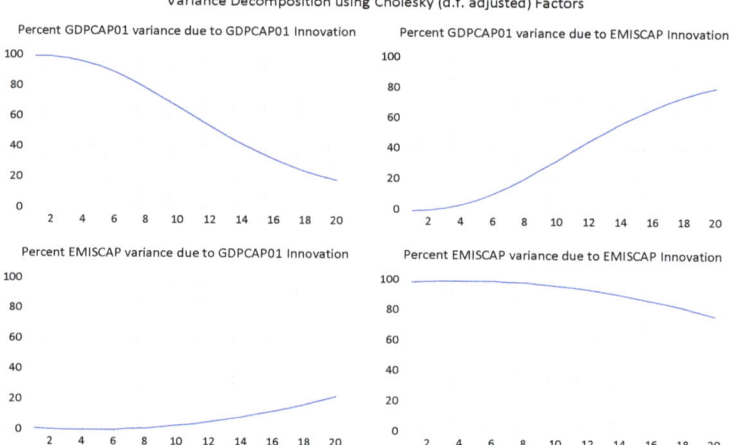

Figure 3. Variance Decomposition Analysis analysis of the variables employed.

Based on our findings an innovation on greening interprets the income variability with an increasing rate and reaches 80% after of 20 periods validating the slow process through which greening entrepreneurship may affect income volatility. On the other hand the rate is even slower to interpret the greening variance attributed to income innovation reaching 20% of the total variance. This result is indicative that other than income motivation could promote the adoption of greening practices. Last but not least the MAE = 0.098 and RMSE = 0.118 validating a good forecast ability.

4. Conclusions

Green or sustainable practices in the agro-food sector have become common in modern societies. Especially in the EU, this trend has been imposed on different stages of the agro-food industry including the farm-to-fork strategy in line with the SDG strategy, which aims to deliver nutritious and affordable food for a growing world. Actually, in EU, the particular strategy aims' to make food systems fair, healthy and environmentally friendly. The present work has employed the BVAR methodology to identify the interlinkage among emissions per capita generated by the agro-food sector as a proxy for the greening of the agro-food sector and GDP per capita. Our findings confirm that greening is far from being successful since the effort to reduce carbon emissions is not accompanied by economic efficiency. To synopsize, more steps should be taken in order for ecoefficiency to become an achievable objective in the agro-food sector.

Author Contributions: Conceptualisation, E.Z. and G.A.; methodology, E.Z., S.G. and S.T.; software, E.Z.; validation, G.T., S.T. and S.G.; formal analysis, E.Z.; investigation, E.Z.; resources, S.T.; data curation, G.T.; writing—original draft preparation, E.Z., G.A. and G.T.; writing—review and editing, E.Z., G.A. and G.T.; supervision, E.Z.; project administration, E.Z. All authors have read and agreed to the published version of the manuscript.

Funding: This research received no external funding.

Institutional Review Board Statement: Not applicable.

Informed Consent Statement: Not applicable.

Data Availability Statement: The dataset is available at www.FAOSTAT.org. (accessed on 1 June 2023).

Conflicts of Interest: The authors declare no conflict of interest.

References

1. Preiss, M.; Vogt, H.M.J.; Dreher, C.; Schreiner, M. Trends Shaping Western European Agrifood Systems of the Future. *Sustainability* **2022**, *14*, 3976. [CrossRef]
2. Hurduzeu, G.; Pânzaru, L.R.; Medelete, M.D.; Ciobanu, A.; Enea, C. The Development of Sustainable Agriculture in EU Countries and the Potential Achievement of Sustainable Development Goals Specific Targets (SDG 2). *Sustainability* **2022**, *14*, 5798. [CrossRef]
3. Narayan, P.K.; Popp, S. Size and power properties of structural break unit root tests. *Appl. Econ.* **2013**, *45*, 721–728. [CrossRef]
4. Bloor, C.; Matheson, T. Real-time conditional forecasts with Bayesian VARs: An application to New Zealand. *N. Am. J. Econ. Financ.* **2011**, *22*, 26–42. [CrossRef]
5. Meredith, S.; Allen, B.; Kollenda, E.; Maréchal, A.; Hart, K.; Hulot, J.F.; Frelih-Larsen, A.; Wunder, S. *European Food and Agriculture in a New Paradigm: Can [67]global Challenges Like Climate Change Be Addressed through a Farm to Forkapproach?* Institute for European Environmental Policy and the Ecologic Institute: Berlin, Germany, 2021.
6. Chygryn, O. Green entrepreneurship: EU experience and Ukraine perspectives. *CSEI Work. Pap. Ser.* **2017**, *6*, 6–13.
7. Kabiraj, S.; Topkar, V.; Walke, R.C. Going Green: A Holistic Approach to Transform Business. *Int. J. Manag. Inf. Technol.* **2010**, *2*, 22–31. [CrossRef]
8. Lütkepohl, H. Vector autoregressive models. In *Handbook of Research Methods and Applications in Empirical Macroeconomics*; Edward Elgar Publishing: Cheltenham, UK, 2013.
9. Ivanov, V.; Kilian, L. A practitioner's guide to lag order selection for VAR impulse response analysis. *Stud. Nonlinear Dyn. Econom.* **2005**, *9*, 2. [CrossRef]
10. Giannone, D.; Reichlin, L.; Sala, L. VARs, common factors, and the empirical validation of equilibrium business cycle models. *J. Econom.* **2006**, *132*, 257–279. [CrossRef]
11. Kang, S.H.; Islam, F.; Tiwari, A.K. The dynamic relationships among CO_2 emissions, renewable and non-renewable energy sources, and economic growth in India: Evidence from time-varying Bayesian VAR model. *Struct. Change Econ. Dyn.* **2019**, *50*, 90–101. [CrossRef]
12. Pesaran, H.H.; Shin, Y. Generalized impulse response analysis in linear multivariate models. *Econ. Lett.* **1998**, *58*, 17–29. [CrossRef]
13. Sarantis, N.; Stewart, C. Structural, VAR and BVAR models of exchange rate determination: A comparison of their forecasting performance. *J. Forecast.* **1995**, *14*, 201–215. [CrossRef]
14. Tsioptsia, K.A.; Zafeiriou, E.; Niklis, D.; Sariannidis, N.; Zopounidis, C. The Corporate Economic Performance of Environmentally Eligible Firms Nexus Climate Change: An Empirical Research in a Bayesian VAR Framework. *Energies* **2022**, *15*, 7266. [CrossRef]
15. Purcel, A.A. New insights into the environmental Kuznets curve hypothesis in developing and transition economies: A literature survey. *Environ. Econ. Policy Stud.* **2020**, *22*, 585–631. [CrossRef]
16. Yan, H.; Xiao, W.; Deng, Q.; Xiong, S. Analysis of the Impact of US Trade Policy Uncertainty on China Based on Bayesian VAR Model. *J. Math.* **2022**, 7124997.
17. Brahmasrene, T.; Huang, J.C.; Sissoko, Y. Crude oil prices and exchange rates: Causality, variance decomposition and impulse response. *Energy Econ.* **2014**, *44*, 407–412. [CrossRef]
18. Jakada, A.H.; Mahmood, S.; Ahmad, A.U.; Garba Muhammad, I.; Aliyu Danmaraya, I.; Sani Yahaya, N. Driving Forces of CO_2 Emissions Based on Impulse Response Function and Variance Decomposition: A Case of the Main African Countries. *Environ. Health Eng. Manag. J.* **2022**, *9*, 223–232. [CrossRef]
19. Gorodnichenko, Y.; Lee, B. Forecast error variance decompositions with local projections. *J. Bus. Econ. Stat.* **2020**, *38*, 921–933. [CrossRef]

Disclaimer/Publisher's Note: The statements, opinions and data contained in all publications are solely those of the individual author(s) and contributor(s) and not of MDPI and/or the editor(s). MDPI and/or the editor(s) disclaim responsibility for any injury to people or property resulting from any ideas, methods, instructions or products referred to in the content.

Proceeding Paper

Partial Substitution of Fresh Microalgae with Baker's Yeast (*Saccharomyces cerevisiae*) Enhances the Growth of Juvenile *Ostrea edulis* and *Ruditapes decussatus* †

Dimitrios K. Papadopoulos [1], Ioannis Georgoulis [1], Athanasios Lattos [1], Konstantinos Feidantsis [1], Basile Michaelidis [1] and Ioannis A. Giantsis [2,*]

[1] Laboratory of Animal Physiology, Department of Zoology, Faculty of Science, School of Biology, Aristotle University of Thessaloniki, 54124 Thessaloniki, Greece; dkpapado@bio.auth.gr (D.K.P.); georgoim1707@gmail.com (I.G.); lattosad@bio.auth.gr (A.L.); kfeidant@upatras.gr (K.F.); michaeli@bio.auth.gr (B.M.)

[2] Department of Animal Science, Faculty of Agricultural Sciences, University of Western Macedonia, 53100 Florina, Greece

* Correspondence: igiantsis@uowm.gr

† Presented at the 17th International Conference of the Hellenic Association of Agricultural Economists, Thessaloniki, Greece, 2–3 November 2023.

Abstract: The hatchery culture of bivalve mollusks depends on feeding with fresh microalgae which represent up to 50% of the production costs. We investigated the growth performance of juvenile *Ostrea edulis* and *Ruditapes decussatus* under 15% and 30% replacement of microalgae with *Saccharomyces cerevisiae*. Metabolic indices were measured along with weight-specific growth rate and condition index for 28 days. 15% substitution led to great results, whereas 30% yeast-fed treatments displayed poor growth and a depressed metabolism.

Keywords: aquaculture; bivalves; yeast; microalgae substitution

1. Introduction

Microalgae production is the main limiting factor impeding the industrial growth of the bivalve aquaculture industry since it corresponds to 30–50% hatchery production's operating costs [1,2]. Diets aiming to substitute live microalgae have been implemented in the early stages of shellfish culture, with varying outcomes [3–5]. Yeast cells possess the capability for mass production, are highly stable in water, have an appropriate size for consumption, and high levels and quality of protein. All these favorable characteristics indicate yeast as a promising substitute for live algal feeds [1,6]. Despite their advantageous aspects, yeast cells present low digestibility and contain limited amounts of polyunsaturated fatty acids [4,7]. Therefore, yeast should be provided to bivalves accompanied by live microalgae, which contain highly unsaturated fatty acids [7]. This study investigated the effects of a partial microalgae replacement (15% and 30%) with baker's yeast on the feeds of juvenile *Ostrea edulis* and *Ruditapes decussatus* by assessing the activities of two key metabolic enzymes (citrate synthase and hydroxyacylCoA dehydrogenase) and the functioning of the respiratory chain through the activity of the electron transport system (ETS). Moreover, the specific growth rate (SGR) of weight and the condition index (CI) were measured.

2. Materials and Methods

Wild juvenile *Ostrea edulis* and *Ruditapes decussatus*, weighting approximately 2 and 4 g, respectively, were placed in rectangular aquaria (50 L) containing natural seawater. Bivalves were fed a live microalgae diet consisting of the marine flagellates *Tisochrysis lutea* (CCAP 927/14) and *Tetraselmis* spp. (Mediterranean strain) as well as the diatom *Chaetoceros calcitrans* (CCAP 1085/3) at a 2:1:1 dry weight ratio. Yeast cells, *Saccharomyces cerevisiae* NCPF 3191 (Sigma-Aldrich, St. Louis, MO, USA), were cultured in a YPD medium and included in two treatments so as to represent 15% and 30% substitution of the microalgae. Treatments were tested in triplicates. Four samplings were performed in a 28-day period on day 1, day 4, day 12, and day 28. At each sampling, the mantle tissue from 6 animals from each treatment was dissected for biochemical analyses. Twelve specimens at the beginning and another twelve at the end of the experiment were used for the condition index calculation, as described by Irisarri et al. [8]. SGR and CI were calculated as follows:

- (SGR) = 100 × ((lnW_2 − lnW_1)/t), where W_1 and W_2 are the initial and final weights (g) of the bivalves and t is the number of feeding days;
- (CI) = (flesh dry weight/shell dry weight) × 100.

The activities of the metabolic enzymes citrate synthase (CS, EC 4.1.3.7) and hydroxyacylCoA dehydrogenase (HOAD, EC 1.1.1.35) were assessed spectrophotometrically based on well-established protocols [9], while ETS activity was determined according to Haider et al. [10]. The results of all the above indices were expressed as means ± standard deviation. One-way analysis of variance (ANOVA) was applied, followed by Tukey's HSD post hoc comparisons to define the statistically significant differences at $p < 0.05$.

3. Results

The condition of Ostrea edulis was similar to all treatments after 28 days, while Ruditapes decussatus fed on 30% yeast exhibited statistically significant lower conditions compared to the 0% and 15% treatments (Table 1). The growth rate of both species was significantly higher in treatments fed on 15% yeast and lower in replicates subjected to a 30% substitution of algae (Table 1).

Table 1. Condition index (CI) and specific growth rate of weight (SGR).

	Ostrea edulis		Ruditapes decussatus	
Condition index	mean	SD	mean	SD
Initial CI	1.86 [a]	0.24	18.35 [a]	1.84
Final CI—0% yeast	2.05 [a]	0.21	18.46 [a]	1.78
Final CI—15% yeast	1.92 [a]	0.12	18.31 [a]	1.43
Final CI—30% yeast	1.94 [a]	0.17	16.68 [b]	0.98
SGR of weight	mean	SD	mean	SD
SGRw 0% yeast	0.102 [a]	0.011	0.07 [a]	0.005
SGRw 15% yeast	0.12 [b]	0.008	0.108 [b]	0.013
SGRw 30% yeast	0.085 [c]	0.007	0.05 [c]	0.006

[a,b,c] Depict statistically different means by ANOVA ($p < 0.05$).

CS and HOAD in Ruditapes decussatus displayed generally similar activities among all treatments until day 4. On days 12 and 28, the activities of both enzymes significantly increased in both yeast-fed replicates, where the activity of these enzymes was similar (Figure 1A,B). Ostrea edulis demonstrated minor differences in the activity of CS at days 1 and 4, regardless of the feed composition. At day 12, both yeast-fed treatments presented a statistically significant increase in activity. At day 28, the 15% treatment had significantly greater activity, and the 30% treatment exhibited decreased activity compared to the control (Figure 2A). Concerning the HOAD in Ostrea edulis, 15% replacement of microalgae resulted in similar or greater control activities, while the 30% replacement led to a generally significantly reduced activity (Figure 2B).

The ETS activity displayed a clear pattern in both bivalves. When fed on 15% yeast, the two species exhibited similar to the control treatment activity of the electron transport system, but when fed on 30% yeast, the activity was significantly reduced (Figures 1C and 2C).

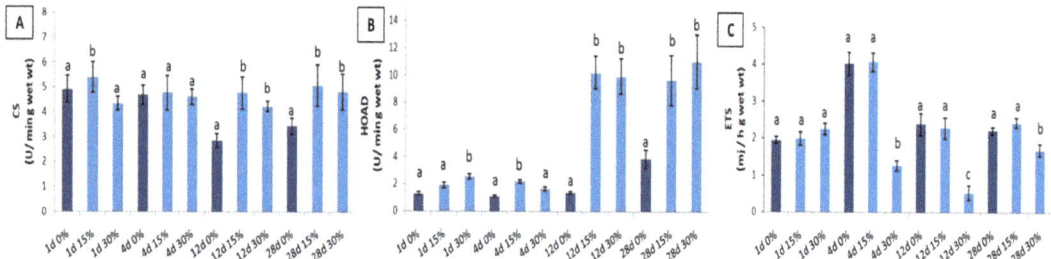

Figure 1. Activity of citrate synthase (**A**), hydroxyacylCoA dehydrogenase (**B**), and electron transport system (**C**) in the mantle tissue of *Ruditapes decussatus*. Dark blue indicates the control treatment. Values are means ± SD. Lower-case letters depict statistically significant differences ($p < 0.05$).

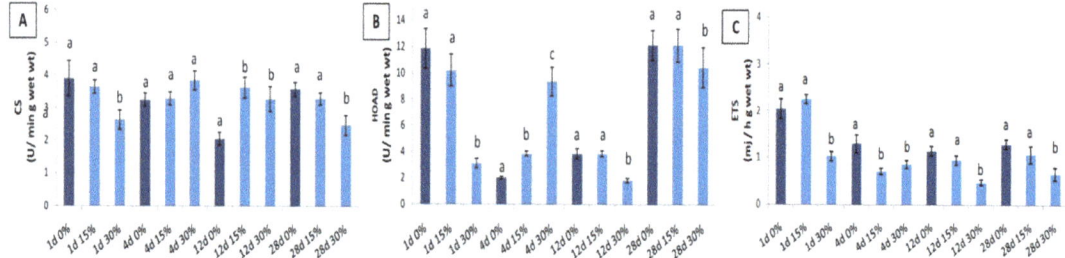

Figure 2. Activity of citrate synthase (**A**), hydroxyacylCoA dehydrogenase (**B**), and electron transport system (**C**) in the mantle tissue of *Ostrea edulis*. Dark blue indicates the control treatment. Values are means ± SD. Lower-case letters depict statistically significant differences ($p < 0.05$).

4. Discussion

Citrate synthase is a key metabolic enzyme that is associated with an organism's capacity for energy production (ATP generation) [11], while hydroxyacylCoA dehydrogenase is involved in fatty acid metabolic processes. Feeding regulates ETS activity, which can be used as an instantaneous index of oyster metabolism [12]. An increased ETS activity is also indicative of a higher rate of ATP production. On the other hand, ETS activity may decrease to conserve cellular resources, which might have happened in the case of 30% algae substitution in both species.

The highest growth rate of weight was detected at 15% yeast-fed replicates at both bivalves. Moreover, the same condition index as well as the similar or increased metabolic intensity in comparison to 100% algae-fed treatment indicate that a 15% substitution of algae enhances the growth of *Ruditapes decussatus* and *Ostrea edulis* juveniles. 30% replacement resulted in depressed ETS activity in both species, decreased activities of CS and HOAD in *Ostrea edulis*, and a lower condition index in *Ruditapes decussatus*. Encouraging results have been reported by many authors when using manipulated yeasts as an algal substitute [3,6] or by using efficiently digested mutant yeast cells [13], reaching substitution percentages of 50–80%.

5. Conclusions

A 15% percentage of fresh microalgae substitution with baker's yeast could be applied in the nursery stages of *Ruditapes decussatus* and *Ostrea edulis* to enhance their growth and eliminate production costs. Longer experiments, which will also include intermediate percentages of algae replacement (e.g., 20% and 25%) and/or different microalgae species, are necessary to assess the highest level of fresh microalgae substitution that can be achieved with *Saccharomyces cerevisiae* for these bivalve mollusks.

Author Contributions: Conceptualization, D.K.P. and A.L.; methodology, D.K.P. and I.G.; software, D.K.P.; validation, K.F., I.A.G.; formal analysis, D.K.P. and I.G.; investigation, D.K.P.; resources, B.M.; data curation, D.K.P. and I.G.; writing—original draft preparation, D.K.P.; writing—review and editing, I.G., A.L., K.F. and I.A.G.; visualization, D.K.P.; supervision, B.M.; project administration, I.A.G.; funding acquisition, B.M. All authors have read and agreed to the published version of the manuscript.

Funding: This research was funded by the 'Innovation Investment Schemes' in the framework of the Operational Program of Central Macedonia 2014–2020, co-funded by the European Social Fund through the National Strategic Reference Framework. Project code KMP6-0078456, MIS: 5136453.

Institutional Review Board Statement: Not applicable, the study only investigates invertebrates.

Informed Consent Statement: Not applicable.

Data Availability Statement: All data of this research are available after communication with the corresponding author.

Conflicts of Interest: The authors declare no conflicts of interest.

References

1. Coutteau, P.; Sorgeloos, P. Substitute diets for live algae in the intensive rearing of bivalve mollusks: A state of the art report. *World Aquacult.* **1993**, *24*, 45–52.
2. Willer, D.F.; Aldridge, D.C. Microencapsulated diets to improve bivalve shellfish aquaculture for global food security. *Glob. Food Secur.* **2019**, *23*, 64–73. [CrossRef]
3. Coutteau, P.; Hadley, N.H.; Manzi, J.J.; Sorgeloos, P. Effect of algal ration and substitution of algae by manipulated yeast diets on the growth of juvenile *Mercenaria mercenaria*. *Aquaculture* **1994**, *120*, 135–150. [CrossRef]
4. Tanyaros, S.; Sujarit, C.; Jansri, N.; Tarangkoon, W. Baker's yeast as a substitute for microalgae in the hatchery rearing of larval and juvenile tropical oyster (*Crassostrea belcheri*, Sowerby 1871). *J. Appl. Aquacult.* **2016**, *28*, 35–46. [CrossRef]
5. Supono, S.; Mugica, M.; Spreitzenbarth, S.; Jeffs, A. Potential for Concentrated Microalgae as Replacement Diets for Juvenile Green-Lipped Mussels, Perna canaliculus. *Aquacult. Res.* **2023**, *2023*, 9841172. [CrossRef]
6. Nell, J.A.; Sheridan, A.K.; Smith, I.R. Progress in a Sydney rock oyster, Saccostrea commercialis (Iredale and Roughley), breeding program. *Aquaculture* **1996**, *144*, 295–302. [CrossRef]
7. Brown, M.R.; Barrett, S.M.; Volkman, J.K.; Nearhos, S.P.; Nell, J.A.; Allan, G.L. Biochemical composition of new yeasts and bacteria evaluated as food for bivalve aquaculture. *Aquaculture* **1996**, *143*, 341–360. [CrossRef]
8. Irisarri, J.; Fernández-Reiriz, M.J.; Labarta, U. Temporal and spatial variations in proximate composition and Condition Index of mussels *Mytilus galloprovincialis* cultured in suspension in a shellfish farm. *Aquaculture* **2015**, *435*, 207–216. [CrossRef]
9. Speers-Roesch, B.; Callaghan, N.I.; MacCormack, T.J.; Lamarre, S.G.; Sykes, A.V.; Driedzic, W.R. Enzymatic capacities of metabolic fuel use in cuttlefish (*Sepia officinalis*) and responses to food deprivation: Insight into the metabolic organization and starvation survival strategy of cephalopods. *J. Comp. Physiol. B* **2016**, *186*, 711–725. [CrossRef]
10. Haider, F.; Sokolov, E.P.; Sokolova, I.M. Effects of mechanical disturbance and salinity stress on bioenergetics and burrowing behavior of the soft-shell clam Mya arenaria. *J. Exp. Biol.* **2018**, *221*, jeb172643. [CrossRef]
11. Kolditz, C.; Borthaire, M.; Richard, N.; Corraze, G.; Panserat, S.; Vachot, C.; Lefèvre, F.; Médale, F. Liver and muscle metabolic changes induced by dietary energy content and genetic selection in rainbow trout (*Oncorhynchus mykiss*). *Am. J. Physiol. Regul. Integr. Comp. Physiol.* **2008**, *294*, R1154–R1164. [CrossRef] [PubMed]
12. García-Esquivel, Z.; Bricelj, V.M.; González-Gómez, M.A. Physiological basis for energy demands and early postlarval mortality in the Pacific oyster, *Crassostrea gigas*. *J. Exp. Mar. Biol. Ecol.* **2001**, *263*, 77–103. [CrossRef]
13. Loor, A.; Bossier, P.; Nevejan, N. Dietary substitution of microalgae with the Saccharomyces cerevisiae mutant, Δmnn9, for feeding Pacific oyster (*Crassostrea gigas*) juveniles. *Aquaculture* **2021**, *534*, 736253. [CrossRef]

Disclaimer/Publisher's Note: The statements, opinions and data contained in all publications are solely those of the individual author(s) and contributor(s) and not of MDPI and/or the editor(s). MDPI and/or the editor(s) disclaim responsibility for any injury to people or property resulting from any ideas, methods, instructions or products referred to in the content.

Proceeding Paper

The Agricultural Knowledge and Innovation System (AKIS) in a Changing Environment in Greece [†]

Epistimi Amerani * and Anastasios Michailidis

Department of Agricultural Economics, Aristotle University of Thessaloniki, 541 24 Thessaloniki, Greece; tassosm@agro.auth.gr
* Correspondence: epistimi@agro.auth.gr
[†] Presented at the 17th International Conference of the Hellenic Association of Agricultural Economists, Thessaloniki, Greece, 2–3 November 2023.

Abstract: The aim of this paper is to answer the question of whether the Greek AKIS system can contribute to the different requirements of the new trends in agriculture according to its main functions. A SWOT analysis has been applied to examine the internal and external environment. Data were collected from 61 experts/representatives of organizations (policy, education, research, consulting, agricultural cooperatives, credit, private companies, and farmers). The data were analysed using Excel spreadsheets and the Statical Package for Social Sciences (SPSS V.28). Based on this method, dominant strengths and weaknesses as well as opportunities and threats of AKIS were identified as a starting point, as well as useful guidance for decision makers, local authorities, and the other actors in Greece.

Keywords: agricultural sector; AKIS; SWOT analysis

1. Introduction

In our era, the agri-food sector has faced a huge challenge: to boost production with increasing demands and constraints placed on it [1]. In the future, feeding nine billion people with continuous pressure on the Earth's natural resources, health, climate, and welfare for both humans and animals is a big challenge for sustainable agriculture. There is an increasing demand for innovative solutions through the continuous renewal of products, processes, and services [2].

The goals related to innovation are increasing their emphasis on encouraging healthy, high-quality products, and environmentally sustainable production methods, including organic production, renewable materials, and biodiversity protection [3]. New social, technical, and economic solutions are needed for farming and rural areas [4]. Innovation is considered one of the key drivers for competitive and sustainable agriculture [5]. In the conventional view, innovation is mainly embodied in technological artifacts (new knowledge and equipment technologies, improved seeds, vaccines, breeding techniques, fertilizers and pesticides, and other agricultural inputs), and its successful application is related to the capacity of the users to learn to 'adopt' them, according to given guidelines. However, in the new network's view, innovation occurs when the network of production changes its way of doing things, so innovation is mainly related to the resulting pattern of interaction between people, tools, and natural resources [4]. Innovation processes are increasingly conceptualized as the outcome of collaborative networks, where information is exchanged and learning processes happen and lead to an expanded knowledge system, including a wide range of stakeholders who innovate and those who benefit (or suffer) from innovation [4]. The combination of technological innovation, improved skills, and an increased capacity of farmers and their organizations [6], and the effective cooperation between the people who produce the knowledge and the end users who utilize it, are optimal solutions for dealing with the above challenges [2].

Citation: Amerani, E.; Michailidis, A. The Agricultural Knowledge and Innovation System (AKIS) in a Changing Environment in Greece. *Proceedings* **2024**, *94*, 29. https://doi.org/10.3390/proceedings2024094029

Academic Editor: Eleni Theodoropoulou

Published: 25 January 2024

Copyright: © 2024 by the authors. Licensee MDPI, Basel, Switzerland. This article is an open access article distributed under the terms and conditions of the Creative Commons Attribution (CC BY) license (https://creativecommons.org/licenses/by/4.0/).

In recent years, AKIS studies agreed on the importance of the direct involvement of farmers in the innovation processes to identify the best response to farm issues and improve innovation effectiveness [7–9]. Direct involvement means an interactive and practical collaboration of all actors (scientific, institutional, business, and civil society) using appropriate tools for the target [10], allowing partners to verify the activity carried out and contribute to the change process. Through the AKIS system, they are given the opportunity to collaborate, share their ideas, and turn existing knowledge and research results into innovative solutions that can be more easily implemented in practice [11].

The main aim of this research is to answer the question of whether the Greek AKIS system can contribute to the different requirements of the new trends in agriculture by evaluating the strengths and weaknesses in terms of its internal environment, as well as the opportunities and threats that come from the external environment.

2. Material and Methods

First, a literature review was carried out with the aim of understanding the internal factors of AKIS operations (strengths and weaknesses), where the participating agencies have a greater capacity for action and control, and then the external elements (opportunities and threats), where their actions are quite limited, but which can significantly influence the situation. SWOT analysis allows an assessment of the parameters of the application of AKIS. To analyse the situation of Greek AKIS, the questionnaire consisted of four sections including strengths (13 factors), weaknesses (11 factors), opportunities (7 factors), and finally threats (8 factors). The surveyed actors were asked to identify if they agreed or disagreed on the typical 5-point Likert scale ranging from 1 (Strongly Disagree) to 5 (Strongly Agree). Data were collected through a survey of 61 expert representatives (mainly senior managers) from all participating bodies (Ministry, Region, Chamber, NGO, ELGO-Dimitra, Research Institutes, Educational Institutions, private consulting companies, supply of inputs, manufacturing companies, cooperatives, credit institutions, and farmers). Data were collected during December 2022 and March 2023 using an online survey tool after an initial phone communication. Descriptive statistics indicators (mean scores, standard deviations, and standard errors) were used to describe and present the main results.

3. Results

Based on the AKIS internal environment evaluation results, the main strength was finding new solutions for agricultural issues (mean: 3.90; SD: 0.98 and SE: 0.12). The findings revealed that the main weakness of AKIS is the ageing population of farmers (mean: 3.84; SD: 1.05 and SE: 0.13). In terms of external opportunities, AKIS has the potential to develop further, due to new opportunities and environmental factors (mean: 4.16; SD: 0.76 and SE: 0.10). However, the most significant threat to AKIS is the complexity of legal and regulatory frameworks (mean: 4.18; SD: 0.82 and SE: 0.11) (Tables 1 and 2)

Table 1. External factors evaluation matrix.

External Factors	Mean	SE	SD
Opportunities			
O1: Farming system to produce high-value products	4.08	0.09	0.69
O2: New market information system	3.95	0.11	0.82
O3: New opportunities and environmental potential to develop agriculture	4.16	0.10	0.76
O4: Strengthen policies in the European Union	3.72	0.11	0.90
O5: Development of programs, institutions, and facilities	3.87	0.12	0.97
O6: Increasing economic growth rate	3.62	0.12	0.97

Table 1. Cont.

External Factors	Mean	SE	SD
Threats			
T1: Complexity of legal and regulatory frameworks	4.18	0.11	0.82
T2: Inadequate balance of supply and demand of products	3.46	0.11	0.87
T3: High fluctuations in prices of inputs and outputs	3.72	0.02	0.93
T4: Adverse environment due to conditions of uncertainty (recession, pandemic, war)	4.10	0.11	0.89
T5: Most innovations are capital-intensive	3.62	0.13	1.00
T6: The lack of financial and government support	3.77	0.14	1.09
T7: Unforeseen environmental changes	3.79	0.13	1.00
T8: Low resilience of agricultural holdings	3.79	0.13	1.02

Table 2. Internal factors evaluation matrix.

Internal Factors	Mean	SE	SD
Strengths			
S1: Strengthening of interactive learning through the sharing of different types of knowledge	3.66	0.15	1.15
S2: Improving farmers' access to a new, diverse, and growing information system	3.74	0.12	0.96
S3: Educating farmers to improve their skills	3.75	0.14	1.10
S4: Boosting productivity and farmers' incomes and subsequently improving their standard of living	3.44	0.14	1.07
S5: Increasing and attracting investment	3.33	0.14	1.08
S6: Finding new solutions for agricultural problems	3.90	0.12	0.98
S7: Enhancing coordination among AKIS actors	3.57	0.14	1.12
S8: Developing each actor's new capacities and skills within AKIS	3.64	0.14	1.10
S9: Changing farmers' knowledge, attitudes, and strengthening of participatory spirit	3.56	0.14	1.10
S10: Improving farmers' access to international markets	3.13	0.12	0.90
S11: Improvement in the responsibility of actors to farmers	3.39	0.13	0.99
S12: Preventing anti-competitive practice	3.05	0.14	1.10
S13: Empowerment of farmers to increase critical thinking skills to be able to analyse situations and determine their main demands	3.43	0.13	1.02
Weaknesses			
W1: Ageing of the agricultural population	3.84	0.13	1.05
W2: Lack of focus in dealing with diverse demands that come from different farmers	3.75	0.10	0.79
W3: Lack of enough development of social capital between farmers	3.80	0.12	0.91
W4: Ignorance of poor and marginal farmers	3.82	0.13	0.99
W5: High costs of advisory service	3.31	0.13	1.02
W6: Lack of enough use of new information and communication technologies	3.39	0.14	1.07
W7: Insufficient opportunities of education and training programs	3.34	0.13	1.03
W8: Inadequate control and evaluation systems by regional authorities	3.82	0.13	1.01
W9: Lack of synergies between actors to co-create the appropriate innovation	3.80	0.13	0.96
W10: Inadequate significant organizational capacity of advisors	3.46	0.12	0.92
W11: Lack of awareness of possibilities to receive advisor services	3.67	0.12	0.89

4. Discussion and Conclusions

This research focuses on the question of whether the Greek AKIS system can contribute to the different requirements of the new trends in agriculture, according to its main functions such as the guidance of search, knowledge development, network formation and knowledge diffusion, entrepreneurial activities, market formation, resource mobilization, and formation of legitimacy [12,13]. The actors supported that the existing AKIS develop new knowledge for solving agricultural problems, mobilize resources for educating farmers to improve their skills, and strengthen the farmers' access to communication information technologies (agreed by 60–75%). The ageing and ignorance of poor and marginal farmers were considered the main inhibiting factors for its operation (agreed by 65%). The existence of agricultural systems such as integrated farming management, organic farming, and precision agriculture were considered opportunities for the development of AKIS (agreed by 84%). On the other hand, the actors support that the complexity of legal and regulatory frameworks is a threat to the system (agreed by 80%). The analysis presents a starting point and useful guidance both for decision makers and the other actors for the enhancement of AKIS.

Author Contributions: A.M. writing—original draft preparation, visualization, supervision, project administration, writing—review and editing; E.A. Conceptualization, methodology, software, validation, investigation, resources, data curation, writing—review and editing. All authors have read and agreed to the published version of the manuscript.

Funding: This research was funded by the Hellenic Foundation for Research and Innovation (HFRI) under the third call for HFRI PhD Fellowships, grant number 6422.

Informed Consent Statement: Informed consent was obtained from all subjects involved in the study.

Data Availability Statement: Data will be made available upon request.

Conflicts of Interest: The authors declare no conflicts of interest.

References

1. Panetto, H.; Lezoche, M.; Hernandez, J.E.; Alemany, M.M.E.; Kacprzyk, J. Special issue on Agri-food 4.0 and digitalization in agriculture supply chains—New directions, challenges, and applications. *Comput. Ind.* **2020**, *116*, 103188. [CrossRef]
2. European Union-Standing Committee on Agricultural Research (EU SCAR). *Preparing for Future AKIS in Europe*; European Commission: Brussels, Belgium, 2019.
3. European Council Presidency Conclusions–Goteborg European Council (15–16 June 2001). Available online: http://www.consilium.europa.eu/uedocs/cms_data/docs/pressdata/en/ec/00200-r1.en1.pdf (accessed on 18 September 2023).
4. Knickel, K.; Brunori, G.; Rand, S.; Proost, J. Towards a Better Conceptual Framework for Innovation Processes in Agriculture and Rural Development: From Linear Models to Systemic Approaches. *J. Agric. Educ. Ext.* **2009**, *15*, 131–146. [CrossRef]
5. Arzeni, A.; Ascioue, E.; Borsotto, P.; Carta, V.; Castelloti, T.; Vagnozzi, A. Analysis of farms characteristics related to innovation needs: A proposal for supporting the public decision-making process. *Land Use Policy* **2021**, *100*, 104892. [CrossRef]
6. Food and Agriculture Organization (FAO). *Agricultural Knowledge and Innovation Systems for Rural Development (AKIS/RD): Strategic Vision and Guiding Principles*; FAO: Rome, Italy, 2000.
7. Botha, N.; Turner, J.A.; Fielke, S.; Klerkx, L. Using a co-innovation approach to support innovation and learning: Cross-cutting observations from different settings and emergent issues. *Outlook Agric.* **2017**, *46*, 87–91. [CrossRef]
8. Fielke, S.; Nelson, T.; Blackett, P.; Bewsell, D.; Bayne, K.; Park, N.; Rijswijk, K.; Small, B. Hitting the bullseye: Learning to become a reflexive monitor in New Zealand. *Outlook Agric.* **2017**, *46*, 117–124. [CrossRef]
9. Ingram, J.; Dwyer, J.; Gaskell, P.; Mills, J.; Wolf, P. Reconceptualising translation in agricultural innovation: A co-translation approach to bring research knowledge and practice closer together. *Land Use Policy* **2018**, *70*, 38–51. [CrossRef]
10. Barcellini, F.; Prost, L.; Cerf, M. Designers' and users' roles in participatory design: What is actually co-designed by participants? *Appl. Ergon.* **2015**, *50*, 31–40. [CrossRef] [PubMed]
11. Feo, E.; Mareen, H.; Burssens, S.; Spangle, P. The Relevance of Videos as a Practical Tool for Communication and Dissemination in Horizon2020 Thematic Networks. *Sustainability* **2021**, *13*, 13116. [CrossRef]
12. Hermans, F.; Klerkx, L.; Roep, D. Structural conditions for collaboration and learning in innovation networks: Using an innovation system performance lens to analyze agricultural knowledge systems. *J. Agric. Educ. Ext.* **2015**, *21*, 35–54. [CrossRef]
13. Zahran, Y.; Kassem, H.S.; Naba, S.M.; Alotaibi, B.A. Shifting from Fragmentation to Integration: A Proposed Framework for Strengthening Agricultural Knowledge and Innovation System in Egypt. *Sustainability* **2020**, *12*, 5131. [CrossRef]

Disclaimer/Publisher's Note: The statements, opinions and data contained in all publications are solely those of the individual author(s) and contributor(s) and not of MDPI and/or the editor(s). MDPI and/or the editor(s) disclaim responsibility for any injury to people or property resulting from any ideas, methods, instructions or products referred to in the content.

Proceeding Paper

Sustainability Assessment of Highly Biodiversified Farming Systems: Multicriteria Assessment of Greek Arable Crops [†]

Andreas Michalitsis [1], Ferdaous Rezgui [2], Fatima Lambarraa-Lehnhardt [2], Paschalis Papakaloudis [1], Maria Laskari [1], Efstratios Deligiannis [1] and Christos Dordas [1,*]

[1] Laboratory of Agronomy, School of Agriculture, Aristotle University of Thessaloniki, 54124 Thessaloniki, Greece; amichalits@agro.auth.gr (A.M.); papakalp@agro.auth.gr (P.P.); marialaskari00@gmail.com (M.L.); deliefst@agro.auth.gr (E.D.)

[2] Leibniz Centre for Agricultural Landscape Research (ZALF), Eberswalder Str. 84, 15374 Müncheberg, Germany; ferdaous.rezgui@zalf.de (F.R.); fatima.lehnhardt@zalf.de (F.L.-L.)

* Correspondence: chdordas@agro.auth.gr

[†] Presented at the 17th International Conference of the Hellenic Association of Agricultural Economists, Thessaloniki, Greece, 2–3 November 2023.

Abstract: The intensive agriculture that is used in many countries has led to a reduction in biodiversity and the deterioration of the environment. Therefore, it is important to increase the adoption of cropping systems with high biodiversity. The objectives of the present study were the following: 1. assess the performance and sustainability of novel highly diversified production systems compared to the current traditional system and 2. provide quantitative economic and ecosystem service information for farmers, extension workers, and policy makers in order to support the development of sustainable and resilient high species cultivar/landrace diversification (HSD) production systems. The rotation of wheat–pea–barley was a system with low energy inputs and high outputs, significantly increasing the energy efficiency. Also, the same system demonstrated better economic and environmental indices, making it a suitable cropping system for Mediterranean areas.

Keywords: crop rotation; intercropping; pea; co-design; wheat

1. Introduction

The wide use of intensive agriculture in many countries has had many adverse consequences as it caused an increase in soil salinity and the deterioration of plant growth environments [1]. The deterioration in plant growth environments is also exacerbated by climate change, such as the increases in temperature and changes in rainfall, which will make agricultural production even more vulnerable in the future [2,3]. To alleviate these challenges, it is necessary to use sustainable agricultural systems and increase the biodiversity of cropping systems.

The diversification of agricultural production systems implies forfeiting the economies of scale by increasing expenses per unit of output, reducing the efficiency of machinery, and applying less specialised knowledge and labour division [4]. The ecological benefits of diversified farming systems were found to be insufficient to outbalance the economic costs in the short term [5], even though many examples showed that diversified farming practices have the potential to lead to higher and more stable yields [6], increase profitability, and reduce risks in the long term [5]. Therefore, research on diversified systems requires short- and long-term economic analyses to identify efficient policy support measures.

The objectives of the present study were the following:

Assess the performance of novel highly diversified production systems compared to the current traditional system.

Provide quantitative economic and ecosystem service information for farmers, extension workers, and policy makers in order to support the development of sustainable and resilient HSD production systems.

2. Materials and Methods

The approach that was followed was an integrated approach that incorporated stakeholder expertise, analysis of empirical data, and quantitative modelling of the economic and agro-environmental performance of novel production systems. Furthermore, the quantitative data that were used for the modelling process of this study were obtained from previous, similar experiments of the laboratory in the same place and using the methodology followed by Rezgui et al. (2023, under review) [7]. This combination of sources allowed us to capture the short- and long-term effects of diversified production systems. The work focused on HSD arable rotations used in Mediterranean areas and especially in Greece. The work was organised by generating and assessing diversified crop rotations with the cropping system assessment framework. During the co-design process, three systems were developed: (i) Diversified system 1 (DIV1) was a wheat–oilseed rape–barley rotation; (ii) Diversified system 2 (DIV2) was a rotation of wheat–pea–barley; and the third diversified system (DIV3) was a wheat-intercropping of barley with common vetch–barley rotation. The three diversified systems were compared to a typical sole cropping system in the region of wheat and barley monoculture.

The indicators used to evaluate the four systems included energy efficiency, total renewable and non-renewable input energy per system, and pesticide load indicator, along with three sub-indicators (health load, ecotoxicity load, and fate load), and the economic performances of the four systems (farming profit, farming income, and farming cost).

3. Results

Energy-use efficiency was determined for the four systems, and it was found that DIV2 is the most energy-use efficient system, followed by DIV3 (Figure 1). These results were probably observed because the pea crop was more energy efficient due to a high grain energy output and low energy inputs when compared to the RS than the vetch–barley intercrop as well as the rapeseed crop (DIV1).

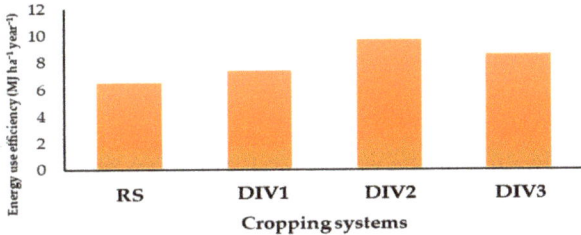

Figure 1. Energy-use efficiency of the four systems that were co-designed with the stakeholders of the agri-food chain.

The total renewable and non-renewable input energy per system was calculated and it was found that DIV1 was the one with the highest non-renewable energy input, followed by the RS. DIV2 and DIV3 were the ones with the highest renewable energy input (Figure 2).

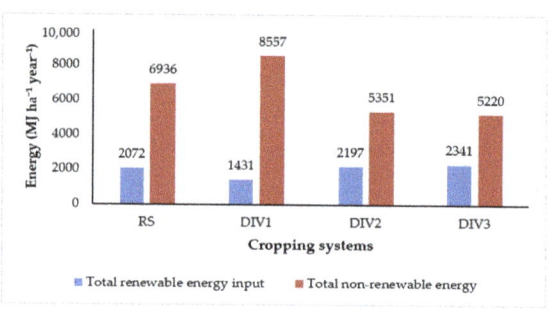

Figure 2. Total renewable and non-renewable input energy per system.

From the four systems that were assessed the system, the RS had the highest pesticide health load. This means that the pesticides which were used for this system type were the most toxic to humans compared with the pesticides used for other systems. In addition, DIV1 had the most toxic effect on mammals, birds, fish, daphnia, algae, aquatic plants, earthworms, and bees (ecotoxicity load). The fate load of the pesticides used for the four systems was relatively similar, with DIV2 and DIV3 having the lowest averages (less pesticides for more crops). The pesticide load of the RS was the highest, indicating that the pesticides used for this system were the most dangerous in terms of quantity and toxicity (Figure 3).

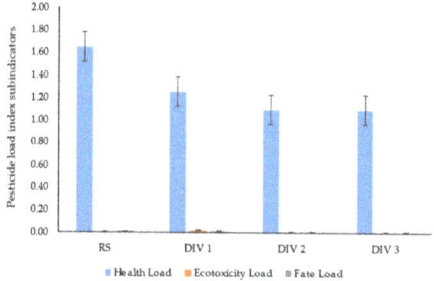

Figure 3. Average pesticide-load sub-indicators per system health load, ecotoxicity load, and fate load.

There was a 67% increase in total costs in DIV1, a 55% increase in DIV2, and a 32% increase in DIV3 compared with the reference system (RS). In addition, there was a 71% increase in income with DIV3, followed by a 48% increase with DIV2 and a 28% increase with DIV1 compared to the reference system (Figure 4).

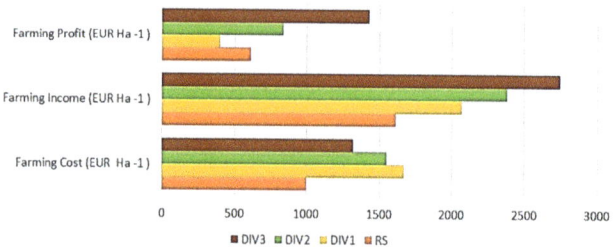

Figure 4. Economic indices of the four systems that were assessed.

4. Discussion

Based on the results, when legumes are incorporated in the cropping system, the result is that we have better environmental indices and higher farming profits. Similar results were reported in other studies, where the inclusion of legumes reduced the inputs and decreased the environmental impact of cropping systems [6,8]. However, the data are limited to Mediterranean cropping systems.

5. Conclusions

The four cropping systems that were evaluated gave interesting data that can be used to design more sustainable cropping systems. DIV3 is a system with low energy inputs and high outputs, significantly increasing the energy efficiency. Also, the same system has better economic and environmental indices than the other three systems, promising a sustainable cropping system for the Mediterranean areas.

Author Contributions: C.D., F.R. and F.L.-L.: conceptualisation and methodology; A.M., P.P., M.L. and E.D.: field measurement and data curation. C.D.: writing—original draft preparation. A.M., P.P., M.L. and E.D.: visualisation. F.R. and F.L.-L.: writing—review and editing. C.D.: supervision. All authors have read and agreed to the published version of the manuscript.

Funding: This project Biodiversify (Boost ecosystem services through high Biodiversity-based Mediterranean Farming systems) is funded by the General Secretariat for Research and Technology of the Ministry of Development and Investments under the PRIMA Programme. PRIMA is an Art.185 initiative supported and co-funded under Horizon 2020, the European Union's Programme for Research and Innovation.

Institutional Review Board Statement: There is no institutional review board statement.

Informed Consent Statement: Not applicable.

Data Availability Statement: The data presented in this study are available on request from the corresponding author.

Conflicts of Interest: The authors declare no conflicts of interest.

References

1. Abberton, M.; Batley, J.; Bentley, A.; Bryant, J.; Cai, H.; Cockram, J.; Yano, M. Global agricultural intensification during climate change: A role for genomics. *Plant Biotechnol. J.* **2016**, *14*, 1095–1098. [CrossRef] [PubMed]
2. Bodner, G.; Nakhforoosh, A.; Kaul, H.P. Management of crop water under drought: A review. *Agron. Sustain. Dev.* **2015**, *35*, 401–442. [CrossRef]
3. McKersie, B. Planning for food security in a changing climate. *J. Exp. Bot.* **2015**, *66*, 3435–3450. [CrossRef] [PubMed]
4. Klasen, S.; Meyer, K.M.; Dislich, C.; Euler, M.; Faust, H.; Gatto, M.; Hettig, E.; Melati, D.N.; Jaya, N.S.; Otten, F.; et al. Economic and ecological trade-offs of agricultural specialization at different spatial scales. *Ecol. Econ.* **2016**, *122*, 111–120. [CrossRef]
5. Rosa-Schleich, J.; Loos, J.; Musshoff, O.; Tscharntke, T. Ecological-economic trade-offs of Diversified Farming Systems—A review. *Ecol. Econ.* **2019**, *160*, 251–263. [CrossRef]
6. Reckling, M.; Albertsson, J.; Topp, C.F.; Vermue, A.; Carlsson, G.; Watson, C.; Jensen, E.S. Does cropping system diversification with legumes lead to higher yield stability? Diverging evidence from long-term experiments across Europe. In Proceedings of the European Conference on Crop Diversification, Budapest, Hungary, 18–21 September 2019; pp. 18–21.
7. Rezgui, F.; Rosati, A.; Lambarra-Lehnhardt, F.; Paul, C.; Reckling, M. Sustainability assessment of Mediterranean farming systems: The case of olive agroforestry in central Italy. *Eur. J. Agron.* **2024**, *152*, 127012. [CrossRef]
8. Uthes, S.; Sattler, C.; Zander, P.; Piorr, A.; Matzdorf, B.; Damgaard, M.; Sahrbacher, A.; Schuler, J.; Kjeldsen, C.; Heinrich, U.; et al. Modeling a farm population to estimate on-farm compliance costs and environmental effects of a grassland extensification scheme at the regional scale. *Agric. Syst.* **2010**, *103*, 282–293. [CrossRef]

Disclaimer/Publisher's Note: The statements, opinions and data contained in all publications are solely those of the individual author(s) and contributor(s) and not of MDPI and/or the editor(s). MDPI and/or the editor(s) disclaim responsibility for any injury to people or property resulting from any ideas, methods, instructions or products referred to in the content.

Proceeding Paper

Important Parameters Connected to Farmers' Networking and Training That Give Added Value to "Fasolia Vanilies Feneou" and "Fava Feneou" Products †

Elissavet Ninou [1,*], Fokion Papathanasiou [2], Anthoula Tsipi [3], Anastasia Kargiotidou [4], Georgia Vasiligianni [5], Konstantinos Koutis [6] and Ioannis Mylonas [3,*]

1. Department of Agriculture, International Hellenic University, 57400 Sindos, Greece
2. Department of Agriculture, University of Western Macedonia, 50100 Koila Kozanis, Greece; fpapathana-siou@uowm.gr
3. Greece Institute of Plant Breeding and Genetic Resources, ELGO DEMETER, Kourtidou 56–58 & Nirvana, 11145 Athens, Greece; tsipianti@gmail.com
4. Institute of Industrial & Forage Crops, ELGO DEMETER, Kourtidou 56–58 & Nirvana, 11145 Athens, Greece; kargiotidou@elgo.gr
5. KIATO UNION IKE, Kleisthenous 4, 20200 Kiato Korinthias, Greece; gewrgia_vasil@hotmail.com
6. AEGILOPS—Network for Biodiversity and Ecology in Greece Ano Lechonia, 37300 Agria Volos, Greece; koutisresfarm@gmail.com

* Correspondence: lisaninou@gmail.com (E.N.); ioanmylonas@yahoo.com (I.M.)
† Presented at the 17th International Conference of the Hellenic Association of Agricultural Economists, Thessaloniki, Greece, 2–3 November 2023.

Abstract: The official designation of the bean "Fasolia Vanilies Feneou" and grass pea "Fava Feneou" as Protected Geographical Indication (PGI) products do not extend protection to their cultivated genetic material due to their non-inclusion in the National Catalog of Varieties [EC 2008/62/EK (official Greek Gazette) FEK 165/30-Juanuary-2014] as recognized traditional cultivars. This omission poses a significant risk to the genetic diversity of these varieties, potentially leading to the loss of their distinct characteristics, decreased yields, and compromised quality. The primary objective of this project is to ensure the preservation of these local varieties through a comprehensive study of their genetic variability. Additionally, it aims to adhere to official protocols for describing and subsequently registering these varieties in the National List of Varieties. This registration will enhance the product's value and secure its unique identity. The experimentation phase of the project focuses on evaluating the landrace to select plants that demonstrate improved productivity and quality. This work presents the parameters connected with the description of the unique identity of this product; its origin, traceability, and local agricultural practices; and specific product characteristics that will contribute to this. The product will be utilized by Kiato Union IKE and, at the same time, farmers will be trained in the excellent seed reproduction and production of the product. This initiative promises several benefits for the agricultural cooperative and producers in Feneos.

Keywords: *Phaseolus vulgare* L.; *Lathyrus* sp.; landraces; added value; biodiversity protection

1. Introduction

The common bean is a globally regarded legume that is a significant supplier of top-notch proteins, carbohydrates, vitamins, minerals, dietary fiber, phytonutrients, and antioxidants for human consumption. Many of these substances have substantial beneficial effects on human well-being. Consequently, the common bean could be considered a promising functional food [1,2]. Furthermore, recent studies have unveiled the health-promoting nutraceutical potential of grass peas [3]. Common bean and lathyrus cultivation have traditionally formed an integral part of rural economies in Greece [4]. Production has traditionally revolved around local landraces cultivated by small-scale farmers employing low-input production systems. As members of the Fabaceae family, common beans and lathyrus also contribute significantly to sustainable agriculture through their capacity to fix atmospheric nitrogen, thereby reducing the dependence on fertilizer applications. Moreover, in contemporary agricultural contexts, landraces have the potential to assume a pivotal role in low-input cultivation systems, serving as a valuable reservoir of genetic material. This genetic diversity can enhance tolerance to both abiotic and biotic stresses and can facilitate the adaptation of modern cultivars to the challenges posed by climate change [5].

The dedication of the local agricultural cooperative KIATO UNION to cultivating common beans—"Fasolia Vanilies Feneou"—and grass pea—"Fava Feneou"—is paramount for preserving the cultural heritage of Feneos. Project M16SYN2-00320 is geared towards securing and conserving the indigenous variety that serves as the source of the PDO product under consideration. This goal will be realized by applying established official description protocols and capitalizing on existing knowledge to characterize and describe the genetic material.

2. Material and Methods

The initial phase of the project's first year involved conducting a pilot study to evaluate the landraces' genetic diversity and identify plants demonstrating enhanced productivity and quality while preserving all inherent plant characteristics for product production. The genetic material employed in this study consisted of common bean—"Fasolia Vanilies Feneou"—and grass pea—"Fava Feneou"—varieties of seeds sourced from the Kiato Union. The experiment was conducted during the 2022–23 growing season on the Institute of Genetic Improvement and Plant Genetic Resources' farm in Thessaloniki Therme (for grass peas) and on the farm of the University of Western Macedonia-Department of Agriculture (for common beans) (Figure 1). Six hundred plant positions were established for each landrace with low plant density. All observations were made at the individual plant level and pertained to various agronomic traits. Moreover, a field experiment was established in the area of Feneos, ~1 ha, for both landraces to study their genetic variation (Figure 1). Three research Institutions (ELGO-Dimitra, University of Western Macedonia, and International Hellenic University), the local Agricultural Community represented by the agricultural cooperative of Florina, an NGO Aegilops, and an Advisor (Tsipi Anthoula) cooperated under the PAA M16.1–16.2 project (M16SYN2-00320) to achieve this goal.

(a) (b)

Figure 1. Field experiments: (**a**) in the area of Feneos; (**b**) in the area of the University of Western Macedonia-Department of Agriculture.

3. Results and Discussion

The official designation of vanilla bean and fava bean as Protected Geographical Indication (PGI) products does not extend protection to their cultivated genetic material due to their non-inclusion in the National Catalog of Varieties [EC 2008/62/EK (official Greek Gazette) FEK 165/30-January-2014] as recognized traditional cultivars. This omission poses a significant risk to the genetic diversity of these varieties, potentially leading to the loss of their distinct characteristics, decreased yields, and compromised quality. The primary objective of this project is to ensure the preservation of these local varieties through a comprehensive study of their genetic variability.

For beans, the yield ranged from 10 to 400 g per plant, the number of pods ranged from 7 to 330 per plant, flowering ranged from 42 to 56 days, the growth type was dwarf, and the biological cycle was late. Similar variability was observed for lathyrus.

These crops are staples of the agricultural landscape and represent a rich and enduring part of the local tradition and identity. However, the continued cultivation of these crops is under threat due to the risk of genetic erosion. Common beans, such as "vanilla", and grass peas, such as "fava", are grown on approximately 400 hectares in the Feneos region. This expansive cultivation area demonstrates these crops' substantial role within the local agricultural community. Moreover, they contribute substantially to the local economy, estimated at a significant figure. Protective interventions are urgently needed to ensure the ongoing cultivation of common beans—"Fasolia Vanilies Feneou"—and grass peas—"Fava Feneou"—and to preserve their cultural and economic significance. These interventions should focus on maintaining genetic diversity, implementing sustainable farming practices, and raising awareness about the critical importance of these crops locally and beyond. By taking these measures, we can safeguard the future of these crops and ensure they continue to be vital elements of Feneos's agricultural and cultural legacy.

4. Conclusions

Currently, vanilla beans and fava beans are cultivated on approximately 4000 hectares, yielding around 3 tons per hectare, with a market price of roughly EUR 2.8 per kilogram. This cultivation plays a pivotal role in the local agricultural community of Feneos, contributing significantly to the economy; the anticipated benefits of implementing this program are expected to yield approximately a 25% annual profit increase. This increase will be derived from a combination of factors, including improved productivity due to the utilization of enhanced genetic material, improved consulting services, and increased market value due to the authentication of the product's origin and quality.

The comprehensive strategy of project M16SYN2-00320, funded within the Agricultural Development Program 2014–2020 (Measure 16), specifically Sub-Measure 16.1–16.2, aims to enhance the preservation and subsequent utilization of this valuable resource by completing the following:

1. Register and identify the landrace with new legislation and EU directives using the provided description protocol.
2. Define the protected variety and apply for its registration in the National List of Varieties.
3. Establish and implement an innovative framework/process to disseminate the best conservation and seed production practices for the landrace within the region of origin, ensuring certification and seed purity. This initiative will be executed through collaboration between KIATO UNION and the support of research institutes and Agricultural University researchers.
4. Authenticate the landrace through morphological and qualitative characteristics and DNA techniques.
5. Provide consulting services to improve farming techniques for farmers, including field schools, e-learning, online applications, and networking via an online platform.
6. Document reduced product inputs.

Author Contributions: Conceptualization, I.M., F.P. and E.N.; methodology, I.M., F.P., E.N., A.T., A.K., G.V. and K.K.; validation, I.M., F.P., E.N. and A.T.; formal analysis, I.M., F.P., E.N., A.T., A.K., G.V. and K.K.; investigation, I.M., F.P., E.N., A.T., A.K., G.V. and K.K.; resources, I.M., F.P., E.N. and A.T.; data curation, I.M., F.P. and E.N.; writing—original draft preparation, I.M., F.P., E.N., A.T., A.K., G.V. and K.K.; writing—review and editing, I.M., F.P., E.N., A.T., A.K., G.V. and K.K.; visualization,

I.M., F.P. and E.N.; supervision E.N., I.M. and F.P.; project administration, E.N. and I.M. All authors have read and agreed to the published version of the manuscript.

Funding: This work was funded in the context of the Agricultural Development Program 2014–2020 (Measure 16), and in particular Sub-Measure 16.1–16.2. Establishment and operation of Operational Groups of the European Innovation Partnership for the productivity and sustainability of agriculture (Project No M16SYN2-00320).

Institutional Review Board Statement: Not applicable.

Informed Consent Statement: Not applicable.

Data Availability Statement: Data are only available on request due to privacy restrictions.

Conflicts of Interest: The authors declare no conflicts of interest.

References

1. Caradoc-Martínez, A.; Loarca-Piña, G.; Oomah, B.D. Antioxidant Activity in Common Beans (*Phaseolus vulgaris* L.). *J. Agric. Food Chem.* **2002**, *50*, 6975–6980. [CrossRef] [PubMed]
2. Chávez-Mendoza, C.; Sánchez, E. Bioactive Compounds from Mexican Varieties of the Common Bean (*Phaseolus vulgaris*): Implications for Health. *Molecules* **2017**, *22*, 1360. [CrossRef] [PubMed]
3. Lambein, F.; Travella, S.; Kuo, Y.-H.; Van Montagu, M.; Heijde, M. Grass Pea (*Lathyrus sativus* L.): Orphan Crop, Nutraceutical or Just Plain Food? *Planta* **2019**, *250*, 821–838. [CrossRef] [PubMed]
4. Koutsika-Sotiriou, M.; Mylonas, I.G.; Ninou, E.; Traka-Mavrona, E. The Cultivation Revival of a Landrace: Pedigree and Analytical Breeding. *Euphytica* **2010**, *176*, 15–24. [CrossRef]
5. Korpetis, E.; Ninou, E.; Mylonas, I.; Ouzounidou, G.; Xynias, I.N.; Mavromatis, A.G. Bread Wheat Landraces Adaptability to Low-Input Agriculture. *Plants* **2023**, *12*, 2561. [CrossRef] [PubMed]

Disclaimer/Publisher's Note: The statements, opinions and data contained in all publications are solely those of the individual author(s) and contributor(s) and not of MDPI and/or the editor(s). MDPI and/or the editor(s) disclaim responsibility for any injury to people or property resulting from any ideas, methods, instructions or products referred to in the content.

Proceeding Paper

Co-Design and Co-Evaluation of Traditional and Highly Biodiversity-Based Cropping Systems in the Mediterranean Area [†]

Paschalis Papakaloudis [1], Andreas Michalitsis [1], Maria Laskari [1], Efstratios Deligiannis [1], Fatima Lambarraa-Lehnhardt [2] and Christos Dordas [1,*]

1. Laboratory of Agronomy, School of Agriculture, Aristotle University of Thessaloniki, 54124 Thessaloniki, Greece; papakalp@agro.auth.gr (P.P.); amichalits@agro.auth.gr (A.M.); marialaskari00@gmail.com (M.L.); deliefst@agro.auth.gr (E.D.)
2. Leibniz Centre for Agricultural Landscape Research (ZALF), Eberswalder Str. 84, 15374 Müncheberg, Germany; fatima.lehnhardt@zalf.de
* Correspondence: chdordas@agro.auth.gr
† Presented at the 17th International Conference of the Hellenic Association of Agricultural Economists, Thessaloniki, Greece, 2–3 November 2023.

Abstract: Intensive agriculture has created several problems in cropping systems that threaten the sustainability of agricultural production. In order to design new cropping systems, a new approach is emerging to support the transition toward sustainable agriculture: a co-design and co-evaluation process that involves stakeholders in the agrifood chain. The present work therefore describes the co-design and co-evaluation process that was followed to design a highly diversified cropping system in a Mediterranean environment. The different systems that were co-designed include the reference system, with wheat and barley in rotation, as well as three diversified systems that were also proposed and co-evaluated: the rotation of wheat, oil seed rape, and barley (DIV1); the rotation of wheat, pea, and barley (DIV2); and the rotation of wheat, intercrops of barley-common vetch, and barley (DIV3). The best system that was selected from the different stakeholders was the DIV3, as it had the highest evaluation of the stakeholders using agronomic, environmental, and socio-economic criteria.

Keywords: intercropping; crop rotation; reference system; diversification; co-design; co-evaluation

1. Introduction

The Mediterranean basin is characterized by a high dependence on agricultural imports, especially cereals and legumes. Over the past 30 years, policies aimed at intensifying agricultural production have led to trajectories that have generally increased the incomes and market orientation of agricultural systems for farm households. However, the resulting economic pressure has encouraged specialization, leading to monocultures that have caused environmental degradation, such as a loss of biodiversity, which threatens the provision of ecosystem services (ES) [1]. Moreover, there is a strong need to develop modern and sustainable agriculture in the Southern Mediterranean countries to stabilize rural populations by providing them with real economic prospects and better social conditions. One way to achieve this is to increase the biodiversity of cropping systems.

Developing highly diversified-based agriculture often requires more than just efficiency or substitution strategies; it requires farming systems to be redesigned [2]. This is a knowledge-intensive approach that potentially empowers farmers and advisors in the quest for agricultural innovations [3,4]. Moreover, biodiversity-based agriculture is highly context-dependent, as designing highly diversified innovative systems requires combining locally relevant empirical knowledge with scientific process-based knowledge [4]. Therefore, a participatory approach is the most relevant way to hybridize scientific information and the expert knowledge of actors [5], acknowledging and taking advantage of the fact

that farmers are also designers [1,6]. In such innovation processes, researchers act as partners in the overall approach [1,4], with one of their main roles being to structure and steer the design process [7].

The objective of the present work is to co-design locally promising innovations based on diversified cropping and farming systems using participatory methods and to co-evaluate the effects of the co-designed diversified cropping and farming systems compared to current ones.

2. Materials and Methods

The approach followed in the present study involved the participation of stakeholders from the agrifood chain. Two workshops were held on the university farm, the first being a co-design workshop aimed at finding innovative and highly species-diversified (HSD) options adapted to the Thessaloniki region case study. The second workshop took place one and a half years later and was aimed at evaluating the HSD systems suggested using a number of indicators belonging to different agronomic, environmental, and socio-economic dimensions (Table 1). Local stakeholders evaluated and ranked potential innovations in dedicated meetings based on data derived from a multi-criteria ex ante assessment, which led to a fine-tuning of the co-designed systems in an iterative manner.

Table 1. Indicators used to evaluate cropping system diversification, ordered within three categories: agronomic, environmental, and socio-economic.

Agronomic	Environmental	Socio-Economic
Grain yield (t ha^{-1})	N losses (volatilization, leaching)	Farming profit
Grain protein concentration	Soil CO_2 sequestered/emitted	Economic independence (from fuel and mineral N)
Yield variability	Energy use efficiency	Economic cost
Soil organic carbon content	Renewable energy input	Material additional cost
Soil erosion	Non-renewable energy input	Workload
Soil N mineralized		Employment of workers
Pest control (weeds, pests, and diseases)		

During the co-design process, four cropping systems were developed and evaluated. The first system served as the reference (RS) and involved a two-year rotation of wheat and barley. In this system, wheat is sown during the first growing season, while barley is sown during the second growing season. The second system (DIV1) consisted of a three-year rotation of wheat the first year, oilseed rape the second year, and barley the third year. The third system (DIV2) included a three-year rotation of wheat, pea, and barley. For these two diversified systems, wheat is cultivated for the first year, the following crop is oilseed rape for DIV1, and pea for DIV2, and finally, in the third year, both systems include barley. Finally, the fourth system (DIV3) involved a three-year rotation with wheat during the first growing season, intercropping of barley with common vetch for the second growing season, and lastly, barley for the third growing season.

3. Results and Discussion

The different stakeholders that were involved were as follows: 20 agricultural students that are also farmers, 1 seed producer and supplier of agricultural supplies, e.g., pesticides, fertilizers, etc., 13 farmers, and 5 researchers. All the participants were involved in cropping systems in Central Macedonia (Figure 1).

Most of the stakeholders indicated that the socio-economic aspects are more important, with 37.8%, followed by 35.4% of the agronomic and 26.8% of the environmental (Figure 2).

The systems that were better according to the stakeholders were DIV3 and DIV,2 as they had the best evaluation regarding the indicators that were used, such as agronomic, environmental, and socio-economic (Figure 3).

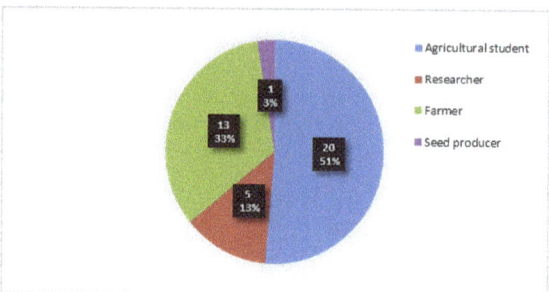

Figure 1. Stakeholders who attended the workshop.

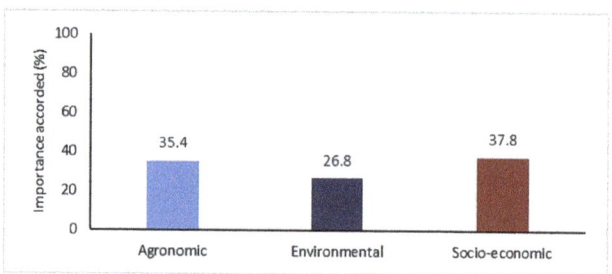

Figure 2. Importance according to each dimension by the stakeholders.

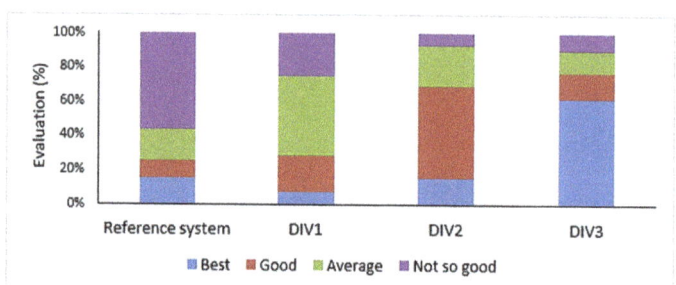

Figure 3. Evaluation of the four cropping systems designed in 2021.

4. Discussion

Based on the results, cropping systems that included legume species, either as sole crops or as intercrops with cereal, were found to be more preferable by the stakeholders [1,4]. Additionally, the farmers' main concerns were related to their final income, which is associated with socio-economic factors [5,7]. Similar responses have been reported in other studies, and it was found that it is better when legumes are included in the cropping system in a rotation or with intercropping, and more stakeholders recognized the need to develop highly diversifying cropping systems; however, the data are limited for Mediterranean cropping systems [1,3–5].

5. Conclusions

The main conclusions of the present study are that agronomic and socio-economic dimensions were the most important for the participants (over 70% combined). Furthermore, DIV2 and DIV3 were selected as the most satisfactory alternative cropping systems. Finally, throughout the discussion, it was obtained that the farmers were more concerned about the socio-economic dimension regarding the final profit.

Author Contributions: C.D. conceptualization and methodology; A.M., P.P., M.L. and E.D., field measurement and data curation; C.D., writing—original draft preparation; A.M., P.P., M.L. and E.D., visualization; F.L.-L., writing—review and editing; C.D., supervision. All authors have read and agreed to the published version of the manuscript.

Funding: This project Biodiversify (Boost ecosystem services through high Biodiversity-based Mediterranean Farming systems) is funded by the General Secretariat for Research and Technology of the Ministry of Development and Investments under the PRIMA Programme. PRIMA is an Art.185 initiative supported and co-funded under Horizon 2020, the European Union's Programme for Research and Innovation.

Institutional Review Board Statement: There is no institutional review board statement.

Informed Consent Statement: Not applicable.

Data Availability Statement: The data presented in this study are available on request from the corresponding author.

Acknowledgments: We are grateful to Anastasios Lithourgidis and the personnel of the University Farm of the Aristotle University of Thessaloniki for assistance with the field experiments. Also, we are grateful to George Menexes for assistance with the statistical analysis of the data.

Conflicts of Interest: The authors declare no conflicts of interest.

References

1. Salembier, C.; Aare, A.K.; Bedoussac, L.; Raj Chongtham, I.; de Buck, A.; Dhamala, N.R.; Dordas, C.; Renate Finckh, M.; Hauggaard-Nielsen, H.; Krysztoforski, M.; et al. Exploring the inner workings of design-support experiments: Lessons from 11 multi-actor experimental networks for intercrop design. *Eur. J. Agron.* **2023**, *144*, 126729. [CrossRef]
2. Hill, J. Breeding components for mixture performance. *Euphytica* **1996**, *92*, 135–138. [CrossRef]
3. Horlings, L.G.; Marsden, T.K. Towards the real green revolution? Exploring the conceptual dimensions of a new ecological modernisation of agriculture that could 'feed the world' May 2011. *Glob. Environ. Chang.* **2011**, *21*, 441–452. [CrossRef]
4. Klerkx, L.; van Mierlo, B.; Leeuwis, C. Evolution of systems approaches to agricultural innovation: Concepts, analysis and interventions. In *Farming Systems Research into the 21st Century: The New Dynamic*; Darnhofer, I., Gibbon, D., Dedieu, B., Eds.; Springer: Dordrecht, The Netherlands, 2012. [CrossRef]
5. Duru, M.; Therond, O.; Martin, G.; Martin-Clouaire, R.; Magne, M.A.; Justes, E.; Journet, E.P.; Aubertot, J.N.; Savary, S.; Bergez, J.E.; et al. How to implement biodiversity-based agriculture to enhance ecosystem services: A review. *Agron. Sustain. Dev.* **2015**, *35*, 1259–1281. [CrossRef]
6. Salembier, C.; Segrestin, B.; Berthet, E.; Weil, B.; Meynard, J.M. Genealogy of design reasoning in agronomy: Lessons for supporting the design of agricultural systems. *Agric. Syst.* **2018**, *164*, 277–290. [CrossRef]
7. Martin, G. A conceptual framework to support adaptation of farming systems–Development and application with Forage Rummy. *Agric. Syst.* **2015**, *132*, 52–61. [CrossRef]

Disclaimer/Publisher's Note: The statements, opinions and data contained in all publications are solely those of the individual author(s) and contributor(s) and not of MDPI and/or the editor(s). MDPI and/or the editor(s) disclaim responsibility for any injury to people or property resulting from any ideas, methods, instructions or products referred to in the content.

Proceeding Paper

Building Advisors and Researchers' Capacity to Support Agricultural Knowledge and Innovation Systems in Europe: The Case of the I2CONNECT Summer School †

Eleni Zarokosta * and Alex Koutsouris

Department of Agricultural Economics & Rural Development, Agricultural University of Athens, 11855 Athens, Greece; koutsouris@aua.gr
* Correspondence: elenazarokosta@aua.gr; Tel.: +30-2105294816
† Presented at the 17th International Conference of the Hellenic Association of Agricultural Economists, Thessaloniki, Greece, 2–3 November 2023.

Abstract: The I2CONNECT Horizon project introduced summer school training, aiming at strengthening the capacity of future advisors and researchers to support interactive innovations. The training consisted of two online sessions and a four-day face-to-face course, covering basic concepts and various methodological tools for stimulating active participation and strengthening innovation networks. The findings indicate the effectiveness of interactive training in cultivating skills and attitudes that enable innovations and also imply the need for the integration of participatory learning and methodological knowledge on interactive processes into university curricula. Modifying traditional university education in this direction could enhance the design and implementation of interactive projects, facilitating actors' navigation through innovative ecosystems.

Keywords: interactive training; networks; AKIS; education; innovation support services

Citation: Zarokosta, E.; Koutsouris, A. Building Advisors and Researchers' Capacity to Support Agricultural Knowledge and Innovation Systems in Europe: The Case of the I2CONNECT Summer School. *Proceedings* **2024**, *94*, 33. https://doi.org/10.3390/proceedings2024094033

Academic Editor: Eleni Theodoropoulou

Published: 25 January 2024

Copyright: © 2024 by the authors. Licensee MDPI, Basel, Switzerland. This article is an open access article distributed under the terms and conditions of the Creative Commons Attribution (CC BY) license (https://creativecommons.org/licenses/by/4.0/).

1. Introduction

According to the Agricultural [Knowledge and] Innovation Systems (A[K]IS) thinking, innovations are complex processes in which new ideas are developed and implemented by networks of multiple actors. In innovation networks, actors engage in social learning and adaptive experimentation to achieve desirable outcomes. Currently, such multi-actor/interactive approaches are gaining ground, being a key component in policy interventions and initiatives, such as the Strategic Working Group on Agricultural Knowledge and Innovation Systems of the Standing Committee on Agricultural Research of the EU (SWG SCAR-AKIS), the European Innovation Partnership for Agricultural Productivity and Sustainability (EIP-AGRI) and Horizon 2020 projects.

These approaches embrace actors' meaningful interaction throughout the entire innovation process since all relevant actors are considered the owners of the same complex problem, though from different angles. At the same time, these actors are considered sources of complementary knowledge, values, interests [1] and practices, which potentially—if put together—lead to a viable solution(s). The recognition that complex problems require the full engagement of diverse actors in networks leads to the need for new ways of actors' mobilization and coordination that facilitate knowledge co-creation and social learning. In this framework, within the emerging pluralistic advisory landscapes, a new set of Innovation Support Service's (ISS) [2] functions emerges as compared to that of 'conventional' advisory services, including access to knowledge; advisory, consultancy and backstopping; marketing and demand articulation; networking facilitation and brokerage; capacity building; access to resources; institutional support for niche innovation; and scaling mechanisms stimulation [3]. In this respect, a major role of ISS is that of the co-learning facilitator, aiming at the development of common meaning and language between dialogue partners in order to encourage change and develop innovative solutions. Therefore, the advisors involved in interactive innovations need new knowledge and skills as well as a methodological toolkit to successfully shape and deliver advisory services tailored to clients' needs.

The i2connect project (https://i2connect-h2020.eu/, accessed on 19 September 2023) identified basic concepts and modes of learning relevant to the qualification of advisors and facilitators engaged in interactive innovation [4] in order to set up three summer schools in the period 2022–2024. On this basis, it adopted a non-directive, participant and problem-solving-oriented training approach [5,6] to support trainees in their own learning about concepts and methods appropriate for interactive innovation. The purpose of this work is to present the experience of the first i2connect summer school and the lessons learned, facilitating similar future efforts.

2. Materials and Methods

The first summer school was organized by the Agricultural University of Athens with the close collaboration of trainers from the University of Hohenheim, the University College Dublin, the Széchenyi István University and the Berner Fachhochschule. Twenty-six (26) MSc and PhD students from universities from eleven European countries participated. The summer school was carried out in 3 stages, including 2 two-hour online meetings and a course with physical presence for four full days. The first online meeting (28 June 2022) aimed at familiarizing participants with each other, the objectives, the structure and the basic concepts of the training, as well as assigning them the task of studying an interactive project from their country—mainly through EIP-AGRI.

The face-to-face course took place in the period from 23 to 29 July 2022 at the Mediterranean Agronomic Institute of Chania (MAICH), Crete. The course covered basic concepts of (interactive) innovation and network facilitation (Table 1). The trainers/facilitators utilized a variety of interactive exercises to encourage trainees'/participants' active engagement and trigger their creativity (e.g., cross the river, guiding the blind, AKIS analysis, controlled dialogue, egg dropping, role-playing in facilitation, walk and talk, etc.). This way, the trainees were sensitized and learned the roles undertaken and the competencies needed for successfully delivering interactive advisory services. In this respect, a variety of methodological tools were also used, such as the Spiral of Initiatives/Innovations and the Circle of Coherence [7]. Furthermore, the trainees participated in a farm visit to interview local actors engaged in an ongoing innovation project and put the tools they had learned into practice. At the end, the trainees evaluated the course with a questionnaire comprising 34 Likert-type questions and 4 open questions regarding (a) what they liked best about the training, (b) which topics were covered insufficiently, (c) suggestions for improvements, and (d) feedback about their personal learning.

Table 1. Overview of the structure and topics covered in the summer school.

Daily Sessions	Monday	Tuesday	Wednesday	Thursday	Friday
Morning sessions		Types of advisory approaches The AKIS concept	Spiral of innovation Cold & warm processes The role of advisors in innovation process	Debriefing of field visit conclusions Facilitation	Networking My own role as an advisor Evaluation
Afternoon Sessions	Introduction	Interactive approaches Competencies of advisors Communication	Farm visit -Preparation -Field trip	Facilitation exercises Debriefing Conclusions	

In the second online meeting (2 November 2022), the trainees reflected on their learnings and further strengthened their network. The participants were invited to exchange experiences in small groups based on questions such as the following: Q1: What feelings occur when thinking of the summer school in Chania? Q2: What is most prominent in my mind related to the summer school? Q3: In what way did the learnings of the summer school change my way of thinking/worldview? and Q4: What have I put into practice so far? When? How did I feel?

3. Results

The quantitative analysis of the questionnaires showed that the course exceeded the expectations of the trainees, rating their overall satisfaction at 3.83 out of 4, with the rates of the majority of the questionnaire items rated over 4 (and half of the items over 4.5/5). Specifically, the course was found to be well planned and organized (rated at 4.7 out of 5) and the content of the training quite comprehensive (4.17); the training was adjusted to the current capabilities of the trainees (4.09), while the involvement of trainees with different backgrounds was also very positive (4.67). The teaching aids used were helpful (4.74), and the methods used made the understanding of the tools easy (4.46), thus increasing trainees' confidence in their future use (4.54). In addition, the trainees were particularly satisfied with their cooperation with the trainers (4.78), who were found to be knowledgeable about the training topics (4.65), able to explain concepts and tools clearly (4.35) and were supportive and helpful when needed (4.91). The trainers were found to be excellent at encouraging active participation and interaction among trainees (4.91) and creating a constructive working atmosphere (4.91). This, along with trainees' good cooperation (4.75), positively influenced peer-to-peer interactions, resulting in increasing trainees' collaborative attitude (4.83) and their motivation to pursue further learning (4.65). As a result, they found that the course was useful, particularly as regards their professional growth (4.42).

The qualitative evaluation confirmed these findings. According to the trainees' comments, they enjoyed a dynamic and inspiring environment when working in small groups, which helped them to keep their energy high throughout the training. The training "was fun", "interesting" and "a true co-learning experience". Trainees got "good knowledge", and although "the training was pretty different from what I expected ... it was better this way". They particularly liked "the practical side of learning (no boring lectures in a classical classroom)" and that "I was able to learn about myself and gain confidence in what I am capable of". Another trainee said: "I built a network of people that study and work in the sector; learned about innovation approaches; used new tools and expressed myself at the facilitation exercise".

An indication of students' appetite "to learn even more" was their suggestions that the next summer school should cover topics such as the facilitation of farmers' discussion groups, project management, conflict management, etc. Their recommendations included the course extension to five full days and the enrichment of thematic areas (e.g., AKIS) with more detailed knowledge. More time should be devoted to examples and developing facilitation competencies as well as to outdoor activities, including the field visit.

The follow-up online meeting confirmed that the experience of the summer school continued to induce feelings of excitement among the trainees. Communication skills, particularly active listening and exercising patience, better understanding of networking and exercising self-confidence were stated as the most prominent learnings. Moreover, certain facilitation activities and tools were put into practice after the course, indicating the impact and effectiveness of interactive training in cultivating skills and attitudes that enable interactive innovations.

4. Discussion and Conclusions

The assessment provides empirical evidence suggesting changes to the traditional (top-down/ex-cathedra) agronomic education offered to future advisors in Higher Education Institutes, especially with a view to the emerging 'paradigm' of interactive innovations. Participatory learning emerges as an essential pillar of advisors' education, indicating the integration of methodological knowledge about shaping and facilitating interactive processes into university curricula. Such knowledge is useful not only for advisors but academics/researchers as well, helping with tasks such as analyzing innovation networks, identifying the roles of relevant actors and navigating through innovation ecosystems. The expected benefits include more effective integration of actors and better design/implementation of interactive projects (e.g., HORIZON and EIP-AGRI).

Author Contributions: Conceptualization E.Z. and A.K., methodology E.Z. and A.K., formal analysis E.Z., Data curation A.K., writing-original draft preparation E.Z., writing- review and editing A.K., funding acquisition A.K. All authors have read and agreed to the published version of the manuscript.

Funding: This research was funded by the EU (H2020) Project I2CONNECT (Grand Agreement Number 863039).

Institutional Review Board Statement: Not applicable.

Informed Consent Statement: Informed consent was obtained from all subjects involved in the study.

Data Availability Statement: Data are contained within the article.

Acknowledgments: The authors wish to express their gratitude to the training team (Maria Gerster-Bentaya, University of Hohenheim, Germany; Stefan Dubach, Berner Fachhochschule, Switzerland; Andras Ver, Széchenyi István University, Hungary; and Monica Gorman, University College of Dublin, Ireland, as well as to all the participants in the 1st i2connect summer school.

Conflicts of Interest: The authors declare no conflicts of interest.

References

1. Beers, P.J.; Sol, A.J. Guiding multi-actor innovation and education projects. In *Theory and Practice of Advisory Work in a Time of Turbulences, Proceedings of the XIX ESEE, Assisi, Perugia, Italy, 15–19 September 2009*; Paffarini, C., Santucci, F.M., Eds.; INEA: Perugia, Italy, 2009; pp. 223–229. Available online: https://edepot.wur.nl/12566 (accessed on 19 September 2023).
2. Knierim, A.; Labarthe, P.; Laurent, C.; Prager, K.; Kania, J.; Madureira, L.; Ndah, T. Pluralism of agricultural advisory service providers—Facts and insights. *J. Rural. Stud.* **2017**, *55*, 45–58. [CrossRef]
3. Faure, G.; Knierim, A.; Koutsouris, A.; Ndah, T.; Audouin, S.; Zarokosta, E.; Wielinga, E.; Triomphe, B.; Mathé, S.; Temple, L.; et al. How to strengthen innovation support services in agriculture with regards to multi-stakeholders approaches. *J. Innov. Econ. Manag.* **2019**, *1*, 145–169. [CrossRef]
4. Hoffmann, V.; Gerster-Bentaya, M. *Rural Extension Handbook, Volume III: Training Concepts and Tools*; Margraf Publishers: Weikersheim, Germany, 2011.
5. Hoffmann, V.; Gerster-Bentaya, M.; Christinck, A.; Lemma, M. *Rural Extension Handbook, Volume I: Basic Issues and Concepts*; Margraf Publishers: Weikersheim, Germany, 2009.
6. Debruyne, L.; Lybaert, S. I2CONNECT Deliverable 1.4: Repository of Required Competencies of an Innovation Advisor. Available online: https://i2connect-h2020.eu/wp-content/uploads/2021/09/Deliverable-1-4-1.pdf (accessed on 19 September 2023).
7. Wielinga, E.; Sjoerd, R. *Energizing Networks. Tools for Co-Creation*; Wageningen Academic Publishers: Wageningen, The Netherlands, 2020.

Disclaimer/Publisher's Note: The statements, opinions and data contained in all publications are solely those of the individual author(s) and contributor(s) and not of MDPI and/or the editor(s). MDPI and/or the editor(s) disclaim responsibility for any injury to people or property resulting from any ideas, methods, instructions or products referred to in the content.

Proceeding Paper

The Use of Digital Media in Equestrian Clubs in Croatia [†]

Maja Gregić *, Tina Bobić, Ranko Gantner and Vesna Gantner

Department of Animal Production and Biotechnology, Faculty of Agrobiotechnical Sciences Osijek, University of J. J. Strossmayer Osijek, V. Preloga 1, 31000 Osijek, Croatia; tbobic@fazos.hr (T.B.); rgantner@fazos.hr (R.G.); vgantner@fazos.hr (V.G.)
* Correspondence: mgregic@fazos.hr
[†] Presented at the 17th International Conference of the Hellenic Association of Agricultural Economists, Thessaloniki, Greece, 2–3 November 2023.

Abstract: This paper aimed to analyze the use of digital media, and the research was conducted via a web survey sent by e-mail to equestrian clubs in Croatia. Social media has significantly altered the way of communication and the availability of information in all segments of life and work, including horse breeding. Within digital media, an extremely large amount of information is available that is not necessarily relevant and true. To prevent the use of inadequate information, in 60% of equestrian clubs in Croatia, certain persons are responsible for the content. Less than 50% of the respondents follow influencer posts. Furthermore, 90% of the respondents believe that digital media is an excellent tool that can help in the work of equestrian clubs, while 80% of the respondents believe that it is currently underutilized.

Keywords: equestrian clubs; digital media; information exchange

1. Introduction

New media habits in the era of digitization are challenging previous understandings of who and what receive media attention. Definitions, interpretations, and understanding of social media have been the subject of debate and frequent corrections over the past 20 years. Ref. [1] claims that social media is a dynamic and contextual concept and understanding social media is temporally, spatially, and technologically sensitive. Furthermore, the content on social media is variable and does not objectivity represent one time period or locality in its entirety. Social media has dramatically changed the dominant way of communication, and today everyone can share content beyond the "gatekeeping" function of traditional media and with the help of new, relatively cheap technologies [2]. In recent years, the influence of social media on the spread and quality of knowledge has become increasingly evident. The development of social media represents a challenge for traditional sources of knowledge and raises questions about how we should interpret and evaluate available knowledge. Social media platforms provide enormous opportunities for people to communicate with each other and allow for the distribution of misinformation to flourish [3]. The presence of fake news in the media has become a global problem, being especially prominent during the coronavirus pandemic, but it is not present in all countries to the same extent. For example, Croatia is a country whose inhabitants know how to recognize and avoid fake news. According to the research of [4], the most reliable sources of information in Croatia are television and radio, while the Internet (excluding social networks) is in second place. Although Croats are considered to be successful in identifying fake news, their actual ability to recognize it does not reflect this. Furthermore, social media has become a well-established way of communication; for example, football clubs from the UEFA League communicate with their fans through the Facebook application [5]. Ref. [6], in their research, used the Facebook application as a tool to conduct a survey related to horse welfare. Furthermore, a relatively new term related to social media is influencer. An influencer is considered to be a person who has an influence on other people and thus can

Citation: Gregić, M.; Bobić, T.; Gantner, R.; Gantner, V. The Use of Digital Media in Equestrian Clubs in Croatia. *Proceedings* **2024**, *94*, 34. https://doi.org/10.3390/proceedings2024094034

Academic Editor: Eleni Theodoropoulou

Published: 25 January 2024

Copyright: © 2024 by the authors. Licensee MDPI, Basel, Switzerland. This article is an open access article distributed under the terms and conditions of the Creative Commons Attribution (CC BY) license (https://creativecommons.org/licenses/by/4.0/).

influence their views on various things. Descriptions of influencers usually focus on what influencers do. Ref. [7] emphasizes that influencers on social networks "gather" followers through blogs, vlogs, and similar content, presenting their everyday life in text and images and promoting various products and services through advertisements [8]. Ref. [9] states that influencers play an important role in consumption processes. Digital media users perceive themselves to have a close relationship with influencers, so they are more likely to be motivated by social media marketing than traditional advertising. Furthermore, influencer marketing is often not perceived as advertising but as a recommendation from a friend [9]. According to Radmann et al. [10], influencers in horse breeding mainly focus their communication on issues related to horses, and their (and their followers') love for horses creates the intimacy necessary to form a basis for other messages (knowledge and advertising). Intimacy and authenticity are strengthened in the interaction between followers and influencers, and they receive positive feedback for their way of working with their horses. The feeling gained from this kind of communication is likely to make commercial recommendations less visible and more compelling for consumption. Croatia, with 70 registered equestrian clubs in the Croatian Equestrian Federation [11], is not a large market for digital media. The size of the Croatian market is particularly evident in comparison with that of Sweden, which has 900 horse-related clubs, of which about half are riding schools; or Norway, whose Norwegian Equestrian Federation (NRYF) is the 13th largest sports federation in the world, with 30,000 active members in 340 registered clubs [10,12,13]; and Germany, where equestrian sport ranks 8th in terms of the number of members in equestrian clubs [14]. The interest in riding and equestrian sports is not only expressed by the number of members of a country's equestrian associations but also by activities on social networks. Numerous blogs as well as accounts on different applications (Facebook, Instagram, Twitter, and YouTube) are dedicated to horses. Analysis of the content on social media shows that young riders analyze and harmonize their qualities in working with horses in relation to the content published by their role models (influencers) on social media [15]. Furthermore, social media often presents an ideal image of the interaction between man and horse, which creates pressure and discomfort in the follower. Considering the increasing importance of digital media in all aspects of life and work, the aim of this paper was to analyze the use of digital media in equestrian clubs in Croatia.

2. Methods

This research was conducted via a web survey sent by e-mail to equestrian clubs registered in Croatia. The questionnaire included questions related to the use of digital media in the work of equestrian clubs, means of communication with their members, and their creation and monitoring of digital media. All clubs were members of the Croatian Equestrian Association, and the questionnaire was filled out by the presidents of the equestrian clubs. In terms of age, 55% of survey respondents were over 41 years old. The web survey was conducted over 14 days, in the period from 31 January 2022 to 13 February 2022. The results of the web survey are presented graphically (MS Excel).

3. Results and Discussion

Members of equestrian clubs in Croatia combine the use of various digital tools for exchanging information, among which the most represented are WhatsApp (80%), calling (75%), and Viber, while those with less representation include publications on websites (40%) and sending SMSs (40%) (Figure 1). Changes in communication with members of equestrian clubs have occurred in parallel with the development of technology. The same trends in changes in the type of communication are happening all around the world, and WhatsApp is currently one of the most popular communication platforms [16].

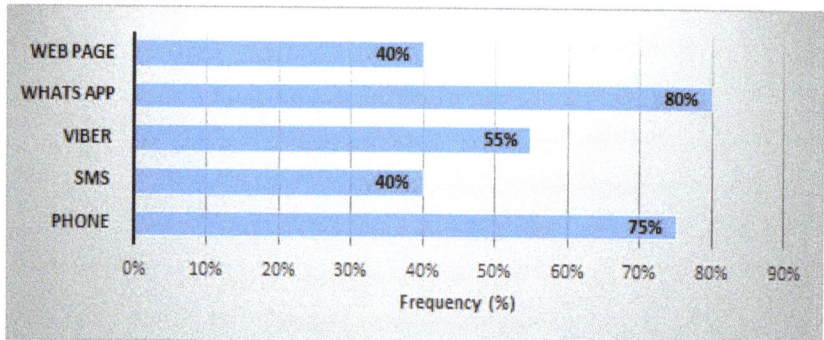

Figure 1. Frequency of use of certain communication tools among members of equestrian clubs in Croatia.

Monitoring and managing digital platforms are common practices in all countries, given that the visibility of posts has no limits and information reaches users quickly, but on the other hand, there are no reviews, so the credibility is questionable. The most popular digital media is Facebook (100%), which was used by all the studied equestrian clubs in Croatia. It was followed by Instagram (65%), while blogs and YouTube appeared sporadically in 5% of cases (Figure 2). Equestrian clubs in Croatia follow the trend of representation on Facebook (Vale and Fernandes, 2017), which contributes to the welfare of horses [6].

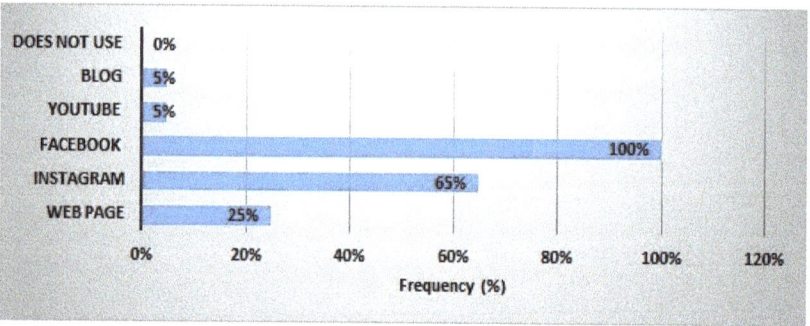

Figure 2. Frequency of use of digital media (networks) used by equestrian clubs in Croatia.

The accuracy of posts on official digital platforms should be satisfactory considering that 60% of the equestrian clubs assigned a person to create and publish content on them, while 35% of the clubs assigned multiple persons. Furthermore, in 10% of the equestrian clubs, content on digital media could be created by all club members (Figure 3). In reducing the number of content creators on digital media, the possibility of spreading fake news among equestrian club members also decreases [4,10,17].

In the case of particular problems at work (health, training, equipment, etc.), 100% of respondents said they would ask for the help of a professional, but in 35% of cases, respondents would check and find information on the Internet, while 20% of respondents would use books and 5% scientific papers. The development of social networks poses new challenges to traditional sources of knowledge and raises the question of how we should interpret the available information. Based on the conducted research, it is evident that expertise is valued in equestrian clubs, given that in the case certain problems arise, members first contact experts in the reference area. The above indicates that there is no disruption in the understanding of information [3]. Influencers in the equestrian world in more developed countries are well positioned [10], while in Croatia, 45% of respondents said they follow them constantly, 5% occasionally, and 50% of respondents do not follow influencers' posts at all. In addition, 10% of the respondents were considered influencers in equestrian clubs in Croatia. Furthermore, 90% of the respondents believed that digital media is a helpful tool and that its use improves their work in equestrian clubs. A total of 80% believed that digital media is not used enough in the "equestrian world". Equestrian clubs in Croatia follow world trends in horse breeding in 15% of cases. It is believed that the younger population is more inclined towards social media usage, more technically skilled, and uses digital media more often [18–20].

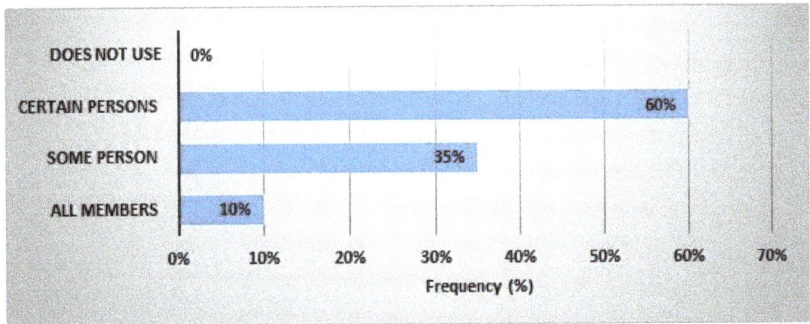

Figure 3. Creation of content on digital media under the authority of equestrian clubs in Croatia.

4. Conclusions

At a time when digital media has almost completely taken over most means of communication and is becoming one of the main sources of information in all segments of life and work, equestrian clubs in Croatia are also following this trend. The complexity of running an equestrian club is greatly facilitated by the use of digital media, which club managers have recognized. Furthermore, false data and unverified information represent major global problems to which special attention should be paid, and systematic work should be carried out to reduce their spread. To increase the accuracy of information, 60% of equestrian clubs in Croatia assign specific persons to the creation of content. Less than 50% follow influencer posts. Also, digital media represents an excellent tool in the work of equestrian clubs in Croatia (in the opinion of 90% of respondents) if used rationally and in a controlled manner, but currently, it is not used enough (in the opinion of 80% of respondents).

Author Contributions: Conceptualization, M.G.; methodology, V.G.; software, V.G.; validation, M.G. and T.B.; formal analysis, M.G.; investigation, M.G. and T.B.; resources, V.G. and R.G.; data curation, M.G. and T.B.; writing—original draft preparation, M.G.; writing—review and editing, V.G.; visualization, M.G. and R.G.; supervision, M.G.; project administration, R.G. and T.B.; funding acquisition, M.G. All authors have read and agreed to the published version of the manuscript.

Funding: This research and its dissemination were supported by the Fund for Bilateral Relations within the Financial Mechanism of the European Economic Area and Norwegian Financial Mechanism for the period 2014-2021 (grant number: 04-UBS-U-0031/23-14).

Institutional Review Board Statement: This study did not require ethical approval.

Informed Consent Statement: Not applicable.

Data Availability Statement: The data are unavailable due to privacy restrictions.

Conflicts of Interest: The authors declare no conflicts of interest.

References

1. Papacharissi, Z. We have always been social. *Soc. Media Soc.* **2015**, *1*, 1–2. [CrossRef]
2. Bruce, T. *Terra Ludus: A Novel about Media, Gender and Sport*; Sense Publishers: Rotterdam, The Netherlands, 2016. [CrossRef]

3. Wardle, C.; Understanding Information Dissorder. First Draft. 2019. Available online: https://firstdraftnews.org (accessed on 1 March 2022).
4. Krelja Kurelović, E.; Tomac, F.; Polić, T. Načini informiranja i prepoznavanje lažnih vijesti kod studenata u hrvatskoj tijekom COVID-19 pandemije. *Zb. Veleučilišta Rijeci* **2021**, *9*, 119–130. [CrossRef]
5. Voigt, M.A.; Hiney, K.; Richardson, J.C.; Waite, K.; Borron, A.; Brady, C.M. Show horse welfare: Horse show competitors' understanding, awareness, and perceptions of equine welfare. *J. Appl. Anim. Welf. Sci.* **2016**, *19*, 335–352. [CrossRef] [PubMed]
6. Vale, L.; Fernandes, T. Social media and sports: Driving fan engagement with football clubs on Facebook. *J. Strateg. Mark.* **2017**, *26*, 37–55. [CrossRef]
7. Abidin, C. Communicative intimacies: Influencers and perceived interconnectedness. *ADA J. Gender N. Media Technol.* **2015**, *8*. [CrossRef]
8. Pöyry, E.; Pelkonen, M.; Naumanen, E.; Laaksonen, S.M. A call for authenticity: Audience responses to social media influencer endorsements. *Int. J. Strat. Commun.* **2019**, *13*, 336–351. [CrossRef]
9. Stubb, C.; Colliander, J. This is not sponsored content"—The effects of impartiality disclosure and e-commerce landing pages on consumer responses to social media influencer posts. *Comput. Hum. Behav.* **2019**, *98*, 210–222. [CrossRef]
10. Radmann, A.; Hedenborg, S.; Broms, L. Social Media Influencers in Equestrian Sport. *Front. Sports Act. Living* **2012**, 1–13. [CrossRef] [PubMed]
11. Hrvatski Konjički Savez, HKS. 2022. Available online: https://www.konjicki-savez.hr/ (accessed on 1 March 2022).
12. Svenska Ridsport Főrbundet, SRF. 2021. Available online: https://www.ridsport.se (accessed on 1 March 2022).
13. Hatlevoll, M. Flest Privateide Rideskoler. 2019. Available online: https://www.hest.no/article.html?news.nid=12629 (accessed on 1 March 2022).
14. Eurostat. 2019. Available online: https://ec.europa.eu/eurostat (accessed on 30 March 2021).
15. Broms, L.; Bentzen, M.; Radmann, A.; Hedenborg, S. *Stable Cultures in Cyberspace: A Study About Equestrians' Use of Social Media as Knowledge Platforms*; Scandinavian Sport Studies Forum: Malmö, Sweden, 2021.
16. Kootbodien, A.; Prasad, N.; Ali, M.S.S. Trends and Impact of WhatsApp as a Mode of Communication among Abu Dhabi Students. *Media Watch* **2018**, *9*, 257–266. [CrossRef]
17. Hine, C. Headlice eradication as everyday engagement with science: An analysis of online parenting discussions. *Public Understand. Sci.* **2014**, *23*, 574–591. [CrossRef] [PubMed]
18. Kosić, S. Online društvene mreže i društveno umrežavanje kod učenika osnovne škole: Navike facebook generacije. *Život škola* **2010**, *56*, 103–125.
19. Stipetić, L.; Benazić, D. Ružić, E. Ponašanje korisnika društvenih mreža za razmjenu video sadržaja. *CroDiM* **2021**, *4*, 93–106.
20. Wang, Y.; McKee, M.; Torbica, A.; Stuckler, D. Systematic literature review on the spread of health-related misinformation on social media. *Soc. Sci. Med.* **2019**, *240*, 112552. [CrossRef] [PubMed]

Disclaimer/Publisher's Note: The statements, opinions and data contained in all publications are solely those of the individual author(s) and contributor(s) and not of MDPI and/or the editor(s). MDPI and/or the editor(s) disclaim responsibility for any injury to people or property resulting from any ideas, methods, instructions or products referred to in the content.

Proceeding Paper

An Investigation of the Digital Presence of Agricultural Stores in Greece [†]

Konstantinos Demestichas *, Antonis Vlandis, Maria Ntaliani * and Constantina Costopoulou *

Informatics Laboratory, Agricultural University of Athens, 75 Iera Odos St., 11855 Athens, Greece; antonisvlandis@gmail.com
* Correspondence: cdemest@aua.gr (K.D.); ntaliani@aua.gr (M.N.); tina@aua.gr (C.C.)
[†] Presented at the 17th International Conference of the Hellenic Association of Agricultural Economists, Thessaloniki, Greece, 2–3 November 2023.

Abstract: Websites are one of the most important digital marketing tools for businesses, through which they interact with users and establish their online presence. A well-designed website is effective in attracting and retaining customers and increasing sales. Automated website evaluation tools are a quick and easy solution for assessing a website, offering immediate results and suggestions for its improvement. In this study, the characteristics of the digital presence of agricultural stores in Greece during 2021–2022 were investigated, using Website Grader and Google Lighthouse tools for a sample of 282 websites. This work shows potential improvements of agricultural store websites over time and can also be used to improve evaluation tools.

Keywords: digital marketing; website performance; website assessment; agricultural stores; Greece

1. Introduction

Electronic markets for agricultural products existed long before the advent of the Internet, since as early as the mid-1970s, some US agricultural industries supported electronic trading mechanisms [1]. There are many indications that electronic commerce can reduce the cost and increase the demand of agri-food products [2].

In today's competitive digital economy, all large businesses and organizations have their own website, which is considered one of the most important components of their operation and an integral part of their business activities [3]. A well-designed website can assist in an increase in sales and business profit; however, websites that are not functional and do not offer user interaction capabilities are off-putting and work negatively for both the user and the company itself. The problem of evaluating websites becomes evident, in order to determine measures or indicators that will assess whether a website is performing its function properly, i.e., retaining existing customers and attracting new ones.

A lot of agricultural stores in Greece have websites, through which they provide information about their products and services to farmers and the general public. The purpose of this research study is to investigate the characteristics of the digital presence of agricultural stores in Greece. A more specific goal is the evaluation of the websites of agricultural stores (as a means of their digital presence) using automated evaluation tools.

2. Methods

The research was carried out for the years 2021 and 2022. For the purposes of the research, an internet search was carried out through the Google and Microsoft search engines to identify the agricultural stores that maintain a website. An agricultural store was defined as any online store that: (a) is located within the Greek territory, (b) provides agricultural/livestock/zootechnical supplies to farmers and (c) has a website through which it carries out online sales. The search was based on a number of keywords in Greek, that here are translated into English for ease of reference: "agricultural store",

"farm department store", "farm center", "farm produce store", "farm supply store", "farm supplies", "agricultural equipment", "agricultural machinery", "agricultural services", "animal feed", etc.

The search initially resulted in 282 electronic addresses (URLs), which were checked for their validity and the fulfillment of the three aforementioned criteria for inclusion in this research. Websites that corresponded to online agricultural stores were included, while unverified URLs, social media pages (e.g., Facebook), blogs, pet accessory stores, sites with hosting problems, and online directory listings were excluded.

The sites that were included in the sample of the present research amounted to 239 different URLs. The majority of these shops (210) were farm supply stores. For each store, the collected information included a series of metadata, such as location, year of establishment, supported languages, and social media presence. As part of the research, the digital presence of agricultural stores in Greece was evaluated using automated website evaluation tools. Website Grader was used as the primary assessment tool for both years, while Google Lighthouse was only applied for the 2022 assessment. Both tools evaluate a website with a score from 0 to 100.

Website Grader assesses four key metrics [4], namely, Performance: overall appearance of a website (rating 0 to 30); Search Engine Optimization (SEO): ranking of the website by search engine users (rating 0 to 30); Mobile Readiness: capacity to view the website on a mobile device (rating 0 to 30); and Security: existence of a security certificate (rating 0 to 10). Google Lighthouse assesses each of the following five metrics on a scale of 0–100, namely, Performance: speed of website loading; Accessibility: ease of usage by persons with disabilities; Best Practices: implementation of security aspects and standards of web development; SEO: capability of crawling by search engines; and Progressive Web Application (PWA): audits of operation. It has to be noted that from October 2022, the PWA metric is assessed using a binary system instead of a point-based system.

3. Results

A statistical analysis was undertaken for the obtained data. For the year 2021 (Figure 1), the Website Grader results for websites regarding the mean score and standard deviation are as follows: 13.96 ± 7.01 for Performance; 26.59 ± 5.16 for SEO; 19.35 ± 10.09 for Mobile Readiness; and 4.86 ± 3.73 for Security.

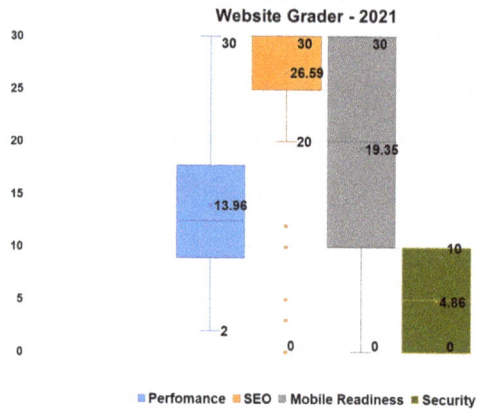

Figure 1. Boxplots of Website Grader results for Performance, SEO, Mobile Readiness and Security (Year = 2021).

For the year 2022 (Figure 2), the Website Grader results for websites regarding the mean score and standard deviation are as follows: 13.62 ± 6.94 for Performance; 27.41 ± 2.82 for SEO; 20.46 ± 9.27 for Mobile Readiness; and 5.71 ± 3.57 for Security.

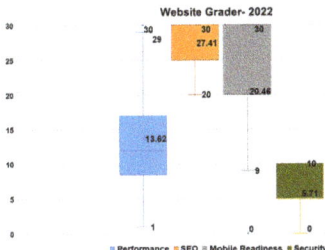

Figure 2. Boxplots of Website Grader results for Performance, SEO, Mobile Readiness and Security (Year = 2022).

For the year 2022 (Figure 3), the Google Lighthouse results for websites regarding the mean score and standard deviation are as follows: 49.71 ± 21.39 for Performance; 81.32 ± 12.11 for Accessibility; 79.02 ± 12.50 for Best Practices; 84.59 ± 10.42 for SEO; and 37.29 ± 8.72 for PWA.

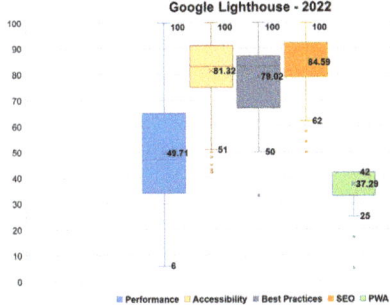

Figure 3. Boxplots of Google Lighthouse results for Performance, Accessibility, Best Practices, SEO and PWA.

Table 1 depicts the correlations between the scores of the various control elements of the two evaluation tools. A strong positive correlation (0.653) was recorded between the Performance metric in Website Grader and the Performance metric in Google Lighthouse, as well as between Security in Website Grader and Best Practices in Google Lighthouse (0.603).

Table 1. Pearson coefficients (r) between Website Grader and Google Lighthouse website evaluation tools metrics (significant correlations are shown in bold).

Tool	Checkpoint	Website Grader				Google Lighthouse				
		Performance	SEO	Mob	Security	Performance	Accessibility	Best Practices	SEO	PWA
Website Grader	Performance									
	SEO	−0.103								
	Mobile	−0.157	0.043							
	Security	0.037	−0.221	0.265						
Google Lighthouse	Performance	**0.653**	−0.046	−0.175	−0.053					
	Accessibility	−0.134	0.043	**0.229**	0.191	−0.144				
	Best Practices	0.111	−0.139	**0.347**	**0.603**	0.127	**0.251**			
	SEO	−0.259	**0.457**	**0.467**	0.064	**−0.261**	**0.410**	0.176		
	PWA	0.010	−0.028	**0.254**	**0.389**	−0.044	0.047	**0.243**	0.110	

147

4. Conclusions

Comparing the scores between 2021 and 2022 using Website Grader showed variations in SEO, Mobile Readiness and Security, indicating potential improvements or positive changes that were made by websites over time. In particular, it appears that the overall mean scores across the four metrics do not differ significantly between the two years; however, individual websites varied in their ranking in relation to their overall score, as some of them showed improvement and others decline in performance in terms of specific metrics.

Performance and SEO are metrics used by both tools; however, due to the different rating scale of the two tools, the results are not directly comparable. This is in agreement with other research [5], which evaluated specific websites with different tools and found that websites are evaluated and interpreted differently and receive a different score for metrics with same names by each tool. However, as shown through the correlations that were recorded between the scores of the two tools' metrics, Performance and SEO, these are strongly positively and moderately positively correlated, respectively. This seems to be an indicator of reliability of the provided evaluation results of the two tools.

Author Contributions: Conceptualization, K.D. and A.V.; methodology, K.D. and C.C.; validation, A.V., K.D., M.N. and C.C.; data curation, A.V.; writing—original draft preparation, K.D. and A.V.; writing—review and editing, C.C. and M.N.; visualization, A.V.; supervision, K.D., C.C. and M.N. All authors have read and agreed to the published version of the manuscript.

Funding: This research received no external funding.

Institutional Review Board Statement: Not applicable.

Informed Consent Statement: Not applicable.

Data Availability Statement: Data are available on request.

Conflicts of Interest: The authors declare no conflicts of interest.

References

1. Montealegre, F.; Thompson, S.; Eales, J.S. An empirical analysis of the determinants of success of food and agribusiness e-commerce firms. *Int. Food Agribus. Manag. Rev.* **2007**, *10*, 61–81.
2. Zeng, Y.; Jia, F.; Wan, L.; Guo, H. E-commerce in agri-food sector: A systematic literature review. *Int. Food Agribus. Manag. Rev.* **2017**, *20*, 439–460. [CrossRef]
3. Herrada-Lores, S.; Iniesta-Bonillo, M.Á.; Estrella-Ramón, A. Weaknesses and strengths of online marketing websites. *Span. J. Mark.-ESIC* **2022**, *26*, 189–209. [CrossRef]
4. Costopoulou, C.; Ntaliani, M.; Ntalianis, F. An analysis of social media usage in winery businesses. *Adv. Sci. Technol. Eng. Syst.* **2019**, *4*, 380–387. [CrossRef]
5. Kaur, S.; Kaur, K.; Kaur, P. An empirical performance evaluation of universities website. *Int. J. Comput. Appl.* **2016**, *146*, 10–16. [CrossRef]

Disclaimer/Publisher's Note: The statements, opinions and data contained in all publications are solely those of the individual author(s) and contributor(s) and not of MDPI and/or the editor(s). MDPI and/or the editor(s) disclaim responsibility for any injury to people or property resulting from any ideas, methods, instructions or products referred to in the content.

Proceeding Paper

Factors Connected with the Registration of "Sikali Vevis" as a Geographical Indication Protection (PGI) Product [†]

Ioannis Mylonas [1,*], Fokion Papathanasiou [2], Elissavet Ninou [3,*], Anthoula Tsipi [1], Dimitrios Kostitsis [4], Iosif Sistanis [1], Chrysanthi Pankou [5] and Kostantinos Koutis [6]

[1] Institute of Plant Breeding and Genetic Resources, ELGO DIMITRA, Kourtidou 56–58 & Nirvana, 11145 Athens, Greece; anthitsipi@gmail.com (A.T.); aff00018@uowm.gr (I.S.)
[2] Department of Agriculture, University of Western Macedonia, 50100 Koila Kozanis, Greece; fpapathanasiou@uowm.gr
[3] Department of Agriculture, International Hellenic University, 57400 Sindos, Greece
[4] Agriculture Cooperative of Florina, 7th Noemvriou 26, 53100 Florina, Greece; dkostitsis@gmail.com
[5] Institute of Industrial & Forage Crops, ELGO DIMITRA, Kourtidou 56–58 & Nirvana, 11145 Athens, Greece; cpankou@gmail.com
[6] AEGILOPS—Network for Biodiversity and Ecology in Greece Ano Lechonia, 37300 Agria Volos, Greece; koutisresfarm@gmail.com
* Correspondence: ioanmylonas@yahoo.com (I.M.); lisaninou@agr.teithe.gr or lisaninou@gmail.com (E.N.)
[†] Presented at the 17th International Conference of the Hellenic Association of Agricultural Economists, Thessaloniki, Greece, 2–3 November 2023.

Abstract: The rate of rye consumption is increasing due to its benefits for human health. "Sikali Vevis" is a cultivated traditional rye population of the Vevi area, Florina of Western Macedonia, Greece, which supports the local agricultural community. However, the identity of this traditional population is not yet protected. This work, funded under the Agricultural Development Program 2014–2020 (Measure 16), Sub-Measure 16.1–16.2 (project M16SYN2-00321), will present the parameters connected with the description of the unique identity of this product, its origin, its traceability, local agricultural practices, and specific characteristics that will contribute to the protection of this traditional population.

Keywords: rye; local landrace; PGI product; quality; agricultural development

1. Introduction

Rye (*Secale cereale* L.) is a cereal crop recognized for its robust winter endurance, ability to withstand various environmental and biological challenges, and suitability for nutrient-depleted, sandy soils with low pH levels [1]. The majority of global rye production, exceeding 90%, is concentrated in Europe's northern, eastern, and central regions [2], where it has been cultivated for its grains since the Bronze Age. Within the European Union, rye grains serve diverse purposes, with 41% allocated for human consumption, 32% allocated for animal feed, 12% used as a raw material for bioethanol production, 10% allocated for biogas generation, and 5% allocated for seed multiplication [3]. Rye stands out among the cereals due to its notably high dietary fiber content [4] and its rich assortment of bioactive compounds [5]. Beyond its dietary importance, rye has played a pivotal role in numerous breeding programs by serving as a source of disease resistance in wheat. This includes resistance against diseases like powdery mildew, stripe rust, and stem rust [6].

The European Union first adopted the system for the protection of geographical indications and the designations of origin of agricultural products and foodstuffs (regulations 2081/92 and 2082/92) to define rules on the certificates of specific characteristics for European agricultural products. "Sikali Vevis" is a traditional rye population (*Secale cereale* L.) cultivated in the Vevi area of Florina, Western Macedonia, that supports the local agricultural community. The rate of rye consumption has been increasing recently due to its benefits for human health, so there is an increasing interest in rye consumption as a food. "Sikali Vevis" is cultivated in disadvantaged and remote areas, having particular characteristics and unique qualities; these unique characteristics have not yet been recorded systematically to promote and protect the name of this agricultural product in alignment with the EU and Greek regulations [7]. This will help secure the identity and bring added value to this unique rye product, achieving better market prices and improving the producers' income [8]. PGI emphasizes the relationship between the specific geographic region and the product name, where a particular quality, reputation, or other characteristic is attributable to its geographical origin [9].

An essential target of the project M16SYN2-00321 is to record the agronomic and quality characteristics of this rye population, to investigate the parameters connected with its unique origin, and collect all the required information for this unique product to apply for registration under the scheme of Protected Geographical Indication. This work includes the initial results of this study.

2. Methodology

Extensive research and interviews with the local farmers were conducted to identify the traditional uses of the rye "Sikali Vevis" population; different samples were collected from the area's farmers in cooperation with the Agricultural Cooperative of Florina. A pilot study using six representative samples of the cultivated local rye population "Sikali Vevis" originated from 6 producers located in Vevi, Florina, Western Macedonia, Greece, was established in the Farm of the Department of Agriculture of the University of Western Macedonia (Table 1). The seeds were sown according to the local producers' standard agricultural practices, and seed productivity values and protein content were recorded during growth and agronomic parameters. The productivity and quality were estimated according to standard practices.

Table 1. An indicative range of the "Sikali Vevis" seeds' agronomic and quality characteristics is expressed as mean yield.

"Sikali Vevis"	Seed Yield (kg/ha) [1]	Protein Content (%)
Population 1	1507	12.6
Population 2	1659	12.4
Population 3	1621	12.1
Population 4	1296	13.7
Population 5	1549	12.5
Population 6	1859	11.8

[1] The average yield derived from four different replications.

3. Results and Discussion

The name of the agricultural food product is "Sikali Vevis", and the description of the agricultural product consists of the seeds produced by the local population of rye (*Secale cereale* L.) cultivated in the Vevi area since the 19th century. Vevi belongs to the Prefecture of Western Macedonia, the Municipality of Florina. It is located east of the city of Florina, about 20 km from it, in a southwestern mountain branch of the Voras mountain (Kaimaktsalan), which ends in the narrows of Kirli Derven, near the village of Kleidi, while the southeastern branch ends in the plain of Florina. The production process follows the traditional agricultural practices without any additional inputs; it is essential that, according to the locals, "Sikali Vevis" is cultivated in mountainous and disadvantaged areas and especially in fields of low-level fertility. It is characteristic that "Sikali Vevis" can grow very well in low-fertility-level soils and show good tolerance to frost in subzero environments. The average range of the recorded seed yield and protein content span from 1296 (kg/ha) to 1859 (kg/ha), and the protein content ranges from 11.8% to 13.7%.

The seeds and inflorescence are shown in Figure 1.

Figure 1. The seeds just before harvesting of "Sikali Vevis".

4. Conclusions

This multifaceted approach of the project M16SYN2-00321 funded by the Greek Agricultural Development Program 2014–2020 (Measure 16) and, in particular, Sub-Measure 16.1–16.2, will enhance the safeguarding and subsequent utilization of this valuable resource through the following ways:

- Recognizing and identifying the traditional population through applying for registration on the National List. This ensures production protection across a broader geographical area by gathering the data necessary to submit a dossier as a PGI (Protected Geographical Indication) product.
- Developing improved genotypes suitable for organic environments.
- Establish and implement an innovative procedure for disseminating the best practices in conserving and producing seeds of the preserved variety within its region of origin. This process is tailored to the specific conditions, ensuring seed certification and an adequate purity level. With ELGO researchers' support, the cooperative will lead this effort.
- Authentication of the morphological, qualitative characteristics, and DNA techniques.
- Providing valuable advisory services to farmers, which include field schools, e-learning opportunities, an online application system, and networking through an online platform.
- Documenting the reduction in product inputs, promoting sustainability and resource efficiency.

Author Contributions: Conceptualization, I.M., F.P. and E.N.; methodology, I.M., F.P., E.N., A.T., D.K., I.S., C.P. and K.K.; validation, I.M., F.P., E.N., A.T. and I.S., formal analysis, I.M., F.P., E.N., A.T., D.K., I.S., C.P. and K.K.; investigation, I.M., F.P., E.N., A.T. and I.S.; resources I.M., F.P., E.N., A.T. and I.S.; data curation, I.M., F.P., E.N., A.T. and I.S.; writing—original draft preparation, I.M., F.P., E.N., A.T., D.K., I.S., C.P. and K.K.; writing—review and editing, I.M., F.P., E.N., A.T., D.K., I.S., C.P. and K.K.; visualization, I.M., F.P., E.N., and I.S.; supervision, I.M., F.P., E.N.; project administration, I.M., F.P. and E.N.; funding acquisition, I.M., F.P. and E.N. All authors have read and agreed to the published version of the manuscript.

Funding: This work was funded in the context of the Agricultural Development Program 2014–2020 (Measure 16), and in particular Sub-Measure 16.1–16.2. Establishment and operation of Operational Groups of the European Innovation Partnership for the productivity and sustainability of agriculture (Project No M16SYN2-00321).

Institutional Review Board Statement: Not applicable.

Informed Consent Statement: Not applicable.

Data Availability Statement: Data are only available on request due to privacy restrictions.

Conflicts of Interest: The authors declare no conflicts of interest.

References

1. Miedaner, T.; Laidig, F. *Advances in Plant Breeding Strategies: Cereals*; Al-Khayri, M., Jain, S., Johnson, D., Eds.; Springer Nature: Cham, Switzerland, 2019; pp. 343–372.
2. The Top 10 Rye Producing Countries of the World. Available online: https://www.worldatlas.com/articles/the-top-10-rye-producing-countries-of-the-world.html (accessed on 1 September 2023).
3. EU Cereals Balance Sheets 2016/17 and 2017/18. Available online: https://ec.europa.eu/agriculture/sites/agriculture/files/cereals/presentations/oilseeds/balance%E2%80%93sheets%E2%80%93and%E2%80%93forecasts_en.pdf (accessed on 1 September 2023).
4. Rakha, A.; Åman, P.; Andersson, R. Characterisation of Dietary Fibre Components in Rye Products. *Food Chem.* **2010**, *119*, 859–867. [CrossRef]
5. Koistinen, V.M.; Mattila, O.; Katina, K.; Poutanen, K.; Aura, A.M.; Hanhineva, K. Metabolic profiling of sourdough fermented wheat and rye bread. *Sci. Rep.* **2018**, *8*, 5684. [CrossRef] [PubMed]
6. Li, J.; Zhao, L.; Lü, B.; Fu, Y.; Zhang, S.; Liu, S.; Yang, Q.; Wu, J.; Li, J.; Chen, X. Development and Characterization of a Novel Common Wheat–Mexico Rye T1DL·1RS Translocation Line with Stripe Rust and Powdery Mildew Resistance. *J. Integr. Agric.* **2023**, *22*, 1291–1307. [CrossRef]
7. Regulation (EU) No 1151/2012 of the European Parliament and of the Council of 21 November 2012 on Quality Schemes for Agricultural Products and Foodstuffs. Available online: https://eur-lex.europa.eu/legal-content/en/ALL/?uri=CELEX:32012R1151 (accessed on 1 September 2023).
8. Protected Geographical Indication (PGI). Available online: https://agriculture.ec.europa.eu/farming/geographical-indications-and-quality-schemes/geographical-indications-and-quality-schemes-explained_en#pgi (accessed on 1 September 2023).
9. Geographical Indications and Quality Schemes Explained. Available online: https://agriculture.ec.europa.eu/farming/geographical-indications-and-quality-schemes/geographical-indications-and-quality-schemes-explained_en (accessed on 1 September 2023).

Disclaimer/Publisher's Note: The statements, opinions and data contained in all publications are solely those of the individual author(s) and contributor(s) and not of MDPI and/or the editor(s). MDPI and/or the editor(s) disclaim responsibility for any injury to people or property resulting from any ideas, methods, instructions or products referred to in the content.

Proceeding Paper

Agricultural Value Added, Farm Business Cycles and Their Relation to the Non-Farm Economy [†]

Christos P. Pappas * and Christos T. Papadas

Department of Agricultural Economics and Rural Development, Agricultural University of Athens, 11855 Athens, Greece; cpap@aua.gr
* Correspondence: christ.pappas@gmail.com
[†] Presented at the 17th International Conference of the Hellenic Association of Agricultural Economists, Thessaloniki, Greece, 2–3 November 2023.

Abstract: This paper investigates the relationship between the gross value added (GVA) of Greece's agricultural sector and the GVAs of the other sectors. The research considers both the relationship between value levels and the cycles of GVAs. Dynamic analysis using ARDL modeling shows that there is no cointegration between agricultural GVA and the other GVAs. However, there is an estimated cointegrating relationship between business cycles of agriculture and those of the rest of the economic sectors, with the cycles of services being the significant variable. Moreover, econometric analysis using NARDL modeling shows that there is a cointegrating relationship between the levels of GVAs as well, when asymmetricity—with respect to GVA changes of the services sector—is introduced.

Keywords: structural transformation; agriculture; business cycles; ARDL and NARDL

1. Introduction

This paper investigates the relationship between agricultural value added (GVA) and the GVAs of the other economic sectors, and the relationship between farm business cycles and the other sectoral business cycles as well. Analysis and estimates refer to the Greek economy and the non-farm sectors, which are classified as industry, construction and services. The results provide information on a significant aspect of linkages between the farm and non-farm economy in the process of growth and transformation, as well as the relationship of their cyclical behavior. Understanding such linkages and relationships is useful to decision makers when sectoral policy measures and growth incentives are devised.

Structural transformation is a prominent feature of economic growth and is regarded as one of the main stylized facts of development [1]. Several studies have been conducted regarding the co-movement of different sectoral outputs [2]. Some investigate such linkages within the context of the Real Business Cycles theory [3]. Empirical research regarding fluctuations in crop output relies heavily on weather shocks and climate changes [4].

However, econometric investigations of the relationship between outputs or the GVA of agriculture and other sectors are limited. Ref. [5] suggests that all economic sectors' outputs are integrated in China, and highlights the prevalence of agriculture in driving other sectors' growth. Ref. [6] supports that Indian sectoral outputs move together and that sectoral growths are interdependent. Regarding the Greek agriculture, ref. [7] argues that linkages between the farm and non-farm sectors are weak. In addition to these studies with variables used at their levels, ref. [8] examined farm business cycles in the U.S. and concluded that they are not correlated with the rest of the economy.

Dynamic analysis shows that even though cointegrating relationships at the variable levels were not found using the ARDL model, when changes in the GVA time series of services are decomposed into positive and negative ones, the estimated NARDL model shows that a significant and negative asymmetric long-term relationship between the GVAs of agriculture and services does exist. In addition, an ARDL model confirms that between the cyclical behavior of GVAs of agriculture and services, there is a significant negative cointegrating relationship.

2. Materials and Methods

We use an annual time series (1960–2020) of sectoral GVAs (at 2015 constant prices) on a logarithmic scale. Deviations of sectoral GVA time-series are taken from the estimation of their trend to estimate their cyclical components, applying the Hodrick–Prescott (HP) filter, despite severe criticisms [9], the Butterworth (BW), the Baxter–King (BK) and the Christiano–Fitzgerald (CF) filters. Results show that the choice of filters does not alter the general picture of cyclical behavior.

The approach initially applied is based on the linear ARDL (p,q) model [10]. We adopted the respective ECM (Error Correction Model) and used it at levels and cyclical components only. In cyclical behavior studies in particular, it has been used to investigate the cointegrating relationship between business cycles of different countries [11]. It is given by Equation (1):

$$\Delta Y_t = \beta_0 + \sum_{j=1}^{p-1} a_j \Delta Y_{t-j} + \sum_{i=1}^{k} \sum_{j=0}^{q-1} \beta_{ij} \Delta X_{it-j} + \theta_0 Y_{t-1} + \sum_{i=1}^{k} \theta_i X_{it-1} + e_t, \quad (1)$$

where Y refers to log GVA of agriculture while X refers to the log GVAs of non-farm sectors.

Asymmetric effects can be searched for and taken under consideration using the NARDL model as per [12]. Decomposing the (kx1) vector of $X_i's$ in positive and negative partial sums of total increases and decreases, that is, $X_t = X_0 + X_t^+ + X_t^-$, with $\sum_{j=1}^{t} \Delta X_j^+ = \sum_{j=1}^{t} \max(\Delta X_j, 0)$; $X_t^- = \sum_{j=1}^{t} \Delta X_j^- = \sum_{j=1}^{t} \min(\Delta X_j, 0)$, we derive the relevant nonlinear ECM. The model can and should be applied only at levels since detrended cycle values cannot be decomposed. NARDL applications in agricultural economics are found in [13].

3. Results

An estimation of the ARDL model rejects the hypothesis of the existence of a linear long-term relationship between the sectoral GVAs, since the F_{PSS}-stat (1.86) is well below the upper and lower bounds' critical values (3.63 and 2.45, respectively). However, when asymmetry is considered and an NARDL model is implemented, a significant cointegrating relationship is found between the variables at their log levels (F_{PSS}-stat = 3.48, p-value (0.05)). Considering that structural change has to do with the continuous growth of services, in accordance with the practice in the literature of selecting a variable as the asymmetric one, the variable treated as such is the GVA of services. The estimated long-run asymmetric equilibrium relationship, derived by the appropriate process after the estimation of the NARDL model, is given by Equation (2):

$$\ln agr_t = 0.79(\ln const_t) + 1.26(\ln ind_t) - 2.15(\ln serv_t^+) - 6.01(\ln serv_t^-) + e_t, \quad (2)$$

where ln denotes the log values of GVA for agriculture ($\ln agr_t$), construction ($\ln const_t$), industry ($\ln ind_t$), and services ($\ln serv_t$), while $\ln serv_t^+$ and $\ln serv_t^-$ are the total increases and decreases in the log GVA up to time t. To check for statistically significant differences between $serv_t^+$ and $serv_t^-$, the Wald test is applied and the results reject the hypothesis of long-term symmetry. This becomes apparent at the dynamic multipliers graph (Figure 1).

In order to investigate the existence of a long-run equilibrium relationship between the cyclical components of agricultural GVA and the other sectoral GVAs, the linear ARDL model is deployed. Four different models are estimated (corresponding to different filters of detrending) and a linear long-term equilibrium relationship is supported, as the model F-values show, which also confirms the negative relationship between the products of the agricultural and services sectors (Table 1).

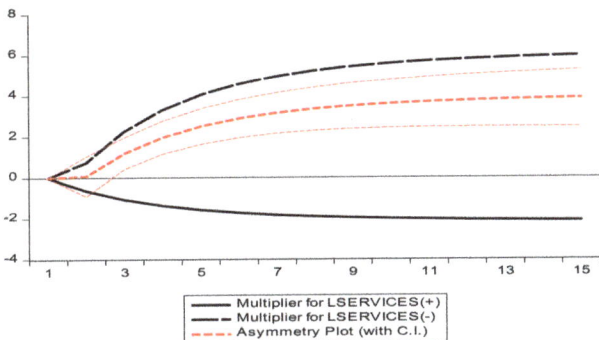

Figure 1. Dynamic multipliers.

Table 1. Dynamic symmetric estimations (cycles) [1].

Variable	BK	BW	CF	HP
$lnagr_{t-1}$	−1.32 (0.00)	−1.65 (0.00)	−1.67 (0.00)	−1.12 (0.00)
lncons	−0.08 (0.37)	−0.07 (0.39)	−0.07 (0.37)	−0.09 (0.35)
lnind	0.51 (0.10)	0.28 (0.37)	0.41 (0.17)	0.56 (0.16)
lnserv	−0.86 (0.05)	−0.67 (0.08)	−0.88 (0.04)	−0.58 (0.11)
$lnserv_{t-1}$	-	-	-	-
$\Delta lnagr_{t-1}$	0.20 (0.16)	0.29 (0.04)	0.37 (0.01)	0.42 (0.01)
$\Delta lnserv$	-	-	-	-
$\Delta lnserv_{t-1}$	-	-	-	-
L_{lncons}	−0.06 (0.38)	−0.04 (0.40)	−0.04 (0.39)	−0.07 (0.36)
L_{lnind}	0.38 (0.14)	0.17 (0.40)	0.25 (0.21)	0.49 (0.08)
L_{serv}	−0.65 (0.05)	−0.37 (0.08)	−0.53 (0.04)	−0.51 (0.10)
R^2	0.63	0.71	0.72	0.62
Adj. R^2	0.60	0.69	0.70	0.60
$x^2_{sc(1)}$	0.10 (0.74)	0.09 (0.76)	0.04 (0.85)	0.43 (0.51)
$F_{sc(1)}$	0.09 (0.76)	0.08 (0.77)	0.03 (0.86)	0.40 (0.52)
$x^2_{sc(2)}$	1.39 (0.50)	4.33 (0.11)	1.64 (0.44)	1.49 (0.48)
$F_{sc(2)}$	0.61 (0.54)	2.06 (0.14)	0.73 (0.48)	0.69 (0.51)
x^2_H	8.8 (0.12)	11.20 (0.05)	3.00 (0.02)	1.62 (0.18)
F_{RESET}	0.22 (0.64)	0.48 (0.49)	0.28 (0.78)	0.17 (0.87)
F_{PSS}	14.32 (0.00)	19.44 (0.00)	22.36 (0.00)	22.52 (0.00)

[1] Dependent variable: $\Delta lnagr$, L are the estimated long-run coefficients, $x^2_{sc(1)}$, $x^2_{sc(2)}$ and F_{sc} (1), F_{sc} (2) denote LM and F tests for serial correlation, x^2_H denotes LM test for homoscedasticity, F_{RESET} denotes LM test for functional form. F_{PSS} is the F-statistic that checks the null hypothesis of non-integration. p-values in parentheses.

4. Discussion

The Greek economy has undergone a structural transformation from an agricultural-based economy to a service-based economy. This shift reflects a negative long-run relationship between the two sectors as the share of agricultural employment and output declines over time. The findings of an asymmetric relationship between the agricultural and service sectors at the level of their values added suggest that there may be additional factors at play. Agriculture seems to be rising faster when economic conditions and services deteriorate, while its reduction in times of service growth follows slower rates.

The service sector is generally considered to be more dynamic and responsive to changes in demand compared to the agricultural sector, which is more dependent on production lags and external factors such as weather conditions and natural disasters. Demand changes in growth periods also reflect more on services. As a result, changes in the service sector may have a greater correlation with economic reverse situations compared to the agricultural sector. During economic downturns or recessions, the service sector may be more vulnerable to declining demand and may experience a more severe contraction than the agricultural sector. Finally, the detrended cyclical behavior of the two sectors and their negative significant relationship is also confirmed.

Future research could focus on the investigated linkages using data from other countries. This would shed more light into relationships of structural changes during the growth processes.

Author Contributions: Conceptualization, C.P.P. and C.T.P.; methodology, C.P.P. and C.T.P.; software, C.P.P.; validation, C.P.P. and C.T.P.; formal analysis, C.P.P. and C.T.P.; investigation, C.P.P. and C.T.P.; resources, C.P.P.; data curation, C.P.P.; writing—original draft preparation, C.P.P.; writing—review and editing, C.P.P. and C.T.P.; visualization, C.P.P.; supervision, C.T.P.; project administration, C.P.P. and C.T.P.; funding acquisition, C.T.P. All authors have read and agreed to the published version of the manuscript.

Funding: This research received no external funding.

Institutional Review Board Statement: Not applicable.

Informed Consent Statement: Not applicable.

Data Availability Statement: Data confirming the reporting results are available at the links: https://economy-finance.ec.europa.eu/economic-research-and-databases/economic-databases_en (accessed on 7 March 2023).

Conflicts of Interest: The authors declare no conflict of interest.

References

1. Bustos, P.; Caprettini, B.; Ponticelli, J. Agricultural Productivity and Structural Transformation: Evidence from Brazil. *Am. Econ. Rev.* **2016**, *106*, 1320–1365. [CrossRef]
2. Carvalho, V.; Gabaix, X. The Great Diversification and Its Undoing. *Am. Econ. Rev.* **2013**, *103*, 1697–1727. [CrossRef]
3. Rebelo, S. Real Business Cycle Models: Past, Present and Future. *Scand. J. Econ.* **2005**, *107*, 217–238. [CrossRef]
4. Deschênes, O.; Greenstone, M. The Economic Impacts of Climate Change: Evidence from Agricultural Output and Random Fluctuations in Weather. *Am. Econ. Rev.* **2012**, *102*, 3761–3773. [CrossRef]
5. Yao, S. Cointegration Analysis of Agriculture and Non-Agricultural Sectors in the Chinese Economy 1952–1992. *Appl. Econ. Lett.* **1994**, *1*, 227–229. [CrossRef]
6. Kanwar, S. Does the Dog Wag the Tail or the Tail the Dog? Cointegration of Indian Agriculture with Nonagriculture. *J. Policy Model.* **2000**, *22*, 533–556. [CrossRef]
7. Kyrkilis, D.; Semasis, S. Greek Agriculture's Failure. The Other Face of a Failed Industrialization. From Accession to EU to the Debt Crisis. *Procedia Econ. Financ.* **2015**, *33*, 64–77. [CrossRef]
8. Da-Rocha, J.M.; Restuccia, D. The Role of Agriculture in Aggregate Business Cycles. *Rev. Econ. Dyn.* **2006**, *9*, 455–482. [CrossRef]
9. Hamilton, J.D. Why You Should Never Use the Hodrick-Prescott Filter. *Rev. Econ. Stat.* **2018**, *100*, 831–843. [CrossRef]
10. Pesaran, M.H.; Shin, Y.; Smith, R.J. Bounds Testing Approaches to the Analysis of Level Relationships. *J. Appl. Econ.* **2001**, *16*, 289–326. [CrossRef]
11. Konstantakopoulou, I.; Tsionas, E.G. Half a Century of Empirical Evidence of Business Cycles in OECD Countries. *J. Policy Model.* **2014**, *36*, 389–409. [CrossRef]
12. Shin, Y.; Yu, B.; Greenwood-Nimmo, M. Modelling Asymmetric Cointegration and Dynamic Multipliers in a Nonlinear ARDL Framework. In *Festschrift in Honor of Peter Schmidt: Econometric Methods and Applications*; Springer: New York, NY, USA, 2014; pp. 281–314. [CrossRef]
13. Rezitis, A.N. Investigating Price Transmission in the Finnish Dairy Sector: An Asymmetric NARDL Approach. *Empir. Econ.* **2019**, *57*, 861–900. [CrossRef]

Disclaimer/Publisher's Note: The statements, opinions and data contained in all publications are solely those of the individual author(s) and contributor(s) and not of MDPI and/or the editor(s). MDPI and/or the editor(s) disclaim responsibility for any injury to people or property resulting from any ideas, methods, instructions or products referred to in the content.

Proceeding Paper

Cultivate Crops or Produce Energy? Factors Affecting the Decision of Farmers to Install Photovoltaics on Their Farmland [†]

Konstantinos Ioannou [1,*], Evangelia Karasmanaki [2], Despoina Sfiri [2], Georgios Tsantopoulos [2] and Kleanthis Xenitidis [2]

1. Hellenic Agricultural Organization Demeter, Forest Research Institute, Vasilika, 57006 Thessaloniki, Greece
2. Department of Forestry and Management of the Environment and Natural Resources, Democritus University of Thrace, 68200 Orestiada, Greece; evkarasm@fmenr.duth.gr (E.K.); ntepisfiri@gmail.com (D.S.); tsantopo@fmenr.duth.gr (G.T.); kxenitid@fmenr.duth.gr (K.X.)
* Correspondence: ioanko@fri.gr
† Presented at the 17th International Conference of the Hellenic Association of Agricultural Economists, Thessaloniki, Greece, 2–3 November 2023.

Abstract: The aim of this study was to examine the factors affecting farmers' willingness to invest in photovoltaics as well as the factors affecting the amount of money they would invest. The study was performed on a representative farmer sample in Northern Greece through the use of structured questionnaires. Two models were developed using categorical regression, with the first model indicating that the willingness to invest was mostly affected by the provision of subsidies and the type of cultivation used for the land in question. The amount of money farmers would invest was mostly affected by the number of hectares of irrigated and dry land that famers had, thereby suggesting that the more farmland they own the more the money they would invest. Results raise policy implications as they show an increased interest in installing renewable systems on farmland which, in turn, raises concerns about the agricultural development of the country.

Keywords: farmers' attitudes; agri-food crisis; willingness to invest; photovoltaics on farmland; factors affecting investments

1. Introduction

Agri-food production is constantly challenged in recent years by various pressures, such as the pandemic and sharp increases in energy prices due to the conflict in Ukraine [1,2]. Despite EU's efforts to tackle the effects of the crisis, food security still relies on a rather volatile environmental and geopolitical context [3,4]. Due to these pressures, a considerable proportion of farmers tend to opt for the installation of photovoltaics on their farmland. This trend, however, may compromise food security and the national agricultural development highlighting the need to dedicate more research on farmers' decision-making. In other words, understanding what affects farmers' decision to install photovoltaics can inform policymaking by pointing at areas that require policy intervention. Hence, this study examines the factors affecting farmers' willingness to install photovoltaics on their farmland, as well as the factors affecting the amount of money farmers would invest in photovoltaics.

2. Methods

The population under study comprised farmer landowners in a typical Greek rural area, the Municipal Unit of Didymoteicho, which is located in Northern Greece. To recruit respondents, the method of simple random sampling was followed with t = 1.96, p = 0.6 and e = 6.3%. Hence, according to the formula of simple random sampling, 233 respondents had to participate in this study in order to achieve a representative sample. Then, respondents were administered structured questionnaires which were completed through personal interviews and, in total, 233 landowners participated in the study. To analyze the collected data, the Statistical Package for the Social Sciences (SPSS) [5] was used and descriptive statistics and categorical regression were specifically applied. Categorical regression was used to build two models, with the first model examining the factors affecting farmers' willingness to install photovoltaics and the second model investigating the factors affecting the amount of money that willing farmers would invest in photovoltaics.

3. Results and Discussion

Regarding respondents' sociodemographic profile, there was an almost equal representation of both genders in the sample, whereby most respondents were married (68.8%) and farming was their main profession (42.9%). As for education level, significant shares of respondents reported being high school graduates (29.8%) and university graduates (14.1%). Respondents reported owning 2272.7 hectares of dry land and 2306.7 hectares of irrigated land. The vast majority of respondents were willing to invest in photovoltaics and they would invest specifically between 10,000 and 20,000 € (17.6%), 2000 and 5000 € (17.2%) and 5000 and 10,000 € (15.9%).

Following descriptive analysis, categorical regression was performed to identify the factors affecting farmers' decision making. Two models were built to explain the dependent variables. In the first model, the dependent variable was "farmers' willingness to invest in photovoltaics on their farmland"; the independent variables can be seen in Table 1. The analysis gave a co-efficient value of multiple determination of $R^2 = 0.310$ and $F = 5.042$, which is statistically important. Taking Figure 1a into account, which displays the transformation plots for the dependent variable, it is indicated that the dependent variable of "farmers' willingness to invest in photovoltaics on their farmland" is mostly affected by the availability of "subsidies for investments in renewables" and a farmer's "level of information about renewable energy investments". Moreover, the dependent variable is affected by the type of crop cultivation, specifically "sugar beet" and "cotton" cultivations, while the number of hectares of dry land that farmers own also exerts a significant effect on the dependent variable. The measures of the relevant importance of the independent variables suggest that "subsidies for investments in renewables" and cultivating "cotton" and "sugar beet" made the highest contribution to the dependent variable.

Table 1. Factors affecting farmers' willingness to invest in photovoltaic systems on their farmland.

Independent Variables	Beta	Std Error	Df	Importance	F
Level of information about renewable energy investments	0.213	0.083	3	0.198	6.656
Subsidies for investments in renewables	0.332	0.063	3	0.330	28.159
The complexity of the licensing process	−0.178	0.067	3	0.087	7.105
Agreement with the installation of solar parks in a location visible from place of residence	0.175	0.112	1	0.110	2.471
Hectares of irrigated land	−0.018	0.110	1	−0.009	0.026
Hectares of dry land	0.166	0.111	1	0.110	2.241
Wheat	−0.078	0.095	1	−0.023	0.668
Cotton	0.239	0.091	1	0.154	6.810
Sunflower	−0.183	0.091	1	−0.035	4.021
Canola	−0.319	0.183	1	−0.079	3.046
Corn	−0.031	0.076	1	−0.006	0.162
Garlic	0.055	0.044	1	0.025	1.624
Sugar beet	0.330	0.174	1	0.138	3.583

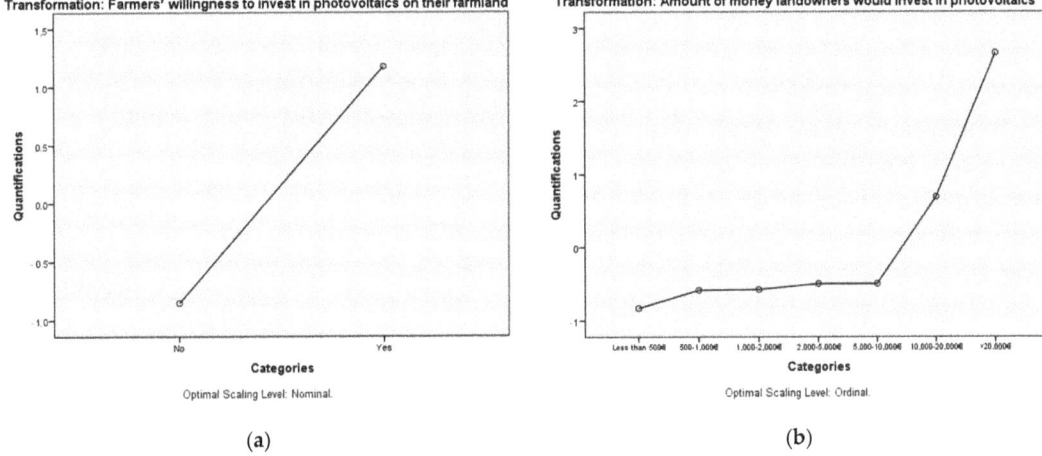

Figure 1. (a) Transformation plots of the independent variable "farmers' willingness to invest in photovoltaics on their farmland"; (b) transformation plots of the independent variable "amount of money farmers are willing to invest in photovoltaics".

In the second model, the dependent variable was "amount of money farmers are willing to invest in photovoltaics"; the independent variables can be seen in Table 2. Regarding this model, analysis gave a value of multiple determination of $R^2 = 0.407$ and $F = 5.003$. Taking Figure 1b into account, which displays the transformation plots for the dependent variable, it is shown that the dependent variable is mostly affected by the variables of "hectares of irrigated land", "hectares of dry land", cultivating "garlic" and the "adoption of pro-environmental behavior". Measuring the relevant importance of the independent variables suggests that the "hectares of irrigated land", "hectares of dry land", cultivating "garlic" and "increasing respect from friends and acquaintances" made the greatest contribution to the dependent variable.

Table 2. Factors affecting the amount of money farmers would invest in photovoltaics.

Independent Variables	Beta	Std Error	Df	Importance	F
Hectares of irrigated land	0.480	0.143	3	0.431	3.901
Hectares of dry land	0.457	0.131	3	0.370	3.909
Wheat	−0.188	0.183	2	−0.081	1.060
Cotton	−0.376	0.170	3	−0.047	1.034
Garlic	0.182	0.155	3	0.113	1.380
Sunflower	−0.012	0.161	1	−0.007	0.005
Canola	0.051	0.078	1	0.024	0.425
Increasing respect from friends and acquaintances	−0.264	0.099	3	0.140	7.091
Adoption of pro-environmental behavior	0.217	0.085	4	0.052	6.478
Occupation	0.054	0.055	1	0.004	0.948

4. Conclusions

The type of cultivation affects the willingness to invest as our results suggest that certain types of land cultivation positively affect this willingness. This suggests that farmers may not be satisfied with the revenues from these crops or that the conditions required for these cultivations may be too demanding. From this perspective, farmers may perceive photovoltaics as a safer and easier solution; however, this points to the risk of replacing crop cultivation with energy production, thereby risking the aggravation of the existing agri-food crisis. Moreover, the availability of subsidies positively affects the willingness to invest and could drive farmers to abandon crop cultivation. Therefore, policymakers should be mindful of being too generous in subsidy schemes but should also try to improve farmer revenues from crop production. Interestingly, the amount of money that farmers would invest was affected by the number of hectares they own. Indeed, the more hectares farmers own the higher the amount of money they are willing to invest in photovoltaics becomes. In other words, ownership of extensive farmland acts as a positive factor for high investments as it allows farmers to continue cultivating their land and to maintain most of their crop cultivations.

Author Contributions: Conceptualization, K.I. and G.T.; methodology, K.I. and G.T.; software, K.I., G.T. and K.X.; validation, K.I., G.T., E.K. and D.S.; formal analysis, K.I. and G.T.; investigation, D.S.; resources, K.I., G.T., E.K. and D.S.; data curation, K.I., G.T., E.K. and D.S.; writing—original draft preparation, K.I., G.T. and E.K.; writing—review and editing, K.I., G.T., E.K. and D.S.; supervision, G.T. All authors have read and agreed to the published version of the manuscript.

Funding: This research received no external funding.

Informed Consent Statement: Not applicable.

Data Availability Statement: No new data were created or analyzed in this study. Data sharing is not applicable to this article.

Conflicts of Interest: The authors declare no conflicts of interest.

References

1. Cavallo, A.; Olivieri, F.M. Sustainable local development and agri-food system in the post Covid crisis: The case of Rome. *Cities* **2022**, *131*, 103994. [CrossRef] [PubMed]
2. Glauben, T.; Svanidze, M.; Götz, L.; Prehn, S.; Jamali Jaghdani, T.; Đurić, I.; Kuhn, L. The War in Ukraine, Agricultural Trade and Risks to Global Food Security. *Intereconomics* **2022**, *57*, 157–163. [CrossRef]
3. Ben Hassen, T.; El Bilali, H. Impacts of the Russia-Ukraine War on Global Food Security: Towards More Sustainable and Resilient Food Systems? *Foods* **2022**, *11*, 2301. [CrossRef] [PubMed]
4. Papadopoulou, C.-I.; Loizou, E.; Melfou, K.; Chatzitheodoridis, F. The Knowledge Based Agricultural Bioeconomy: A Bibliometric Network Analysis. *Energies* **2021**, *14*, 6823. [CrossRef]
5. Šebjan, U.; Tominc, P. Impact of support of teacher and compatibility with needs of study on usefulness of SPSS by students. *Comput. Hum. Behav.* **2015**, *53*, 354–365. [CrossRef]

Disclaimer/Publisher's Note: The statements, opinions and data contained in all publications are solely those of the individual author(s) and contributor(s) and not of MDPI and/or the editor(s). MDPI and/or the editor(s) disclaim responsibility for any injury to people or property resulting from any ideas, methods, instructions or products referred to in the content.

Proceeding Paper

Upgrading Value Chains through Farm Advisory [†]

Maria Spilioti [1,*], Pavlos Karanikolas [1], George Papadomichelakis [2], Konstantinos Tsiboukas [1] and Dimitris Voloudakis [3]

[1] Department of Agricultural Economics and Rural Development, Agricultural University of Athens, 118 55 Athens, Greece; pkaranik@aua.gr (P.K.); tsiboukas@aua.gr (K.T.)
[2] Department of Animal Science and Aquaculture, Agricultural University of Athens, 118 55 Athens, Greece; gpapad@aua.gr
[3] Capacity Building, New Agriculture New Generation, 115 25 Athens, Greece; dimitris@generationag.org
* Correspondence: spimaria1@gmail.com
[†] Presented at the 17th International Conference of the Hellenic Association of Agricultural Economists, Thessaloniki, Greece, 2–3 November 2023.

Abstract: The article discusses the benefits of an integrated farm advisory program on sheep farms, focusing on improving their economic performance. The program involves a team of experts providing advice on animal nutrition and farm management, and conducting a thorough techno-economic analysis before and after recommendations. The economic impact is assessed using a partial budget tool. Results show increased yields, decreased production costs, and increased gross value added. The program requires a cohesive group of experts, trusting relationships between farmers and consultants, and funding. Implementing this program on a large scale can upgrade the relevant value chain.

Keywords: agricultural; counseling; livestock; farmers; management

1. Introduction

With the global human population rising at an alarming rate and climate change posing the greatest threat to food security, it becomes more critical than ever to upgrade the agri-food value chain (VC), which should be in line with the principles of sustainability [1,2]. Particular emphasis must be placed on improving the quality of agricultural products as a way of upgrading production systems, using environmentally friendly methods and ensuring a sufficient farm income, while at the same time promoting social cohesion [3–5]. However, specialised knowledge is required to achieve the above, which farmers often do not possess [6,7]. The EU has identified this lack and created the Agricultural Knowledge and Innovation System (AKIS), which aims to improve European agriculture's efficiency, competitiveness, and sustainability, by providing techniques and financial and environmental advice [8]. In Greece, a professional advisory support program for livestock farmers has been developed and applied under the scientific guidance of professors and advisors of the Agricultural University of Athens [9]; the program is financed by the 'New Agriculture New Generation' organisation through the founding donation of the Stavros Niarchos Foundation. The management of livestock and the feeding of farm animals is the main focus of the program.

The article addresses the practical aspects of this particular farm advisory initiative, including the benefits and the challenges encountered in this effort.

2. Methods

In 2020, the university's team of experts provided six advisories in Thessaly; the same group also offered, from 2021 to 2023, twenty advisories in the same area and six advisories in Crete. Of the above advisories, fifteen are still in progress. The farm advisory (FA) structure includes technical advice, which develops after three visits. The counsellor recorded the current techno-economic situation on the first visit and identified the farmer's

needs. Customised advice was designed based on each farmer's needs between the first and second visit. The second visit was dedicated to presenting the proposed solution to the farmer. Before the third and final visit, the counsellor verified the advice's effectiveness and made adjustments if necessary. The third visit recorded the farm's new techno-economic status following the implementation of the advice (Figure 1).

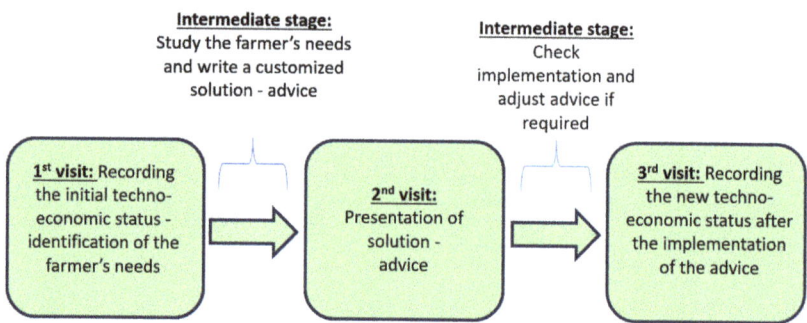

Figure 1. The stages of the advisory program.

Finally, depending on the type of advice, an economic tool is chosen to evaluate and compare the livestock farm's initial and final economic performance. The initial and final state of the livestock farm is compared using a partial budget, or through the calculation of the main economic results.

3. Results and Discussion

The agricultural advisory process can be complex, and farmers must rely heavily on counselors to help them make informed decisions. The basis for such an endeavour is integrating technical and economic expertise at the farm level and a team of experts with internal cohesion. Consultants must provide clear explanations of technical concepts. Yet, some farmers are hesitant to implement the advice; so, the consultant must show them how they will benefit from it upfront. The initiative works in a context of effective communication and trust between the consultant and the farmer. Most of the advice was about improving the nutrition of farm animals, while the rest of the advice related to the implementation of Artificial Rearing of Lambs (ARL), the establishment of small cottage industries, and the purchase of some equipment related to animal husbandry (Figure 2). Nutrition advice prevails because sheep feeding corresponds to 60–70% of the total variable costs of a livestock farm [10]. Most farmers provide an unbalanced ratio, unjustly wasting feed [11,12]. Therefore, the general financial situation of the farm can improve by reducing the cost of feeding the animals. A balanced ration can increase milk and meat yields. The modification of the ration is easy to implement by the farmers without requiring a large waste of financial resources. Regarding the establishment of small cottage industries, small–medium farmers often face difficulties in processing the raw materials they produce, since they do not have access to appropriate food processing equipment, losing a part of the added value produced during processing [2,13]. Thus, offering this kind of advice helps to deal with the problematic situation. In recent years, the rise in the price of sheep's milk makes the farmers seek to increase the marketed amount of milk. One way to achieve the above is the ARL. Through this advice, the lambs will reach the desired weight faster, limiting the waste of resources. The purchase of the appropriate equipment, such as a milking machine, can improve the efficiency of the farm [14].

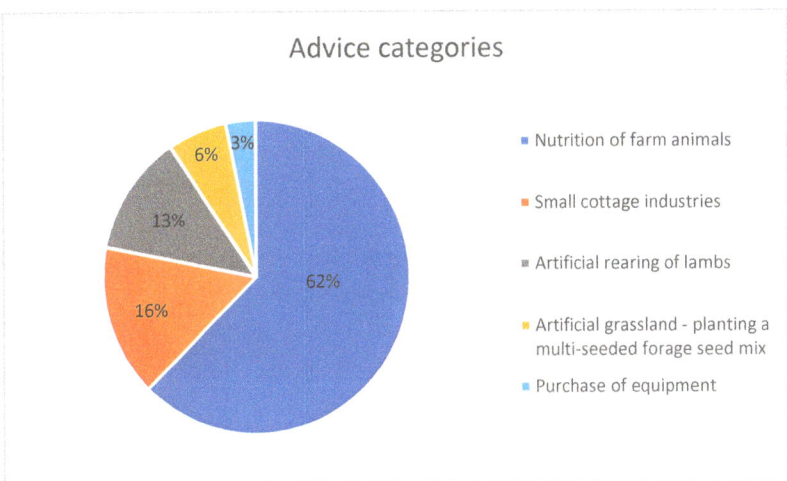

Figure 2. Advice categories.

In Table 1, various specific advice that have been implemented are shown indicatively.

Table 1. FA case studies.

Brief Description of Advice	Capacity of Animals	Total Impact of Advice	Impact of Advice per Ewe (Euros)
Improvement of Existing Ration (IER)	Ewes: 220	Increase of the Gross Added Value (IGAV): 18,121 euros	IGAV: 82.4 euros per ewe
IER, after previously grouping the dairy ewes according to body weight	Ewes: 240	IGAV: 10,055 euros	IGAV: 41.9 euros per ewe
ARL	Ewes: 500 Lamb: 750	IGAV: 10,235 euros	IGAV: 20.5 euros per ewe
Expansion of distribution channels and addition of new products to the existing ones in cottage cheese production	Ewes: 100 Goats: 30	Increase in Revenue (IR) by 9698 euros	IR: 97 euros per eve
IER	Ewes: 100 Goats: 30	IGAV: 5300 euros	IGAV: 53 euros per ewe

Reference: Field Research.

4. Conclusions

An advisory program for livestock farmers has been developed and applied in various regions of Greece, focusing on managing livestock farms and feeding farm animals. Farm advisory is crucial for farmers to boost their productivity and reduce expenses, ultimately improving the performance of their farms. This process requires establishing a coherent group of experts and incorporating technical and financial expertise applied at the farm level. The three-visit method is successful, but communication between the counsellor and farmer is critical. Partial budgeting is the primary method to assess the impact of the advisory, which yields highly favourable outcomes. Another factor that contributed to the success of the program was the availability of funding. Advice for agriculture can increase farmers' income and assist in sustaining the local population, thereby preserving the economic and social fabric of rural areas. As a national AKIS system has not yet been established in Greece, this program could be a successful groundwork for such a system.

Author Contributions: Conceptualization, M.S., P.K., G.P., K.T. and D.V.; methodology, M.S., P.K., G.P. and K.T.; validation, M.S., P.K., G.P., K.T. and D.V.; formal analysis, M.S., P.K., G.P. and K.T.; investigation, M.S., G.P. and K.T.; resources, M.S., P.K., G.P. and K.T.; data curation, M.S., P.K., G.P. and K.T.; writing—original draft preparation, M.S., P.K., G.P., K.T. and D.V.; writing—review and editing, M.S., P.K., G.P., K.T. and D.V.; visualization, M.S., P.K., G.P., K.T. and D.V.; supervision, M.S., P.K., G.P. and K.T.; project administration, G.P. and D.V.; funding acquisition, D.V. All authors have read and agreed to the published version of the manuscript.

Funding: This research was funded by Stavros Niarchos Foundation, grant number 04.0310.

Institutional Review Board Statement: Not applicable.

Informed Consent Statement: Not applicable.

Data Availability Statement: Data are contained within the article.

Conflicts of Interest: The authors declare no conflict of interest.

References

1. Manikas, I.; Sundarakani, B.; Anastasiadis, F.; Ali, B. A Framework for Food Security via Resilient Agri-Food Supply Chains: The Case of UAE. *Sustainability* **2022**, *14*, 6375. [CrossRef]
2. Spilioti, M.; Stachtiaris, S.; Kominakis, A.; Karanikolas, P.; Tsiboukas, K. A niche strategy for geographical indication products, by valorising local resources: The Greek cheese Ladotyri Mytilinis. *Int. J. Agric. Resour. Gov. Ecol.* **2022**, *18*, 160. [CrossRef]
3. Bacon, C.M.; Getz, C.; Kraus, S.; Montenegro, M.; Holland, K. The Social Dimensions of Sustainability and Change in Diversified Farming Systems. *Ecol. Soc.* **2012**, *17*, 41. [CrossRef]
4. Ferris, S.; Robbins, P.; Best, R.; Seville, D.; Buxton, A.; Shriver, J.; Wei, E. 'Linking Smallholder Farmers to Markets and the Implications for Extension and Advisory Services' MEAS. 2014. Available online: https://agritech.tnau.ac.in/dmi/2013/pdf/MEAS%20Discussion%20Paper%204%20-%20Linking%20Farmers%20To%20Markets%20-%20May%202014.pdf (accessed on 25 June 2023).
5. Spilioti, M.; Karanikolas, P.; Kominakis, A.; Stachtiaris, S.; Tsiboukas, K. Geographical indication products and the provision of public goods—A Greek case study. *Int. J. Sustain. Agric. Manag. Inform.* **2023**, *9*, 136–158. [CrossRef]
6. Kahan, D. 'The role of the FARM MANAGEMENT SPECIALIST in extension' FAO. 2013. Available online: https://www.fao.org/3/i3232e/i3232e.pdf (accessed on 25 June 2023).
7. Dockès, A.-C.; Chauvat, S.; Correa, P.; Turlot, A.; Nettle, R. Advice and advisory roles about work on farms. A review. *Agron. Sustain. Dev.* **2019**, *39*, 2. [CrossRef]
8. European Commission. Agricultural Knowledge and Innovation Systems (AKIS), Boosting Innovation and Knowledge Flows across Europe. 2022. Available online: https://ec.europa.eu/eip/agriculture/sites/default/files/eipagri_agricultural_knowledge_and_innovation_systems_akis_2021_en_web.pdf (accessed on 12 May 2023).
9. New Agriculture New Generation. Workforce Development. 2023. Available online: https://www.generationag.org/symboyleytikh-yposthriksh-aigoprobatotrofon-karditsas (accessed on 12 May 2023).
10. Makkar, H. Review: Feed demand landscape and implications of food-not feed strategy for food security and climate change. *Animal* **2018**, *12*, 1744–1754. [CrossRef] [PubMed]
11. Stefanakis, A.; Volanis, M.; Zoiopoulos, P.; Hadjigeorgiou, I. Assessing the potential benefits of technical intervention in evolving the semi-intensive dairy-sheep farms in Crete. *Small Rumin. Res.* **2007**, *72*, 66–72. [CrossRef]
12. Gelasakis, I.A.; Valergakis, E.G.; Fortomaris, P.; Arsenos, G. Farm conditions and production methods in Chios sheep flocks. *J. Hell. Vet. Med. Soc.* **2018**, *61*, 111–119. [CrossRef]
13. Garrity, P.D. Agroforestry and the achievement of the Millennium Development Goals. *Agrofor. Syst.* **2004**, *61*, 5–17.
14. Eastwood, R.C.; Chapman, D.; Paine, M. Networks of practice for co-construction of agricultural decision support systems: Case studies of precision dairy farms in Australia. *Agric. Syst.* **2012**, *108*, 10–18. [CrossRef]

Disclaimer/Publisher's Note: The statements, opinions and data contained in all publications are solely those of the individual author(s) and contributor(s) and not of MDPI and/or the editor(s). MDPI and/or the editor(s) disclaim responsibility for any injury to people or property resulting from any ideas, methods, instructions or products referred to in the content.

Proceeding Paper

Veterinary Students' Perceptions of Entrepreneurship Education [†]

Georgia Koutouzidou [1,*], Vagis Samathrakis [2], Athanasios Batzios [3] and Alexandros Theodoridis [4]

1. Department of Agriculture, School of Agricultural Sciences, University of Western Macedonia, 53100 Florina, Greece
2. Department of Accounting and Information Systems, International Hellenic University, 57400 Thessaloniki, Greece; sbagis@ihu.gr
3. Department of Organization and Business Administration, University of Western Macedonia, 51100 Grevena, Greece; thanos.batzios@gmail.com
4. School of Veterinary Medicine, Aristotle University of Thessaloniki, 54124 Thessaloniki, Greece; alextheod@vet.auth.gr
* Correspondence: gkoutouzidou@uowm.gr; Tel.: +30-6974456828
† Presented at the 17th International Conference of the Hellenic Association of Agricultural Economists, Thessaloniki, Greece, 2–3 November 2023.

Abstract: In this study, the opinions and perceptions of students at a school of veterinary medicine regarding the importance of entrepreneurship education in modern higher education are investigated. A Likert-scale questionnaire design was used to record veterinary students' responses on issues related to entrepreneurship education and its impact on their entrepreneurial mindset, as well as on the students' carrier aspirations and on the factors that influence their carrier choices. The survey was conducted in 2022, and in total, 105 graduates completed the questionnaire. The responses were analyzed through a descriptive statistical analysis using IBM SPSS Statistics 28. The present study confirms that there is a significant need for entrepreneurship education in order to start, develop, and successfully realize business ideas.

Keywords: entrepreneurship; education; university; start-up business

Citation: Koutouzidou, G.; Samathrakis, V.; Batzios, A.; Theodoridis, A. Veterinary Students' Perceptions of Entrepreneurship Education. *Proceedings* **2024**, *94*, 40. https://doi.org/10.3390/proceedings2024094040

Academic Editor: Theodoropoulou

Published: 2 February 2024

Copyright: © 2024 by the authors. Licensee MDPI, Basel, Switzerland. This article is an open access article distributed under the terms and conditions of the Creative Commons Attribution (CC BY) license (https://creativecommons.org/licenses/by/4.0/).

1. Introduction

University entrepreneurship education refers to courses or programs that educate students on various aspects of planning, starting, and managing a modern business. Entrepreneurship education provides students with the skills and tools needed to identify and exploit entrepreneurial opportunities in the market [1] and stimulates them to have greater information, knowledge, and encouragement in supporting their creativity to become entrepreneurs and start their own business [2,3]. Gradually, courses of entrepreneurship have been embedded in the curricula of many universities and higher education institutions, developing novel pedagogies to cultivate students' entrepreneurial mindsets [4].

In line with high modern international standards and considering entrepreneurship as part of its strategic mission, in 2017, the School of Veterinary Medicine of Aristotle University of Thessaloniki introduced a mandatory entrepreneurship course, meaning that the course constitutes an integral part of the institution's educational curriculum. The course, entitled "Entrepreneurship and management of veterinary and animal enterprises", teaches the fundamental principles of entrepreneurship and the basic elements of commercial and tax law and provides the knowledge and skills graduate students need in order to start, develop, and successfully realize business ideas in the field of veterinary medicine and animal production. The aim of this study was to explore the views and the opinions of veterinary students on the role and the need of entrepreneurship courses in modern education programs. Through an empirical analysis, the impact of the provided education on the entrepreneurial mindsets of the students was assessed, and their perceptions concerning the educational modules and topics that should be integrated into the curriculum are discussed in this paper.

Studies related to the topic of this one have been carried out previously, mainly dealing with the development of entrepreneurial skills and competencies in secondary and higher education. Arrighetti et al. [5] conducted one of the first studies on the entrepreneurial orientation of university students by using a large sample of students from the University of Parma (Italy) and Sousa [6–8], identifying which skills and competencies the students required to develop through entrepreneurship education.

2. Methodology

A primary survey was conducted in 2022 by disseminating a Likert-scale questionnaire to graduate students at a school of veterinary medicine in Greece. The questionnaire was hosted online through a Google Form and completed by 105 graduate students that attended the course of entrepreneurship education. The questionnaire was structured in three sections: the first recorded the socio-demographic profile of the students, the second recorded their opinions regarding the quality of the provided education on entrepreneurship, and the third section covered issues related to their carrier aspirations and choices. The responses were recorded in an Excel spreadsheet file, and a descriptive statistical analysis was conducted using IBM SPSS Statistics 28 software.

3. Results and Discussion

Our results show that a large percentage of students (from 55.2% to 59%) do not know the individual educational objectives of the course and believe that it is a module mainly related to economic and financial education. However, 62.9% of the students believe that prospective veterinarians should be informed about the basic principles of marketing and financial management, while 85% believe that having a knowledge of the legal frameworks related to the commercial sector, employment, and tax is a prerequisite for a successful business in the field of veterinary medicine.

Regarding the problems that a veterinarian faces as a new professional and entrepreneur, 93% of the respondents stated that the main problem is related to financial issues, which include insufficient funds, limited access to sources of finance, and the absence of financial programs. This was followed by the problems related to the lack of the appropriate knowledge and skills for planning their future business occupation (77.1%).

The opinions of the students regarding the development of a entrepreneurial spirit and culture of entrepreneurship (develop new business ideas, give advice on these issues and provide incentives for initiating business activities) in higher education in general were particularly negative, with percentages greater than 88%. Despite the high percentage of negative views on the development of entrepreneurial culturethe content of the course of entrepreneurial education provided were considered significant in improving students' knowledge on issues related to business attitudes, values and incentives, on the actions that could potentially be required to start a businessand also on strengthening of their understanding of the creation of networks and the recognition of business opportunities. Contrary to the level of knowledge they believe that they possess, they are particularly cautious about starting, operating, and managing a business (percentages greater than 74%), even though their family and friends would support any decisions related to creating a business (rate greater than 91%).

Figure 1 presents the main factors of the internal environment that prevent students from starting a business. These include (i) a lack of capital, (ii) the risk of losing invested capital, (iii) a lack of knowledge/skills, (iv) a lack of contacts with future customers–suppliers, (v) a fear of failure, (vi) a lack of business skills, and (vi) a lack of a business idea.

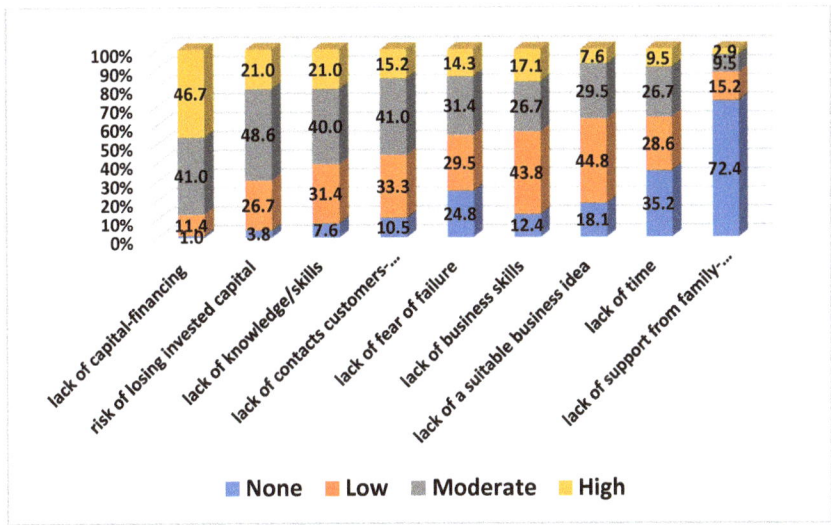

Figure 1. Internal factors that prevent students from starting a business.

Figure 2 presents the external factors that prevent students from starting their own business. The financial crisis and the uncertain economic environment, the volatile tax system, insufficient external financing, the lack of government support for entrepreneurs and of infrastructure, and the lack of knowledge on the legislative and tax framework were indicated by the students as the main problems.

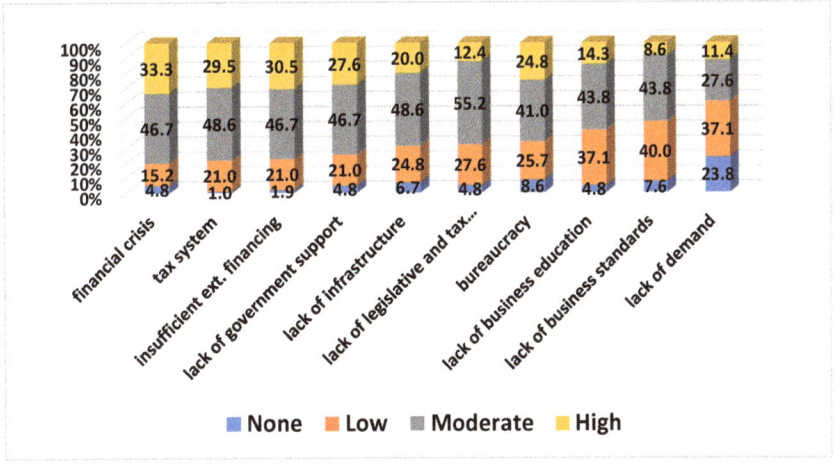

Figure 2. External factors that prevent students from starting a business.

4. Conclusions

This study presents the preliminary results of an empirical survey which was completed by veterinary graduate students in Greece and aimed to investigate and record their perceptions and attitudes concerning entrepreneurship education in universities. The veterinary students believe that, through partaking in the entrepreneurship education course, they could acquire important skills to organize and manage a business, and they are in favor of there being a direct link between their studies and the business environment and the market. Overall, the findings of this study confirm that entrepreneurship education instills entrepreneurial values into university students, benefitting them as they enter the workforce and/or start their own business; hence, entrepreneurship education must be an integral part of modern curricula in higher education.

Author Contributions: All authors (G.K., A.T., V.S. and A.B.) were involved in the analysis of the data and contributed to the final manuscript. V.S. and A.T. designed the survey and collected the data. G.K., A.T. and V.S.: writing—original draft preparation. All authors have read and agreed to the published version of the manuscript.

Funding: This study received no external funding.

Institutional Review Board Statement: Not applicable.

Informed Consent Statement: Informed consent was obtained from all subjects involved in the study.

Data Availability Statement: Data sharing not applicable.

Conflicts of Interest: The authors declare no conflicts of interest.

References

1. Yuan, C.H.; Wang, D.; Mao, C.; Wu, F. An empirical comparison of graduate entrepreneurs and graduate employees based on graduate entrepreneurship education and career development. *Sustainability* **2020**, *12*, 10563. [CrossRef]
2. Jena, R.K. Measuring the impact of business management Student's attitude towards entrepreneurship education on entrepreneurial intention: A case study. *Comput. Hum. Behav.* **2020**, *107*, 106275. [CrossRef]
3. Jiatong, W.; Murad, M.; Bajun, F.; Tufail, M.S.; Mirza, F.; Rafiq, M. Impact of Entrepreneurial Education, Mindset, and Creativity on Entrepreneurial Intention: Mediating Role of Entrepreneurial Self-Efficacy. *Front. Psychol.* **2021**, *12*, 724440. [CrossRef] [PubMed]
4. OECD and Inter-American Development Bank. *Innovative and Entrepreneurial Universities in Latin America*; OECD Skills Studies; OECD Publishing: France, Paris, 2022; pp. 15–70. [CrossRef]
5. Arrighetti, A.; Lasagni, A. Assessing the Determinants of High-Growth Manufacturing Firms in Italy. *Int. J. Econ. Bus.* **2013**, *20*, 245–267. [CrossRef]
6. Sousa, M.J. Entrepreneurship Skills Development in Higher Education Courses for Teams Leaders. *Adm. Sci.* **2018**, *8*, 18. [CrossRef]
7. Linton, G.; Klinton, M. University entrepreneurship education: A design thinking approach to learning. *J. Innov. Entrep.* **2019**, *8*, 3. [CrossRef]
8. Nadelson, S.L.; Aparna, D.; Nageswaran, P.; Benton, T.; Basnet, R.; Bissonnette, M.; Cantwell, L.; Jouflas, G.; Elliott, E.; Fromm, M.; et al. Developing Next Generation of Innovators: Teaching Entrepreneurial Mindset Elements across Disciplines. *Int. J. High. Educ.* **2018**, *7*, 114–126. [CrossRef]

Disclaimer/Publisher's Note: The statements, opinions and data contained in all publications are solely those of the individual author(s) and contributor(s) and not of MDPI and/or the editor(s). MDPI and/or the editor(s) disclaim responsibility for any injury to people or property resulting from any ideas, methods, instructions or products referred to in the content.

Proceeding Paper

The Role of Cooperative Enterprises in the Promotion of Cultural Heritage: A Case Study of the Petrified Forest of Lesvos †

Eleutheria Plousiou [1], Panagiota Sergaki [2] and Ifigeneia Mylona [3,*]

[1] School of Social Sciences, Hellenic Open University, 26331 Patra, Greece; eplousiou.culture@gmail.com
[2] Department of Agricultural Economics, School of Agriculture, Aristotle University of Thessaloniki, 54124 Thessaloniki, Greece; gsergaki@auth.gr
[3] School of Management Science and Technology, International Hellenic University, 65404 Kavala, Greece
* Correspondence: imylona@mst.ihu.gr; Tel.: +30-2510462796
† Presented at the 17th International Conference of the Hellenic Association of Agricultural Economists, Thessaloniki, Greece, 2–3 November 2023.

Abstract: The purpose of this study is to analyze the relationship between SSE bodies, culture and sustainable development, studying the case of the Petrified Forest of Sigri on the island of Lesvos and its interaction with the cooperative of the neighboring settlement of Eresos. Qualitative research was conducted by reviewing the relevant literature and implementing semi-structured interviews. A SWOT analysis was also conducted. The results showed that the Eresos Agricultural and Livestock Cooperative in Sigri, in collaboration with other agencies, can contribute to the sustainability of the area. The cooperative lacks organized promotion, beyond that attempted by the Natural History Museum based in Sigri.

Keywords: Social and Solidarity Economy (SSE); Agricultural Cooperative of Eresos; culture; social enterprises; communication

Citation: Plousiou, E.; Sergaki, P.; Mylona, I. The Role of Cooperative Enterprises in the Promotion of Cultural Heritage: A Case Study of the Petrified Forest of Lesvos. *Proceedings* 2024, 94, 41. https://doi.org/10.3390/proceedings2024094041

Academic Editor: Eleni Theodoropoulou

Published: 4 February 2024

Copyright: © 2024 by the authors. Licensee MDPI, Basel, Switzerland. This article is an open access article distributed under the terms and conditions of the Creative Commons Attribution (CC BY) license (https://creativecommons.org/licenses/by/4.0/).

1. Introduction

Social and Solidarity Economy (SSE) is characterized by the ability to create possibilities for a sustainable and inclusive society by developing entities, based on the principles of solidarity, cooperation, equality and democracy [1]. The promotion of this policy framework arises through the application of new standards, economic and social, turning the economy towards innovation and sustainability. SSE is an activity between private investment and state mechanisms, as social needs cannot be met either by the private sector—it sets the goal of high profit—or by the state—due to a more general political phobia of possible fiscal shortening. Therefore, not being able to fully cover human needs, SSE comes to perform this task, constituting the so-called third sector—the public and the private are the other two sectors—as a pillar of the economy [2]. In the 21st century, economic crisis has forced Greek governments to turn to SSE. In particular, state welfare mechanisms, unable to cope with continuous needs and failing to find real solutions to emerging issues, promoted movements and initiatives towards the development of SSE [3], with part of the funds for it coming from the E.U.

Geoparks are characterized by UNESCO as wider areas that contain a significant number of geological elements, representative of the geological history of the area and of particular scientific value, rarity or aesthetic beauty, simultaneously including sites of ecological, historical or cultural interest (https://www.unesco.org/en/iggp, accessed on 10 March 2023) [4].

Culture refers to the forms of traditional behavior which are characteristic of a given society, of a group of societies, of a certain race, of a certain area or of a certain period of time [5]. Cultural monuments are a kind of territorial capital or development source which must be experienced and enjoyed not only by tourists but also by the residents of the area, and their aesthetic value is as important as their historic narration [6]. This can generate positive economic developments, both for businesses and for local SSE operators, which is a strong incentive for protecting this dimension. Through this kind of protection and promotion, the bases and prospects for sustainability in the region are created. A three-way relationship emerges between cultural management, social enterprises and sustainable development, which contributes to the well-being of society as a whole.

The purpose of this research is to analyze the relationship between SSE bodies (especially agricultural cooperatives), culture and sustainable development, studying the case of the Petrified Forest of Sigri on the island of Lesvos and its interaction with the cooperative of the neighboring settlement of Eresos. The research questions that will be answered in this research are the following:

- What are the collaborations and interactions of the Eresos Agricultural and Livestock Cooperative, as an SSE body, with the university, museum, local government, cultural associations, civil society and other local cooperatives?
- What is the contribution of these cooperative relations to the sustainability of the region and the cooperative?
- Is there any possibility of networking these SSE bodies in order to promote their goals?

The Petrified Forest highlights the existence of a new type of tourism, so-called geotourism, which is a form of tourism based on the geological environment, focused on the geology and landscape of an area as a basis for promoting sustainable tourism development [7]. The Petrified Forest, like all geoparks worldwide, encourages active collaboration with academic institutions and corresponding communities through on-site scientific research, enriching scientific knowledge about the earth and its processes [8].

2. Methodology

This work was based on qualitative research that was carried out following two paths: those of the extensive literature review and the semi-structured interview. Through the literature review, an analysis of the examined topic is attempted regarding the abovementioned research questions. In the research part, three semi-structured interviews were carried out with a member of the Eresos cooperative, a member of the Mesotopos cooperative and the Deputy Mayor of Tourism Development of the Municipality of Western Lesvos Island. Personal interviews gave additional arguments and fidelity to this work through the empirical knowledge of the inhabitants of the area regarding the promotion of the cooperative's products as well as the degree of cooperation between agencies, according to the quadruple helix model. Still, based on the interviews and the literature review, a SWOT analysis was conducted, which was used in order to list the strong and weak elements of the Petrified Forest and the wider area as well as the opportunities that can emerge and the threats it faces.

3. Results

The Eresos Agricultural and Livestock Cooperative collaborates with the Natural History Museum at the level of presentation of wetlands and in educational actions concerning the local community. In general, there is limited help provided to the head of the museum either from the state or from the local government in order to further promote the Petrified Forest and facilitate networking among the relevant organizations [9]. The cooperatives of Eresos collaborate with both the women's cooperatives of Mesotopos and Agras at the level of exchanging products and raw materials.

Help from the local authorities in the promotion of their products is limited related to participation in exhibitions and local celebrations, since there is a lack of organized advertising. The construction of infrastructure, such as that of Sigri-Kalloni road, is also important.

Regarding promotion, it seems that it is carried out by the municipality through their website, while the museum promotes initiatives by organizing events such as conferences that have attracted people in conjunction with the port that was built. There is a shop in the museum to better promote the cooperative's products. Unfortunately, there are road construction problems, and accessing the area is not easy. The relationship between the local government and cooperatives remains mainly in the form of financial support. The local government is not seen as a supporter of these ventures, and problems have been observed in matters of decentralization. In conclusion, SSE and cultural/geological/natural heritage are guarantees of local sustainability, as long as there is a spirit of cooperation and openness to new ideas from all the actors that can cooperate.

The SWOT analysis (Table 1) reveals that strengths and opportunities prevail over weaknesses and threats.

Table 1. SWOT analysis.

Strengths	Weaknesses
Recognized park by UNESCO	Reduced tourist interest
Geological heritage—huge cultural value	Difficult to access
Modern museum of natural history	Limited interest from local authorities and residents in the promotion of the forest and the island
Leader in research and implementation of educational activities, cooperation and networking with international organizations	
Opportunities	Threats
Potential for boosting interest in tourism due to road and port construction	Lack of cooperative spirit among institutions and residents
Increased interest for alternative tourism	Improper knowledge from residents and tourists about the great value of this museum

4. Conclusions

The Eresos Agricultural and Livestock Cooperative in Sigri, in collaboration with other agencies, can contribute to the sustainability of the area. However, the cooperative lacks organized promotion, beyond that attempted by the Natural History Museum based in Sigri. SSE and local sustainable development are interrelated terms. In this regard, the existence of the Natural History Museum and the strong element of culture, which includes the Petrified Forest of Sigri, can act as guarantors of the stability of local sustainability. The Eresos Cooperative, as well as any newly established SSE project, must help more actively in order to give the impetus that is appropriate to the region and to this rare phenomenon of geomorphological texture. Through the interaction of culture and SSE, Sigri and Eresos can take advantage of the comparative advantages and lead the way to sustainable development.

The Petrified Forest in Sigri could play a significant role for the region as it is a well-known monument, and it attracts a lot of tourists from all over the world. It is a monument that in recent years, through the efforts made on the part of the museum, has become globally recognized.

The museum has taken serious initiatives in the direction of promotion, therefore also promoting products. Events and conferences have attracted people, in combination with the port that was built, and we hope that the road construction project mentioned above also breathes life into the area. The museum generally strives for local development through its actions, including collaborations with the University of the Aegean, organizing educational trips and attempts to develop a type of tourism, so-called "geotourism", which from what we see now is becoming what we call "fashionable". The shop inside the museum, where various foods produced by our cooperative are sold, also plays an important role in the promotion of products. Therefore, we can only see positivity from the actions of the museum and the contributions of the Petrified Forest as a cooperative. Furthermore, it must be emphasized that the Petrified Forest, as a world monument, is not only important for the region locally but for the whole island as a touristic destination.

Cooperatives can arguably be most effective at the local level, where they stimulate local social cohesion through day-to-day interactions with members and non-members. In the same way, the Eresos Cooperative offers its members who are producers a series of possibilities that reduce the cost of production and ensure a satisfactory income for them. A cooperative can breathe life into an area through networking with others who have a similar background, such as those in Meteora and Dadia. However, the areas in question, with their special geological and natural topographies, and the SSE agencies based there have common goals.

The belief that geoparks (Petrified Forest, Meteora and Dadia Forest) alone are able to ensure economic viability and sustainability of the area is obviously a mistake. Networking among the relevant players in the area with common interests and goals could activate funds that will boost sustainability. Fine-tuned actions aimed at promoting alternative tourism should arouse the interest of a portion of people who prefer it and advertise it to other people with similar scientific interests or academic identities.

Author Contributions: Conceptualization, E.P. and P.S.; methodology, E.P. and P.S.; investigation, E.P. and P.S.; data curation, E.P.; formal analysis, E.P.; writing—original draft preparation, E.P., P.S. and I.M.; writing—review and editing, E.P., P.S. and I.M.; project administration, P.S. and I.M. All authors have read and agreed to the published version of the manuscript.

Funding: This research received no external funding.

Institutional Review Board Statement: Not applicable.

Informed Consent Statement: Not applicable.

Data Availability Statement: The data presented in this study are available upon request from the corresponding author. The data are publicly unavailable due to privacy restrictions.

Conflicts of Interest: The authors declare no conflicts of interest.

References

1. Spyridakis, M. Market Economy, Economic Anthropology and Social Policy: Karl Polanyi's Contribution to the Critique of the Economic Formation of Modernity. In *Social Thought and Modernity*; Koniordos, S., Ed.; Gutenberg: Athens, Greece, 2010. (In Greek)
2. Nikolopoulos, T.; Kapogiannis, D. *Introduction to the Social and Solidarity Economy—The Meteoric Step of a Possibility*; Colleagues Publications: Athens, Greece, 2014. (In Greek)
3. Kavoulakos, G.; Gritzas, G. *Alternative Economic and Political Spaces*; Kallipos: Athens, Greece, 2015. (In Greek)
4. Unesco, International Geoscience and Geoparks Programme Preserving our Earth's Resources for Future Generations. Available online: https://www.unesco.org/en/iggp (accessed on 20 March 2023).
5. Brumann, C. Writing for culture: Why a successful concept should not be discarded. *Curr. Anthropol.* **1999**, *40* (Suppl. S1), S1–S27. [CrossRef]
6. Papadaki, E. Mediating mediations of the past: Monu-ments on photographs, postcards and social media. *Punctum* **2020**, *134*, 131–144. [CrossRef]
7. Valiakos, H.; Tsalkitzi, O.; Zouros, N. *Nissiope Petrified Forest Park Guide*; Zouros, N., Ed.; MFIADL: Lesvos, Greece, 2015.
8. Bendana, K.; Valiakos, H.; Kontis, V.; Zouros, N. Environmental Awareness Routes and Educational Activities in the Petrified Forest of Lesvos. 2010. Available online: www.lesvosmuseum.gr (accessed on 30 January 2023).
9. Zouros, N.; Velitzelos, E. *Guide to the Lesvos Petrified Forest Park*; Natural History Museum of the Lesvos Petrified Forest: Lesvos, Greece, 2007.

Disclaimer/Publisher's Note: The statements, opinions and data contained in all publications are solely those of the individual author(s) and contributor(s) and not of MDPI and/or the editor(s). MDPI and/or the editor(s) disclaim responsibility for any injury to people or property resulting from any ideas, methods, instructions or products referred to in the content.

Proceeding Paper

Decision Support Model for Integrating the New Cross-Compliance Rules and Rational Water Management [†]

Asimina Kouriati [1], Christina Moulogianni [2], Evgenia Lialia [1], Angelos Prentzas [1], Anna Tafidou [2], Eleni Dimitriadou [1] and Thomas Bournaris [1,*]

1. Department of Agricultural Economics, Aristotle University of Thessaloniki, 54124 Thessaloniki, Greece; kouriata@agro.auth.gr (A.K.); evlialia@agro.auth.gr (E.L.); aprentzas@agro.auth.gr (A.P.); edimitri@agro.auth.gr (E.D.)
2. Department of Mathematics, Aristotle University of Thessaloniki, 54124 Thessaloniki, Greece; kristin@agro.auth.gr (C.M.); annatafidou@gmail.com (A.T.)
* Correspondence: tbournar@agro.auth.gr
† Presented at the 17th International Conference of the Hellenic Association of Agricultural Economists, Thessaloniki, Greece, 2–3 November 2023.

Abstract: The aim of this study is to change land use by applying a decision support model that will contribute to the assimilation of the new cross-compliance rules, to optimal water management, and to the enhancement of the effectiveness and profitability of the farms. The research objective will be achieved by establishing 50-acre pilot fields for five farmer groups through the optimal allocation of limited economic and land resources. The result extracted will lead to the gradual incorporation of the new directives to reduce production costs and recognize the new cross-compliance rules.

Keywords: Common Agricultural Policy; cross-compliance; water management; decision support model

1. Introduction

The research problem to be solved concerns the adaptation of the producers to the new and increasing cross-compliance requirements as the rules will be tightened for the period 2021–2027 and farmers must be ready for the additional obligations of the Common Agricultural Policy. This is why the present study aims to change the land uses by applying a decision support model to five farmer groups located in Thessaloniki, Serres, Kozani, and Kavala. The model will be configured so that its implementation will initially contribute to the assimilation of the new cross-compliance rules, to optimal water management, and to the enhancement of the effectiveness and profitability of the farms. The application of such a developed decision support model will allow each farm to determine its own optimal production plan based on specific limits, with the main objective of using water in a rational way and strengthening the farm's economic position by further contributing to reducing production and labor costs, increasing gross profit, and achieving environmental sustainability. By implementing the above actions, a twofold benefit will be achieved in addition to economic upgrading and increased competitiveness due to the delimitation of the inputs used; farms will be able to further adapt to the new guidelines of the Common Agricultural Policy (reference period: 2021–2027) gradually. The research objective will be achieved by establishing 50 acres of pilot fields for five farmer groups, and the result extracted will lead to the gradual incorporation of the new directives to reduce production costs and recognize the new cross-compliance rules.

The development of a decision support model is a project with a modular implementation process and multiple aspects. This model is based on an existing structure created by the Laboratory of Informatics in Agriculture, which belongs to the Aristotle University of Thessaloniki, and is adapted to the needs of the producers participating in the research.

At the same time, the laboratory's web-based platform will be used after its adaption to the needs of this research. The platform's function concerns the recording of technical and economic data, useful for drawing appropriate conclusions regarding the farms' economic positions. In addition to the aforementioned actions, producers will be taught and familiarized with the use of the platform. The initial use of the platform by the producers is aimed at further adapting it to the users' needs and highlighting possible errors. Regarding the scientific literature, the development of corresponding models and the use of corresponding platforms in various countries are evident [1–4], especially in Greece [5–9]. In fact, the desire to develop web-based platforms for use in the agricultural sector is particularly evident, as highlighted by the review of the most recent literature [10–13]. The remainder of this paper is structured as follows: (1) First, the Materials and Methods section presents the method used and the research stages (Section 2). (2) Subsequently, the research Expectative Results and the contribution to the agricultural sector are described in (Section 3). (3) Finally, the present study's conclusions and innovation parameters are given in detail (Section 4).

2. Materials and Methods

The development of a decision support model (DSM) for the adaptation to cross-compliance rules and farms' economic efficiency achievement is a project with a modular implementation process and multiple and complex aspects. For the model's development, it is initially necessary to collect a set of farmer groups' relevant data using a special questionnaire that is based on the scientific literature [14–17]. After the data collection, the multicriteria decision-making analysis and, especially, the multicriteria weight goal programming are used as they are also proposed by the relevant literature [14,18–25]. These methods are used to develop the decision support model according to the needs of the five farmer groups and to select the 50-acre pilot fields.

Then, the use of the web-based platform is carried out aiming to record the economic and technical data of the fifty-acre pilot fields. Through the use of the web-based platform, the producers' knowledge regarding the farmer group's sustainable position is actually enhanced [26]. In addition, the use of the online platform aims to create a technical and economic database in order to confirm whether the objectives of this research have been achieved in terms of farmer groups' profitability and production costs. In order to fulfill the above-mentioned aim, an economic and technical analysis of the results will be carried out for the economic evaluation of the study and the evaluation of the possibilities of using the new methodology. Minimizing inputs will also be explored. Finally, dissemination actions will be carried out in order to spread the forthcoming results.

3. Expectative Results and Discussion

The present work essentially aims to transform the Laboratory of Informatics in Agriculture's existing research into an organized framework of rational water use management, with the ultimate goal of reducing production and labor costs, increasing gross profit, and achieving the environmental sustainability of Greek farms [8]. This research aim will essentially be achieved with the optimal allocation of the limited economic and land resources of the agricultural producers.

It should be particularly pointed out that the connection of farmers to the decision support model and the electronic management of their farms has multiple benefits since they are part of the innovative and rational management of water use. The organization and extraction—through the model—of an optimal production plan will create more effective farms, based on the challenges linked to the principles of the new Common Agricultural Policy. This study is also an innovative action as it motivates producers to adopt more effective farming methods. Last but not least, it should be also pointed out that the producers' engagement with the decision support model is continuous as they input data individually into a relative web-based platform and will soon be given the opportunity to simulate valid and numerous production plans.

4. Conclusions

It is worth noting that this study is carried out for the first time on such a large scale with a view of extending it to other areas. It should also be considered innovative as it includes information on the main crops of the regions with the aim of managing entire agricultural areas rather than just a single farm while it is known that alternative crops are limited in the area. The process, after the implementation of the decision support model (DSM), will be considered effective if it motivates the producers—through the integration and assimilation of the new cross-compliance rules—to pursue more efficient crops without eliminating the existing ones and always with the aim of increasing their profitability.

Pilot fields can be considered small production plans. Thus, producers will understand the expected profit by implementing this research process on a larger scale. Finally, the farmers' connection with information technology and, in particular, with the decision support model (DSM) has a two-fold perspective as they will be able to enter personalized data themselves and simulate numerous production plans taking into account the new cross-compliance rules and rational water management.

Author Contributions: Conceptualization, A.K. and A.T.; methodology, A.K. and A.T.; validation, E.L., C.M. and A.P.; formal analysis, A.K. and C.M.; investigation, A.T.; resources, A.K., E.L. and A.T.; data curation, E.D. and A.P.; writing—original draft preparation, A.K. and A.P.; writing—review and editing, E.D. and T.B.; supervision, T.B. All authors have read and agreed to the published version of the manuscript.

Funding: This research was funded by the Rural Development Program (RDP) and is co-financed by the European Agricultural Fund for Rural Development (EAFRD) and Greece, grant number M16ΣYN2-00142.

Institutional Review Board Statement: Not applicable.

Informed Consent Statement: Not applicable.

Data Availability Statement: Data are contained within the article.

Conflicts of Interest: The authors declare no conflicts of interest.

References

1. Tiwari, D.N.; Loof, R.; Paudyal, G.N. Environmental–Economic Decision-Making in Lowland Irrigated Agriculture Using Multi-Criteria Analysis Techniques. *Agric. Syst.* **1999**, *60*, 99–112. [CrossRef]
2. Li, Y.P.; Huang, G.H. Interval-Parameter Two-Stage Stochastic Nonlinear Programming for Water Resources Management under Uncertainty. *Water Resour. Manag.* **2007**, *22*, 681–698. [CrossRef]
3. Rupnik, R.; Kukar, M.; Vračar, P.; Košir, D.; Pevec, D.; Bosnić, Z. AgroDSS: A Decision Support System for Agriculture and Farming. *Comput. Electron. Agric.* **2019**, *161*, 260–271. [CrossRef]
4. Meng, C.; Li, W.; Cheng, R.; Zhou, S. An Improved Inexact Two-Stage Stochastic with Downside Risk-Control Programming Model for Water Resource Allocation under the Dual Constraints of Water Pollution and Water Scarcity in Northern China. *Water* **2021**, *13*, 1318. [CrossRef]
5. Moulogianni, C.; Bournaris, T. Assessing the impacts of rural development plan measures on the sustainability of agricultural holdings using a pmp model. *Land* **2021**, *10*, 446. [CrossRef]
6. Latinopoulos, D. Multicriteria Decision-Making for Efficient Water and Land Resources Allocation in Irrigated Agriculture. *Environ. Dev. Sustain.* **2009**, *11*, 329–343. [CrossRef]
7. Manos, B.; Papathanasiou, J.; Bournaris, T.; Voudouris, K. A Multicriteria Model for Planning Agricultural Regions within a Context of Groundwater Rational Management. *J. Environ. Manag.* **2010**, *91*, 1593–1600. [CrossRef] [PubMed]
8. Bournaris, T.; Papathanasiou, J.; Manos, B.; Kazakis, N.; Voudouris, K. Support of irrigation water use and eco-friendly decision process in agricultural production planning. *Oper. Res.* **2015**, *15*, 289–306. [CrossRef]
9. Papathanasiou, J.; Bournaris, T.; Tsaples, G.; Digkoglou, P.; Manos, B.D. Applications of DSSs in Irrigation and Production Planning in Agriculture. *Int. J. Decis. Support Syst. Technol.* **2021**, *13*, 18–35. [CrossRef]
10. Schut, M.; Kamanda, J.; Gramzow, A.; Dubois, T.; Stoian, D.; Andersson, J.; Lundy, M. Innovation Platforms in Agricultural Research for Development: Ex-ante Appraisal of the Purposes and Conditions under Which Innovation Platforms can Contribute to Agricultural Development Outcomes. *Exp. Agric.* **2019**, *55*, 575–596. [CrossRef]
11. Amiri-Zarandi, M.; Hazrati Fard, M.; Yousefinaghani, S.; Kaviani, M.; Dara, R. A Platform Approach to Smart Farm Information Processing. *Agriculture* **2022**, *12*, 838. [CrossRef]

12. Borrero, J.D.; Mariscal, J. A Case Study of a Digital Data Platform for the Agricultural Sector: A Valuable Decision Support System for Small Farmers. *Agriculture* **2022**, *12*, 767. [CrossRef]
13. Runck, B.C.; Joglekar, A.; Silverstein, K.; Chan-Kang, C.; Pardey, P.; Wilgenbusch, J.C. Digital agriculture platforms: Driving data-enabled agricultural innovation in a world fraught with privacy and security concerns. *Agron. J.* **2022**, *114*, 2635–2643. [CrossRef]
14. Bournaris, T. A Multi-Criteria Model for Investigating the Income, Employment, and Environmental Impacts of Irrigated Agriculture. Master's Thesis, Department of Agriculture, School of Geotechnical Sciences, Aristotle University of Thessaloniki, Thessaloniki, Greece, 2003.
15. Martika-Vakirtzi, M.; Dimitriadou, E. *Accounting in Types of Agricultural Holdings*; Grafima: Thessaloniki, Greece, 2007.
16. Kitsopanidis, G.; Kamenidis, C. *Agricultural Economics*, 3rd ed.; ZHTH: Thessaloniki, Greece, 2003.
17. Kouriati, A.; Dimitriadou, E.; Bournaris, T. Farm accounting for farm decision making: A case study in Greece. *Int. J. Sustain. Agric. Manag. Inform.* **2021**, *7*, 77. [CrossRef]
18. Katsaounis, M. Techno-Economic Analysis and Organization of Agricultural Production in the Area Askio of Kozani. Master's Thesis, School of Agriculture, Aristotle University of Thessaloniki, Thessaloniki, Greece, 2012.
19. Briggs, T.; Kunsch, P.L.; Mareschal, B. Nuclear waste management: An application of the multicriteria PROMETHEE methods. *Eur. J. Oper. Res.* **1990**, *44*, 1–10. [CrossRef]
20. Vaillancourt, K.; Waaub, J.-P. Environmental site evaluation of waste management facilities embedded into EUGÈNE model: A multicriteria approach. *Eur. J. Oper. Res.* **2002**, *139*, 436–448. [CrossRef]
21. Kapepula, K.-M.; Colson, G.; Sabri, K.; Thonart, P. A multiple criteria analysis for household solid waste management in the urban community of Dakar. *Waste Manag.* **2007**, *27*, 1690–1705. [CrossRef]
22. Queiruga, D.; Walther, G.; González-Benito, J.; Spengler, T. Evaluation of sites for the location of WEEE recycling plants in Spain. *Waste Manag.* **2008**, *28*, 181–190. [CrossRef]
23. Vego, G.; Kučar-Dragičević, S.; Koprivanac, N. Application of multi-criteria decision-making on strategic municipal solid waste management in Dalmatia, Croatia. *Waste Manag.* **2008**, *28*, 2192–2201. [CrossRef] [PubMed]
24. Wang, J.J.; Jing, Y.Y.; Zhang, C.F.; Zhao, J.H. Review on multi-criteria decision analysis aid in sustainable energy decision-making. *Renew. Sustain. Energy Rev.* **2009**, *13*, 2263–2278. [CrossRef]
25. Moulogianni, C. Comparison of Selected Mathematical Programming Models Used for Sustainable Land and Farm Management. *Land* **2022**, *11*, 1293. [CrossRef]
26. Bournaris, T. Designing and Development of a Web Portal for E-Government and Farm Management. Ph.D. Thesis, Department of Agricultural Economics, School of Agriculture, Aristotle University of Thessaloniki, Thessaloniki, Greece, 2009.

Disclaimer/Publisher's Note: The statements, opinions and data contained in all publications are solely those of the individual author(s) and contributor(s) and not of MDPI and/or the editor(s). MDPI and/or the editor(s) disclaim responsibility for any injury to people or property resulting from any ideas, methods, instructions or products referred to in the content.

Proceeding Paper

Decision Support Model for Input Minimization and the Optimal Economic Efficiency of Agricultural Holdings [†]

Evgenia Lialia [1], Anna Tafidou [2], Asimina Kouriati [1], Angelos Prentzas [1], Eleni Dimitriadou [1], Christina Moulogianni [1] and Thomas Bournaris [1,*]

[1] Department of Agricultural Economics, Aristotle University of Thessaloniki, 54124 Thessaloniki, Greece; evlialia@agro.auth.gr (E.L.); kouriata@agro.auth.gr (A.K.); aprentzas@agro.auth.gr (A.P.); edimitri@agro.auth.gr (E.D.); kristin@agro.auth.gr (C.M.)

[2] Department of Mathematics, Aristotle University of Thessaloniki, 54124 Thessaloniki, Greece; annatafidou@gmail.com

* Correspondence: tbournar@agro.auth.gr

[†] Presented at the 17th International Conference of the Hellenic Association of Agricultural Economists, Thessaloniki, Greece, 2–3 November 2023.

Abstract: This study aims to change land use by implementing a Decision Support Model (DMS) with the goal of reducing water and fertilizer use. The problem is solved by deriving the necessary results of a set of selected pilot fields that belong to a farmer group located in the region of Central Macedonia. In order to define the pilot farms, the necessary data are collected and then processed using multicriteria weighted goal programming in order to develop a Decision Support Model that is related to the reduction of water and fertilizer use.

Keywords: common agricultural policy; input minimization; decision support model

Citation: Lialia, E.; Tafidou, A.; Kouriati, A.; Prentzas, A.; Dimitriadou, E.; Moulogianni, C.; Bournaris, T. Decision Support Model for Input Minimization and the Optimal Economic Efficiency of Agricultural Holdings. *Proceedings* **2024**, *94*, 43. https://doi.org/10.3390/proceedings2024094043

Academic Editor: Eleni Theodoropoulou

Published: 4 February 2024

Copyright: © 2024 by the authors. Licensee MDPI, Basel, Switzerland. This article is an open access article distributed under the terms and conditions of the Creative Commons Attribution (CC BY) license (https://creativecommons.org/licenses/by/4.0/).

1. Introduction

The Common Agricultural Policy (CAP) is one of the most important policies of the European Union, comprising a set of regulations, directives and laws relating to agricultural production, the marketing of agricultural products and all the interventions applied. A key tool for achieving the objectives of the CAP is the use of cross compliance rules and standards, which were developed in the late 1990s and introduced in 2005 as a European Union policy [1,2]. Cross compliance is defined as the set of regulatory standards that farmers follow that relate to the environmental management of natural resources, the protection of the rural landscape, public health, plant and animal health and the implementation of good agricultural practices [3]. Following the cross compliance rules is mandatory for both direct single payment and coupled payments, while their violation results in the reduction of payments [4].

This paper focuses on the optimal management of the water and fertilizer amounts used by producers as a plethora of reports in the literature highlight the continuous efforts in the optimal management of inputs that have been carried out in recent years [5,6]. This aim will be accomplished through the implementation of a Decision Support Model designed especially for farmer group needs. This model has been designed to contribute both to the assimilation of the new cross compliance rules and to the optimal management of water and fertilizers with the ultimate goal of enhancing the farm's efficiency and profitability. This research's aim will be achieved by setting up pilot fields of 100 acres and extending the results to an actual farm area. The results obtained using the model will be used as the basis for the creation of an electronic platform, which will be extremely useful for the extraction of economic results for the farm. The development of similar models and platforms has been extensively studied in various countries [7–10] and particularly in Greece [11–15]. Finally, as the literature suggests, platforms are sought to be developed for use in the agricultural sector in order to modernize it and achieve maximum economic results [16–18].

2. Materials and Methods

This paper aims to enhance the efficiency and profitability of farms while reducing water and fertilizer use through the implementation and use of a Decision Support Model. The scope will be achieved through the optimal allocation of limited economic resources and the land-producers' resources through the establishment of pilot fields by a farmer group, located in the region of Loudia, Thessaloniki. Each producer defines their own individualized production plan with specific limits on irrigation and fertilization, as well as other inputs, in order to obtain their optimal production plan. This leads to the economic upgrade and profitability of the farm and also to the minimization of land fertilization and water use.

The application of a Decision Support Model is required in order to solve, in the best possible way, the problems of the irrational waste of water and fertilizer use in the agricultural sector. At the same time, it is of a great importance to achieve the goals of the CAP, which in this study will be achieved through the application of the cross compliance rules and standards. As the cross compliance rules will become stricter in the coming years, it is crucial to familiarize producers with them.

In order to achieve the research objective, a pilot field of 100 acres will be established by the selected farmer group through the development of a Decision Support Model. The developed model is based on an existing structure created by the Laboratory of Informatics in Agriculture, which belongs to the Aristotle University of Thessaloniki, and is adjusted to the needs of the producers. This process is achieved using the multicriteria weighted goal programming method after the collection of relative data. These data were collected through the use of a specially designed questionnaire based on the literature [19–21]. At the same time, a web-based platform is used by the producers. This platform is also based on the Laboratory in Agriculture's existing infrastructure, is intended to record a set of technical and economic farm data and is essentially used to determine whether the objective of this study has been achieved.

More specifically, this study is carried out in four stages. Firstly, the preparation of the pilot implementation will be achieved through the selection of the pilot fields, crops, location and activities, which are set through the development of a Decision Support Model adjusted to the needs of the producers. Then, a web-based platform will be used, and at the same time, the producers will be taught how to use it properly. During that time, the producers' comments will be used for the optimization of the platform's protocol. This will lead to the creating of a technical and economic database. Then, a technical and economic analysis of the results will take place where the data are evaluated, and the possibilities of utilizing the new methodology and minimizing the inputs will be explored. Finally, the dissemination of the results will be carried out through a set of actions.

3. Expectative Results and Discussion

Connecting farmers to a Decision Support Model and managing their farms electronically has multiple benefits. The main benefit of this research is through rational water and fertilizer use management to reduce production and labor costs, enhance the income and gross profit of producers and increase environmental sustainability regarding farm size [14].

This study will be carried out for the first time and, in parallel, on such a large scale with the aim of spreading its results to the surrounding regions and subsequently to Greece as it includes information on the main crops of Central Macedonia. The above-mentioned process can motivate producers through the reduction of costs in water and fertilizer use to cultivate more efficiently and profitably without eliminating their existing crops.

4. Conclusions

The organization and derivation of an optimal production plan, using a Decision Support Model, will create more efficient farms, based on the challenges linked to the principles of the new CAP. This study is an innovative action as it can motivate producers to follow more efficient ways of cultivation. This will lead to a modernization of farms, thus reducing costs and increasing gross profits. This is the first time that this kind of research will be implemented on such a scale. This fact may lead to its implementation in the surrounding regions beyond Central Macedonia. The present study is also considered innovative as it includes information on the main crops with the aim of managing a set of entire agricultural areas while it is known that alternative crops are limited in the study area. Lastly, the research process encourages producers, by incorporating the new cross compliance rules, to pursue more efficient crops without eliminating existing ones, always looking to increasing their profitability.

Author Contributions: Conceptualization, A.K. and A.T.; methodology, A.K. and A.T.; validation, E.L., C.M. and A.P.; formal analysis, A.K. and C.M.; investigation, A.T.; resources, A.K., E.L. and A.T.; data curation, E.D. and A.P.; writing—original draft preparation, A.K. and A.P.; writing—review and editing, E.D. and T.B.; supervision, T.B. All authors have read and agreed to the published version of the manuscript.

Funding: This research was funded by the Rural Development Program (RDP) and was co-financed by the European Agricultural Fund for Rural Development (EAFRD) and Greece, grant number M16ΣYN2-00056.

Institutional Review Board Statement: Not applicable.

Informed Consent Statement: Not applicable.

Data Availability Statement: Data are contained within the article.

Conflicts of Interest: The authors declare no conflicts of interest.

References

1. Latacz-Lohmann, U.; Buckwell, A.E. Einige ökonomische Überlegungen zu Cross Compliance. *Ger. J. Agric. Econ./Agrarwirtsch.* **1998**, *47*, 429–431.
2. Mann, S. Different perspectives on cross-compliance. *Environ. Values* **2005**, *14*, 471–482. [CrossRef]
3. Pezaros, P. *The Last CAP Reform and Its Implementation in Greece*; Greek Ministry of Agriculture and Food: Athens, Greece, 2007.
4. Regulation (EU) No 1306/2013 of the European Parliament and of the Council of 17 December 2013 on the financing, management and monitoring of the common agricultural policy and repealing Council Regulations (EEC) No 352/78, (EC) No 165/94, (EC) No 2799/98, (EC) No 814/2000, (EC) No 1290/2005 and (EC) No 485/2008. *Off. J. Eur. Union* **2013**, *347*, 549–607.
5. Bournaris, T.; Vlontzos, G.; Moulogianni, C. Efficiency of vegetables produced in Glasshouses: The Impact of Data Envelopment Analysis (DEA) in land management decision making. *Land* **2019**, *8*, 17. [CrossRef]
6. Prentzas, A.; Nastis, S.A.; Moulogianni, C.; Kouriati, A. Technical and economic analysis of farms cultivating cereals and legumes: A Greek case study. *Int. J. Sustain. Agric. Manag. Inform.* **2022**, *8*, 446–459. [CrossRef]
7. Tiwari, D.N.; Loof, R.; Paudyal, G.N. Environmental–Economic Decision-Making in Lowland Irrigated Agriculture Using Multi-Criteria Analysis Techniques. *Agric. Syst.* **1999**, *60*, 99–112. [CrossRef]
8. Li, Y.P.; Huang, G.H. Interval-Parameter Two-Stage Stochastic Nonlinear Programming for Water Resources Management under Uncertainty. *Water Resour. Manag.* **2007**, *22*, 681–698. [CrossRef]
9. Rupnik, R.; Kukar, M.; Vračar, P.; Košir, D.; Pevec, D.; Bosnić, Z. AgroDSS: A Decision Support System for Agriculture and Farming. *Comput. Electron. Agric.* **2019**, *161*, 260–271. [CrossRef]
10. Meng, C.; Li, W.; Cheng, R.; Zhou, S. An Improved Inexact Two-Stage Stochastic with Downside Risk-Control Programming Model for Water Resource Allocation under the Dual Constraints of Water Pollution and Water Scarcity in Northern China. *Water* **2021**, *13*, 1318. [CrossRef]
11. Latinopoulos, D.; Mylopoulos, Y. A multicriteria approach for sustainable irrigation water management: Application in Loudias River Basin. In Proceedings of the International Conference: Protection and Restoration of the Environment VIII, Chania, Greece, 3–7 July 2006.
12. Latinopoulos, D. Multicriteria Decision-Making for Efficient Water and Land Resources Allocation in Irrigated Agriculture. *Environ. Dev. Sustain.* **2009**, *11*, 329–343. [CrossRef]
13. Manos, B.; Papathanasiou, J.; Bournaris, T.; Voudouris, K. A Multicriteria Model for Planning Agricultural Regions within a Context of Groundwater Rational Management. *J. Environ. Manag.* **2010**, *91*, 1593–1600. [CrossRef] [PubMed]

14. Bournaris, T.; Papathanasiou, J.; Manos, B.; Kazakis, N.; Voudouris, K. Support of irrigation water use and eco-friendly decision process in agricultural production planning. *Oper. Res.* **2015**, *15*, 289–306. [CrossRef]
15. Papathanasiou, J.; Bournaris, T.; Tsaples, G.; Digkoglou, P.; Manos, B.D. Applications of DSSs in Irrigation and Production Planning in Agriculture. *Int. J. Decis. Support Syst. Technol.* **2021**, *13*, 18–35. [CrossRef]
16. Amiri-Zarandi, M.; Hazrati Fard, M.; Yousefinaghani, S.; Kaviani, M.; Dara, R. A Platform Approach to Smart Farm Information Processing. *Agriculture* **2022**, *12*, 838. [CrossRef]
17. Borrero, J.D.; Mariscal, J. A Case Study of a Digital Data Platform for the Agricultural Sector: A Valuable Decision Support System for Small Farmers. *Agriculture* **2022**, *12*, 767. [CrossRef]
18. Runck, B.C.; Joglekar, A.; Silverstein, K.; Chan-Kang, C.; Pardey, P.; Wilgenbusch, J.C. Digital agriculture platforms: Driving data-enabled agricultural innovation in a world fraught with privacy and security concerns. *Agron. J.* **2022**, *114*, 2635–2643. [CrossRef]
19. Georgilas, I.; Moulogianni, C.; Bournaris, T.; Vlontzos, G.; Manos, B. Socioeconomic impact of climate change in rural areas of Greece using a multicriteria decision-making model. *Agronomy* **2021**, *11*, 1779. [CrossRef]
20. Martika-Vakirtzi, M.; Dimitriadou, E. *Accounting in Types of Agricultural Holdings*; Grafima: Thessaloniki, Greece, 2007.
21. Kouriati, A.; Dimitriadou, E.; Bournaris, T. Farm accounting for farm decision making: A case study in Greece. *Int. J. Sustain. Agric. Manag. Inform.* **2021**, *7*, 77. [CrossRef]

Disclaimer/Publisher's Note: The statements, opinions and data contained in all publications are solely those of the individual author(s) and contributor(s) and not of MDPI and/or the editor(s). MDPI and/or the editor(s) disclaim responsibility for any injury to people or property resulting from any ideas, methods, instructions or products referred to in the content.

Proceeding Paper

Using Pollen DNA Metabarcoding to Assess the Foraging Preferences of Honeybees in Kastoria Region, Greece [†]

Maria V. Alvanou [1], Maria Tokamani [2], Athanasios Toros [2], Raphael Sandaltzopoulos [2], Konstantinos Zampakas [1], Chrysoula Tananaki [3], Katerina Melfou [1] and Ioannis A. Giantsis [1,*]

[1] Faculty of Agricultural Sciences, University of Western Macedonia, 53100 Florina, Greece; mariaalvanou7@gmail.com (M.V.A.); zampakaskwnstantinos@hotmail.com (K.Z.); kmelfou@uowm.gr (K.M.)
[2] Department of Molecular Biology and Genetics, Democritus University of Thrace, 68100 Alexandroupolis, Greece; athanasiostoros@gmail.com (A.T.); rmsandal@mbg.duth.gr (R.S.)
[3] Laboratory of Apiculture-Sericulture, Aristotle University of Thessaloniki, 57001 Thermi, Greece; tananaki@agro.auth.gr
* Correspondence: igiants@auth.gr
[†] Presented at the 17th International Conference of the Hellenic Association of Agricultural Economists, Thessaloniki, Greece, 2–3 November 2023.

Abstract: Identification of a plant's pollen components can be used to establish its geographical provenance, while also providing insights into the diet and foraging preferences of the honeybee (*Apis mellifera* L.). The diversity and amount of pollen represent crucial factors for pollinators. Here, we identified plant species visited by honeybees by analyzing the pollen pellets collected from honeybees in Kastoria, Greece. The results indicate that pollen from different periods was identified by means of floral composition. An interesting observation is that all identified plants belonged to different genera. Among the identified plants, native ones, such as the Macedonian pine, *Pinus peuce*, present a distinct foraging profile for local honeybees.

Keywords: honeybee; *Apis*; pollen; Kastoria; foraging preferences

1. Introduction

Aside from their role in biodiversity retention, pollinators contribute to the European Union (EU) agricultural industry [1]. More specifically, pollinators are very important for the reproduction and preservation of a plethora of plants (e.g., medical herbal crops, agricultural crops, horticultural, and wild plants) [2–4]. Foraging behavior among pollinators can be affected from all available plants in the foraging area [5]. As the conservation of pollinators is a highly important issue, the identification of the foraging preferences of pollinators is considered a priority. However, most studies have used morphological data, where pollen identification constitutes a difficult task. In Western Macedonia and particularly in the Kastoria region, there are many endemic plant species [6]. Therefore, honeybees feeding on these species may provide honey with specific organoleptic characteristics, while at the same time informing us about its origin [7]. Thus, alternative methodologies, such as molecular techniques to identify the foraging preferences of honeybees, may be valuable for constructing a plant–honeybee interaction network. The main scope of the present study was to evaluate the preferences of the honeybee for plant species using pollen metabarcoding from the Kastoria region.

2. Materials and Methods

During summer 2021, six incoming forager bees with pollen loads were collected from the Kastoria region, Greece. The pollen pellets were collected from the hind legs of each bee and transferred to a sterile tube using disinfected tools. DNA extraction was performed using the kit Nucleospin tissue (Machenery Nagel, Duren, Germany) following the recommendations of the manufacturer. The concentration and purity of the DNA samples was evaluated using a Q5000 Microvolume spectrophotometer (Quawell Technology Inc., San Jose, CA, USA). PCR amplification of the trnL region was carried out using the primer set proposed by Kraaijeveld et al. 2015. The PCR products were resolved using electrophoresis on 1.5% w/v TBE agarose gels and purified using the NucleoMag kit. Amplicon libraries were prepared using the Ion Plus Fragment Library Kit (Cat. No. A28950, Thermo Fisher Scientific, Wilmington, DE, USA), following the instructions of the manufacturer. Sequencing was conducted using the Ion Torrent S5 system on a 530 chip (Cat No. A27764). The resulting reads were quality controlled, and then denoised into ASVs using the DADA2 plugin from the QIIME2 platform. A final taxonomic assignment was performed using BLAST against sequences of the trnL marker region procured from GenBank in 2018.

3. Results and Discussion

In total, we identified 32 plants at the species level, 10 at the genus level and the remaining 4 at the family level. An interesting observation is that all the identified plants belonged to different genera, while the most common families were Asteraceae, Thymelaeaceae, Fabaceae, Eupteleaceae, and Rosaceae. These results indicate a particularly rich flora diversity, in line with the high biodiversity observed in the Kastoria region. A percentage of up to 16.9% remained unassigned.

4. Conclusions

The above study revealed important information regarding the foraging preferences of honeybees across the Kastoria region from three different sampling dates. The use of DNA metabarcoding led us to identify the pollen composition of the samples, which is a more accurate technique in comparison to traditional microscopy methods. Apart from the foraging preferences of honeybees, polled DNA metabarcoding can provide a powerful tool for rapid surveys on plant biodiversity [8]. For the region of Kastoria, which is a highly significant habitat due to its high biodiversity, there is a need for continuous monitoring. However, although we identified plenty of plant species, in two out of six pollen pellets, a high percentage (40–45%) of plants remained unassigned. From this observation, we can assume that these species may be native to this region, and as a result, they are missing from the databases. To conclude, with the use of pollen meta barcodes, we can assess the foraging preferences of the honeybees inhabiting this area and monitor the biodiversity of the areas, while at the same time, it can help us to provide honeybee colonies with their preferred plants.

Author Contributions: Conceptualization, I.A.G. and C.T.; methodology, M.V.A. and M.T.; software, A.T. and R.S.; validation, R.S.; formal analysis, K.M.; investigation, K.Z.; resources, K.Z.; data curation, R.S.; writing—original draft preparation, M.V.A. and M.T.; writing—review and editing, C.T. and I.A.G.; visualization, K.M.; supervision, I.A.G.; project administration, K.M.; funding acquisition, K.M. All authors have read and agreed to the published version of the manuscript.

Funding: This research is supported by the Administrative Region of Western Macedonia, Greece (Special Account for Research and Funds Project Number: 80753).

Institutional Review Board Statement: Not applicable.

Informed Consent Statement: Not applicable.

Data Availability Statement: All data from this research are available after communication with the corresponding author.

Conflicts of Interest: The authors declare no conflicts of interest.

References

1. European Food Safety Authority. Towards an integrated environmental risk assessment of multiple stressors on bees: Review of research projects in Europe, knowledge gaps and recommendations. *EFSA J.* **2014**, *12*, 3594.

2. Ollerton, J.; Winfree, R.; Tarrant, S. How many flowering plants are pollinated by animals? *Oikos* **2011**, *120*, 321–326. [CrossRef]
3. Shakeel, M.; Ali, H.; Ahmad, S.; Said, F.; Khan, K.A.; Bashir, M.A.; Anjum, S.I.; Islam, W.; Ghramh, H.A.; Ansari, M.J.; et al. Insect pollinators diversity and abundance in *Eruca sativa* Mill. (Arugula) and *Brassica rapa* L. (Field mustard) crops. *Saudi J. Biol. Sci.* **2019**, *26*, 1704–1709. [CrossRef] [PubMed]
4. Latif, A.; Malik, S.A.; Saeed, S.; Zaka, S.M.; Sarwar, Z.M.; Ali, M.; Azhar, M.F.; Javaid, M.; Ishtiaq, M.; Naeem-Ullah, U.; et al. Pollination biology of *Albizia lebbeck* (L.) Benth. (Fabaceae: Mimosoideae) with reference to insect floral visitors. *Saudi J. Biol. Sci.* **2019**, *26*, 1548–1552. [CrossRef] [PubMed]
5. Khan, K.A.; Ghramh, H.A.; Ahmad, Z.; El-Niweiri, M.A.; Mohammed, M.E.A. Honeybee (*Apis mellifera*) preference towards micronutrients and their impact on bee colonies. *Saudi J. Biol. Sci.* **2021**, *28*, 3362–3366. [CrossRef] [PubMed]
6. Papanikolaou, A.; Panitsa, M. Plant species richness and composition of a habitat island within Lake Kastoria and comparison with those of a true island within the protected Pamvotis lake (NW Greece). *Biodivers. Data J.* **2020**, *8*, e48704. [CrossRef] [PubMed]
7. Milla, L.; Sniderman, K.; Lines, R.; Mousavi, G.; Derazmahalleh, M.; Encinas Viso, F. Pollen DNA metabarcoding identifies regional provenance and high plant diversity in Australian honey. *Ecol. Evol.* **2021**, *11*, 8683–8698. [CrossRef] [PubMed]
8. Leontidou, K.; Vokou, D.; Sandionigi, A.; Bruno, A.; Lazarina, M.; De Groeve, J.; Li, M.; Varotto, C.; Girardi, M.; Casiraghi, M.; et al. Plant biodiversity assessment through pollen DNA metabarcoding in Natura 2000 habitats (Italian Alps). *Sci. Rep.* **2021**, *11*, 18226. [CrossRef] [PubMed]

Disclaimer/Publisher's Note: The statements, opinions and data contained in all publications are solely those of the individual author(s) and contributor(s) and not of MDPI and/or the editor(s). MDPI and/or the editor(s) disclaim responsibility for any injury to people or property resulting from any ideas, methods, instructions or products referred to in the content.

Proceeding Paper

Towards a Farmer-Centric Approach to Advise Provision [†]

Alex Koutsouris * and Vasiliki Kanaki

Department of Agricultural Economics & Rural Development, Agricultural University of Athens, 11855 Athens, Greece; vskanaki@gmail.com
* Correspondence: koutsouris@aua.gr; Tel.: +30-210-529-4721
[†] Presented at the 17th International Conference of the Hellenic Association of Agricultural Economists, Thessaloniki, Greece, 2–3 November 2023.

Abstract: The objective of this piece of work is to further the understanding of the roles played by a wide range of advice providers in farmer decision-making. Results show that from the perspective of a farmer, advice provision and advice providers are much more varied than is assumed in common perspectives in policy and research. This, in turn calls for a 'farmer centered advice paradigm' while acknowledging (a) the heterogeneity of farmers' circumstances, and (b) that the term advisor may fit any person who provides advice.

Keywords: advice provision; advisors; farmer-centric paradigm

Citation: Koutsouris, A.; Kanaki, V. Towards a Farmer-Centric Approach to Advise Provision. *Proceedings* **2024**, *94*, 45. https://doi.org/10.3390/proceedings2024094045

Academic Editor: Eleni Theodoropoulou

Published: 5 February 2024

Copyright: © 2024 by the authors. Licensee MDPI, Basel, Switzerland. This article is an open access article distributed under the terms and conditions of the Creative Commons Attribution (CC BY) license (https://creativecommons.org/licenses/by/4.0/).

1. Introduction

In the past, the European Commission EC has shown its interest to facilitate the development of farm advisory systems, confirmed through the latest Common Agricultural Policy (CAP; Reg. (EU) 2021/2115). Indeed, the latter stresses again the need for the provision of agricultural advisory services (Articles 15 and 78) while also emphasizing that the advice given shall be impartial and that advisors have no conflict of interest.

The Horizon 2020 AgriLink project [1] focused on the role that advisors play to help farmers to adopt more sustainable farming practices. One of the objectives of the project was to further the understanding of the roles played by a wide range of advice providers in farmer decision-making. Built on 26 case studies, carried out in 13 project partner countries, one of the main findings was that from the perspective of a farmer, advice provision and advice providers are much more varied than is usually assumed [2]. In this piece of work, based on the farmer surveys from two H2020 projects (INNOSETA [3] and AgroFossilFree [4]) we aim at addressing the AgriLink findings and their consequences for innovation support/advisory services.

2. Methodology

The INNOSETA project dealt, among others, with empirical research on innovation processes related to Spraying, Equipment, Training and Advising (SETA). INNOSETA strived to assess end-user needs and interests, and identify factors influencing adoption and diffusion of SETA technologies. Through targeted surveys, in the 8 partner countries, 348 farmers were interviewed in late autumn 2018 till winter 2019. Farmers were selected according to their (pre-defined) cropping system and farm size class. Both adopters and non-adopters of the SETA technologies were included in the sample.

Similarly, in the AgroFossilFree project a survey, addressing different types of renewable and energy saving technologies/practices, was carried out in 8 European countries. Overall, 470 farmers, in late winter 2020 till spring 2021. Additionally, in the AgroFossilFree project the concept of microAKIS was used [5].

3. Results

3.1. INNOSETA [6]

Farmers' most important source of knowledge/know-how on the use and operation of spraying equipment (Figure 1) are their own experience (34%), manufacturers and dealers (25%) and advisors (private: 9% and public/cooperative: 5%). When the three most important sources of information are taken together (Figure 1) again farmers' own experience (23%) and equipment manufacturers and dealers (21%) predominate followed by advisors (private: 9% and public/cooperative: 5%), other farmers (9% peers and 4% farmer groups) and the Internet (11%).

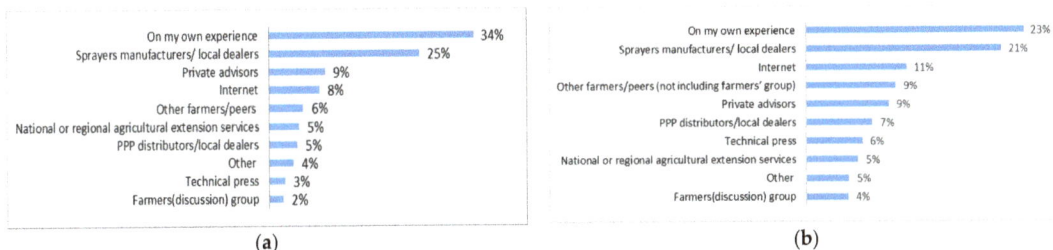

Figure 1. (**a**) Most important source of knowledge/know-how on the use and operation of spraying equipment; (**b**) Three most important sources of knowledge/know-how on the use and operation of spraying equipment.

For adopters, the most important source of information on buying innovative spraying equipment (Figure 2) are sprayers' manufacturers/dealers (29%), farmers' own experience (17%), other farmers (16%) and private advisors (10%). When the three most important information sources are aggregated (Figure 2), sprayers' manufacturers/local dealers (24%) along with other farmers/peers and their own experience (15% each) predominate.

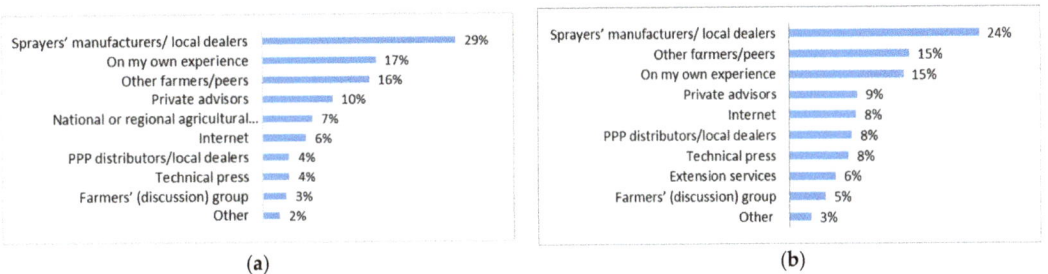

Figure 2. (**a**) Most important farmers' information source on buying innovative spraying equipment; (**b**) Three most important farmers' information sources on buying innovative spraying equipment.

3.2. AgroFossilFree [7]

Farmers' most important source of knowledge/awareness on Renewable Energy Sources (RES) (Figure 3) are the Internet (19.4%), technical press (15.7%), agricultural (public, cooperative) extension/advisory services (15.5%) and their own experience (11.7%). When the three most important sources of information are taken together (Figure 3) the Internet (55.3%) and technical press (41.9%) predominate followed by agricultural extension/advisory services (34.9%). Technology manufacturers/dealers (28.1%), other farmers (24%), farmers' own experience (23.6%) and private advisors (23.4%) also play a role in raising farmers' awareness on RES.

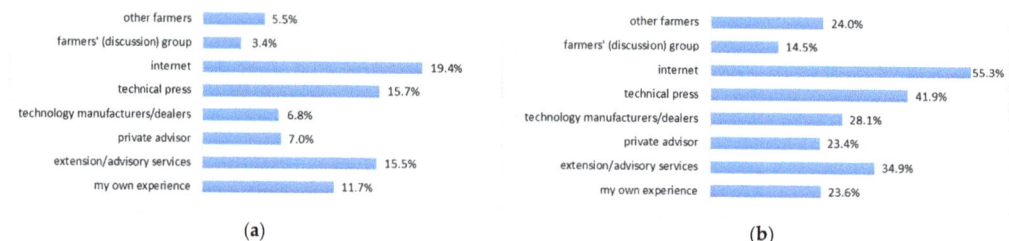

Figure 3. (**a**) Most important source of information on RES; (**b**) Three most important sources of information on RES.

Out of the 438 interviewees (93.2% of the sample) who were aware of RES, 199 (45.4%) use RES on their farms. Among them, the most important source of information/support on the assessment of RES (Figure 4) are farmers' own experience (25.6%), manufacturers/dealers (16.6%), private advisors (15.1%) and agricultural extension services (11.1%). Concerning the three most important sources of information/ support on the assessment of RES (Figure 4) these are manufacturers/ local dealers (58.3%) along with their own experience (43.2%) and private advisors (42.2%). The Internet (34.2%), technical press (26.1%) and agricultural extension services (23.1%) along with other farmers/peers (23.1%) and farmers groups (15.6%) also assist farmers to assess RES.

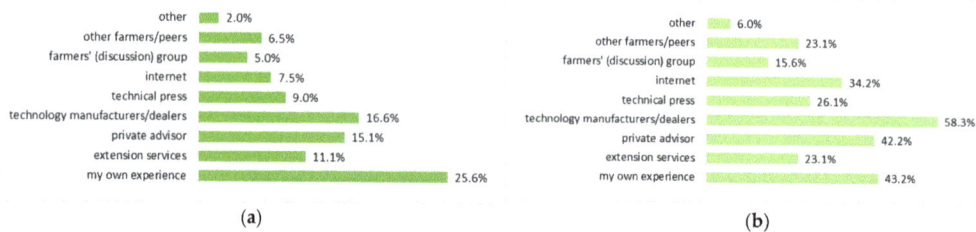

Figure 4. (**a**) Most important source of information/support for RES assessment; (**b**) Three most important sources of information/support for RES assessment.

The most important actors supporting farmers in the establishment and use of RES (Figure 5) are farmers' own experience (31.2%), manufacturers/dealers (23.1%) and private advisors (15.1%). The three most important actors (Figure 5) are manufacturers/ local dealers (61.8%) along with private advisors (43.7%) and their own experience (43.2%). The Internet (26.1%), technical press (24.6%) and national or regional agricultural (public, cooperative) extension services (24.1%) along with other farmers/peers (18.1%) and farmers groups (15.6%) also assist farmers to establish and use RES on their farm.

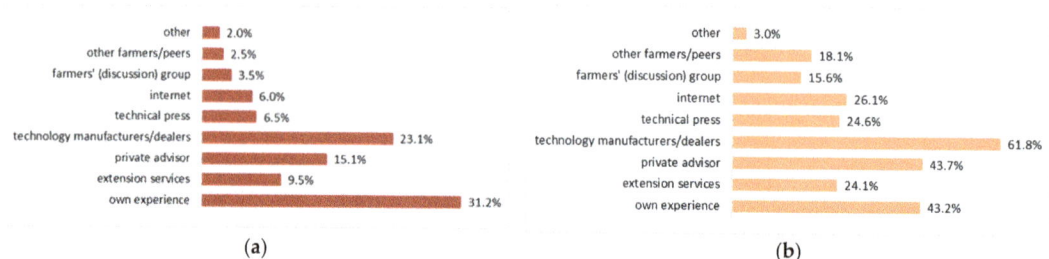

Figure 5. (**a**) Most important source of information/support for RES establishment and use; (**b**) Three most important sources of information/support for RES establishment and use.

4. Discussion and Conclusions

Both projects' findings verify the AgriLink's findings implying that from the perspective of a farmer, advice provision and advice providers are much more varied than is assumed. Therefore, there is a bias in both policy and research in starting from the side of advice provision while having little or no attention for farmers' advice needs. In this respect, countries' AKIS should start taking a closer look at these needs at the micro-level and to try and connect them to advice provision in various AKIS environments (see also [8]). This calls for a 'farmer centred advice paradigm' while also acknowledging (a) the heterogeneity of farmers' microAKISs, and (b) that the term advisor may fit any person who provides advice.

Furthermore, farmers' microAKISs include various sources of advice that are beyond independent influence. Independent advice providers should thus take farmers' reliance on such potentially biased sources as a starting point and help farmers to assess the validity of this type of advice and help them to place their advice needs in a broader context which also includes policy and societal objectives for sustainable development.

Author Contributions: Conceptualization A.K.; methodology A.K. and V.K.; formal analysis A.K.; Data curation A.K.; writing-original draft preparation A.K. and V.K.; writing-review and editing A.K. and V.K.; funding acquisition A.K. All authors have read and agreed to the published version of the manuscript.

Funding: This research was funded by the EU (H2020) as follows: AgriLink, grant agreement number 727577; INNOSETA, grant agreement number 773864; AgroFossilFree, grant agreement number 101000496.

Institutional Review Board Statement: Not applicable.

Informed Consent Statement: Informed consent was obtained from all subjects involved in the study.

Data Availability Statement: Publicly available data were used in this study. This data can be found here: (a) https://www.innoseta.eu/wp-content/uploads/2019/08/D2.2.pdf and (b) https://www.agrofossilfree.eu/wp-content/uploads/2021/10/D1.3.pdf.

Acknowledgments: The authors acknowledge the support of the project coordinators E. Gil and A. Balafoutis and their teams at UPC and CERTH, respectively, as well as both project partners' contribution in carrying out the farmers surveys.

Conflicts of Interest: The authors declare no conflict of interest.

References

1. AgriLink2020. Available online: https://www.agrilink2020.eu/ (accessed on 19 September 2023).
2. Labarthe, P.; Prager, P.; Leloup, H.; Elzen, B.; Collins, K.; Laurent, C.; Redman, M.; Schoorlemmer, H.; Sutherland, L.; Micheloni, C.; et al. Deliverable 5.7 Policy Recommendations Report. Strengthening Farm Advice for Innovation and Sustainability; AgriLink (H2020) Project. 2021. Available online: https://edepot.wur.nl/587603 (accessed on 19 September 2023).
3. INNOSETA. Innovative Spraying Equipment Training Advising. Available online: http://www.innoseta.eu/ (accessed on 19 September 2023).
4. AgroFossilFree. The Path towards a Fossil-Free EU Agriculture. Available online: https://www.agrofossilfree.eu/ (accessed on 19 September 2023).
5. Sutherland, L.A.; Labarthe, P. Introducing 'microAKIS': A farmer-centric approach to understanding the contribution of advice to agricultural innovation. *J. Agric. Educ. Ext.* **2022**, *28*, 525–547. [CrossRef]
6. Koutsouris, A.; Kanaki, V. Deliverable 2.2: Report on Farmers' Needs, Innovative Ideas and Interests. Available online: http://www.innoseta.eu/wp-content/uploads/2019/08/D2.2.pdf (accessed on 19 September 2023).
7. Koutsouris, A.; Kanaki, V. Deliverable 1.3: Report on Farmers' Needs, Innovative Ideas and Interests. Available online: https://www.agrofossilfree.eu/wp-content/uploads/2021/10/D1.3.pdf (accessed on 19 September 2023).
8. Sutherland, L.-A.; Adamsone-Fiskovica, A.; Elzen, B.; Koutsouris, A.; Laurent, C.; Stræte, E.P.; Labarthe, P. Advancing AKIS with Assemblage Thinking. *J. Rural Stud.* **2023**, *97*, 57–69. [CrossRef]

Disclaimer/Publisher's Note: The statements, opinions and data contained in all publications are solely those of the individual author(s) and contributor(s) and not of MDPI and/or the editor(s). MDPI and/or the editor(s) disclaim responsibility for any injury to people or property resulting from any ideas, methods, instructions or products referred to in the content.

Proceeding Paper

Farmers Vocational Education and Training: The Case of Public Institutes of Vocational Training at ELGO-DIMITRA [†]

Vasiliki Bitsopoulou, Eleni Pastrapa, Eleni Zenakou, Despina Sdrali and Eleni Theodoropoulou *

Department of Economics and Sustainable Development, School of Environment, Geography and Applied Economics, Harokopio University of Athens, 17676 Athens, Greece; hp12215210@hua.gr (V.B.); epastrapa@hua.gr (E.P.); elzenakou@hua.gr (E.Z.); dsdrali@hua.gr (D.S.)

* Correspondence: etheodo@hua.gr; Tel.: +30-2109549205
[†] Presented at the 17th International Conference of the Hellenic Association of Agricultural Economists, Thessaloniki, Greece, 2–3 November 2023.

Abstract: The aim of this study was to investigate the current organizational climate of Public Vocational Education and Training Initiatives at ELGO-DIMITRA in Greece. It utilized a SWOT analysis to identify the strengths, weaknesses, opportunities, and threats of the Institute's programs. The findings indicate a need to ensure the quality of Vocational Education and Training Initiatives and enhance the educational services provided to young farmers by adopting a regularly updated framework. This study is crucial for future research, and it is important to replicate it with different focus groups, including trainers, trainees, and graduates, who can provide valuable insight into the sustainability of the Vocational Education and Training Initiatives.

Keywords: ELGO-DIMITRA; young farmers; public vocational education and training

Citation: Bitsopoulou, V.; Pastrapa, E.; Zenakou, E.; Sdrali, D.; Theodoropoulou, E. Farmers Vocational Education and Training: The Case of Public Institutes of Vocational Training at ELGO-DIMITRA. *Proceedings* **2024**, *94*, 46. https://doi.org/10.3390/proceedings2024094046

Academic Editor: Stavriani Koutsou

Published: 4 February 2024

Copyright: © 2024 by the authors. Licensee MDPI, Basel, Switzerland. This article is an open access article distributed under the terms and conditions of the Creative Commons Attribution (CC BY) license (https://creativecommons.org/licenses/by/4.0/).

1. Introduction

The transformation of the agri-food sector is on the threshold of tectonic changes and challenges. The continuous emergence of new agricultural specialties, directly or indirectly related to the agricultural production process, requires investment support measures, and systematic and continuous training of young farmers.

The integration of young people into the agricultural environment of Greece is considered more necessary and timelier than ever, as they possess the appropriate characteristics that will push the agricultural sector towards sustainable development and competitiveness [1]. So, the agricultural sector needs people who fit the following descriptions:

- Receptive to innovation that requires know-how, so people with greater ability to innovate are able to absorb the knowledge at the same time;
- Possess the ability to seize potential opportunities and take entrepreneurial risks.

Moreover, the shift of the new generation to agricultural production should be a conscious choice. Only in this way will the new generation realize its primary role in shaping a favorable framework for innovative agricultural businesses based on extroversion and cooperation (clusters, cooperatives, etc.). This will result to the renewal of local societies that are at risk of desertification, while at the same time giving a perspective and impetus to the Greek agricultural economy.

The Public Institutes of Vocational Training (PIVT) of ELGO-DIMITRA, constitute an important "investment" for the acquisition of upgraded and certified knowledge-skills-abilities in specialized subjects, the adoption of the agricultural entrepreneurial and cooperative culture and the development of extroversion. "Today's farmers are young people of the Renaissance, they must possess the right mix: science, economics, entrepreneurship and environmental awareness to face the challenges of the future" [2].

The object of this study is to investigate the organizational environment of the PIVT of ELGO-DIMITRA [3]. Specifically, we analyze and examine the strengths, weaknesses, opportunities and threats of the six aforementioned Institutes by applying a SWOT analysis, as perceived by the Directors of the Institutes of ELGO-DIMITRA.

The main aim is to draw useful conclusions regarding the critical role they are called upon to play, due to the continuous new data being generated in the agri-food sector.

2. Materials and Methods

In this study a SWOT framework was used to examine the organizational environment of the PIVT of ELGO-DIMITRA in Greece. The research focus group consisted of six Directors of the Institutes located in the regions of Attica, Ioannina, Heraklion, Corinth, Larissa, and Trikala. All six (6) participants were asked to identify the strengths, weaknesses, opportunities and threats that may affect the future and the sustainability of the Greek Vocational Education and Training (VET) Institutes. Data were collected in November and December 2022.

3. Results and Discussion

The collected data are grouped as follows (Table 1).

Table 1. SWOT Analysis of PIVT of ELGO-DIMITRA.

Strengths	Weakness	Opportunities	Threats
Multi-year presence of schools in agricultural education	Limited financial resources	Use of funding provided by the new CAP for agricultural education and training	Continuously changing institutional framework
Free education, housing and food	Deterioration of building facilities, renovation of infrastructure	Upgrading the educational services of the education through targeted initiatives in the organization's new law [4]	High competition with other public and private institutes providing corresponding specializations in the agricultural direction
Appropriate equipped building and laboratory facilities	Lack of permanent staff (educational, administrative and workers)	Creation of new agricultural regional directorates/PIVT	Continuously emerging political—economic—social—environmental challenges
Provision of specific agriculture specializations	Exclusion of young people finishing junior high school	Introducing new specific agriculture specializations	Population aging—desertification agricultural regions
Experienced specialized staff (administrative, educational, technical, workers)	A difficulty observed on the part of the newly admitted vocational high school graduates in the 3rd semester of PIVT in the course: "Gardening Machines and Tools" as they have not been taught it like the DIEK trainees in the first year.	Delivery of distance education due to COVID-19	Years of degradation of the role of school vocational orientation
Educational activities through the ERASMUS+ Program	The training guides were delayed in being approved by the Ministry of Education	New ways of publishing activities and educational projects	Candidates that had previously passed the national examinations did not attend the PIVT
Partnerships with other organizations/collaboration with businesses, local bodies	Greater cooperation with agricultural research institutes of ELGO-DIMITRA	Initiatives and measures to enhance practice support	Delay in activating the evaluation system of educational units of the PIVT
Collaborative climate at all levels	During the studies, a percentage of the students leave their studies at PIVT for specific reasons		

The effectiveness of the quality of agricultural PIVT, as a public and common good "should not be limited to administrative and building criteria, nor to a sterile implementation of decisions taken at a central level" [5,6]. Therefore, they should gather those elements, which strengthen the qualitative performance of their educational work:

- The inseparable link with research;
- Years of educational experience and culture;
- Know-how, extroverted nature;
- Cooperation with other partners and businesses [5].

This is so that they become "an attractive, effective and quality educational policy tool" [7].

ELGO-DIMITRA has another important advantage over other VET providers, with similar agricultural specializations. It has become a central pillar for the direct alignment with the objectives of the new Common Agricultural Policy (CAP), contributing decisively to the connection and consolidation of the tripartite: agricultural Institutes of Vocational Training, Agricultural Entrepreneurship and Innovation for the productive development of the country. ELGO-DIMITRA's dual role in agricultural Institutes of Vocational Training is both in the provision of initial VET and in continuous VET (implementation of training programs).

As a consequence of the undeniable interconnection of these two subsystems and their direct function with the evolving field of the agri-food sector, their integration is imperative [8].

The consolidated SWOT analysis was the basis for the identification of the key components for the formation of a framework for upgrading the quality of agricultural VET through the Public VET Institutes, which should be reviewed regularly and includes (3) groups (Figure 1) coming from the internal and external environment of the Public VET Institutes and also includes priority thematic areas aimed at seeking quality indicators to achieve the objectives:

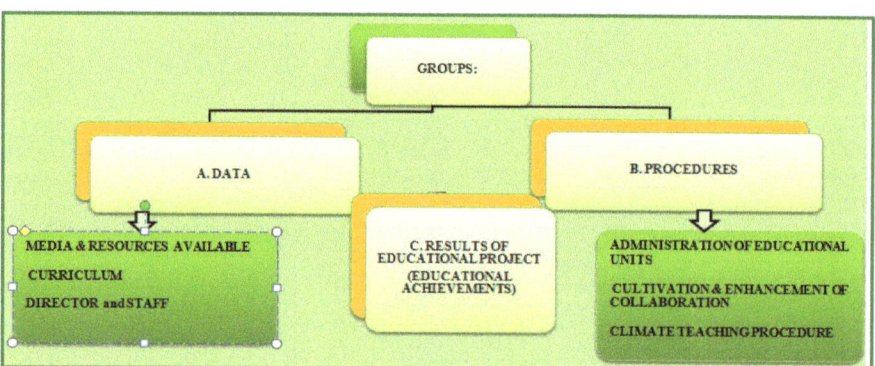

Figure 1. Framework for ensuring and upgrading the quality of educational services of agricultural PIVT [9–11].

Thus, the key to the modernization and sustainable development of farms and businesses is a well-trained agricultural workforce, receptive to innovation, entrepreneurship and partnerships [12]. The driving force will be the agricultural PIVT, as they offer multiple benefits:

- Personal development of trainees through the acquisition of upgraded and certified knowledge, skills and competencies.
- Creating an innovative business ecosystem.
- Socio-economic development and regeneration of local communities.

The research is of particular interest for future research. It is proposed to replicate the current study with different focus groups. A study of trainers and trainees, as well as graduates would give insight into how they perceive the sustainability of the VET Initiatives. The use of quantitative research can also provide an avenue for future research.

4. Conclusions

The PIVT of ELGO-DIMITRA comes to bear a greater extent than before: the education and training of young farmers for the development and strengthening of agricultural entrepreneurship. Therefore, their approach should direct young farmers primarily towards the transmission and implementation of entrepreneurial skills, the search for innovative business ideas and finally, the organization and coordination of an entrepreneurial activity.

Author Contributions: All the authors have contributed equally to all stages of the paper. All authors have read and agreed to the published version of the manuscript.

Funding: This research received no external funding.

Institutional Review Board Statement: Not applicable.

Informed Consent Statement: Informed consent was obtained from all subjects involved in the study.

Data Availability Statement: Data is available under request.

Acknowledgments: The authors thank the six (6) directors of the PIVT of ELGO-DIMITRA for their essential contribution and the two anonymous reviewers for useful comments.

Conflicts of Interest: The authors declare no conflicts of interest.

References

1. Haroutunian, A.S. With an eye on the future. Challenges and prospects for the development of the primary sector in Greece. *Epi Gis Mag.* **2022**, *20*, 16–19. Available online: https://www.piraeusbank.gr/el/agrotes/agrotika-nea-enimerosi/epi-gis#3 (accessed on 7 March 2023). (In Greek).
2. Mergos, G. The promotion of innovation and entrepreneurship. *Epi Gis Mag.* **2016**, *6*, 16. Available online: https://www.piraeusbank.gr/el/agrotes/agrotika-nea-enimerosi/epi-gis#6 (accessed on 7 March 2023). (In Greek).
3. *Regarding the Proportionality Check before the Introduction of New Legalization of Professions (OJ L 173), Ratification of the Agreement between the Government of the Hellenic Republic and the Government of the Federal Republic of Germany on the Hellenic-German Youth Foundation*; Law 4763/2020; FEK 254A′; Official Government Gazette of the Hellenic Republic: Athens, Greece, 2020.
4. *Uniform Regulatory Framework for the Organization and Operation of the Hellenic Agricultural Organization DIMITRA, Establishment and Operation of the Amfissa Traditional Olive Grove Management Body and Other Provisions to Strengthen Rural Development*; Law 5035/2023; FEK 76A′; Official Government Gazette of the Hellenic Republic: Athens, Greece, 2023; p. 16.
5. Foundation for Economic & Industrial Research—IOBE. *Vocational Education and Training in Greece*; IOBE: Athens, Greece, 2021; Available online: http://iobe.gr/docs/research/RES_05_F_27042021_REP_GR.pdf (accessed on 11 March 2023). (In Greek)
6. Hellenic Federation of Enterprises. Vocational Education and Training Reform: Key to a New Productive Model with Better and More Jobs for Young People. *Economy & Business, Special Report The Future of Employment: Initial Vocational Education & Training*. 2020. Available online: https://www.sev.org.gr/ekdoseis/metarrythmisi-tis-epangelmatikis-ekpaidefsis-kai-katartisis-kleidi-gia-ena-neo-paragogiko-protypo-me-kalyteres-kai-perissoteres-douleies-%20gia-tous-neous/ (accessed on 12 March 2023). (In Greek)
7. Delis, V. *Apprenticeship as a Means of Transition to the Modern Work Environment*; Institute for Policy Alternatives: Athens, Greece, 2021; Available online: https://www.enainstitute.org/wp-content/uploads/2021/12/ENA_Paper_Mathiteia-1.pdf (accessed on 12 March 2023). (In Greek)
8. Activities ELGO-DIMITRA/Agricultural Education. Available online: https://www.elgo.gr/index.php?option=com_content&view=category&layout=blog&id=295&Itemid=1352 (accessed on 15 March 2023). (In Greek)
9. Katsarou, E.; Dedouli, M. *Training and Evaluation in the Field of Education*; Ministry of National Education and Religious Affairs, Pedagogical Institute: Athens, Greece, 2008. (In Greek)
10. Sofou, E. Teaching and Education Issues in the Multicultural School: The Self-Evaluation of the School Unit. In *Educational Material Education of Foreign Students*; Katsarou, E., Laikopoulou, M., Eds.; Ministry of National Education and Religious Affairs: Thessaloniki, Greece, 2014. Available online: http://www.diapolis.auth.gr/epimorfotiko_uliko/images/pdf/keimena/yliko/enotita_d/sofoy.pdf (accessed on 12 March 2023). (In Greek)

11. Solomon, I. *Internal Evaluation & Planning of the Educational Project in the School Unit*; Ministry of National Education and Religious Affairs, Pedagogical Institute: Athens, Greece, 1999. Available online: http://users.sch.gr/gkelesidis/images/PDF/AXIOLOGISI/TOMOS1_solomon_MEROS_A.pdf (accessed on 14 March 2023). (In Greek)
12. European Commission. Vocational Education and Training Initiatives. Available online: https://education.ec.europa.eu/el/education-levels/vocational-education-and-training/about-vocational-education-and-training (accessed on 15 March 2023). (In Greek)

Disclaimer/Publisher's Note: The statements, opinions and data contained in all publications are solely those of the individual author(s) and contributor(s) and not of MDPI and/or the editor(s). MDPI and/or the editor(s) disclaim responsibility for any injury to people or property resulting from any ideas, methods, instructions or products referred to in the content.

Proceeding Paper

The Use of Precision Agriculture for Improving the Water Economics of Farms and the Need for Agricultural Advisory †

Georgios Papadavid [1], Georgios Kountios [2,*], Diofantos Hadjimitsis [3] and Maria Tsiouni [2]

1. Agricultural Research Institute, Athalassa, 1516 Nicosia, Cyprus; gpapadavid@ari.moa.gov.cy
2. Department of Agriculture, International Hellenic University, Sindos, 57400 Thessaloniki, Greece; mtsiouni84@yahoo.gr
3. Department of Civil Engineering & Geomatics, Cyprus University of Technology, 31 Archbishop Kyprianos, 3603 Limassol, Cyprus; d.hadjimitsis@cut.ac.cy
* Correspondence: gkountios@ihu.gr
† Presented at the 17th International Conference of the Hellenic Association of Agricultural Economists, Thessaloniki, Greece, 2–3 November 2023.

Abstract: The rational management of water, which is determined by the Framework Directive 2000/60/EC of the EU, is a contractual obligation of the Agricultural Sector of Cyprus, both towards the European Union and the next generations of Cypriot citizens. To make decisions about sustainable water use and improve water use, it is necessary to understand the water use of crops in different water-use areas. Especially in large water projects in Cyprus, there must be a good way to determine the water use of crops so that the correct use of crops can be ensured, thus eliminating problems such as a lack of new information about the crop area and agricultural evaporation, Demand, and water. In most projects, water is managed and supplied based on historical data, and current information is available to determine water demand and availability for large areas. This paper also adds, apart from the clear positive effect of remote sensing and new technologies in crop irrigation, to the emerging need for advisory services for the diffusion of innovation to Cypriot farmers since the estimation of crop water requirements is part of estimating the carbon footprint under the project CARBONICA (EU Funded) for carbon farming.

Keywords: technoeconomic analysis; irrigation water; earth observation; SEBAL method; advisory services

1. Introduction

Prolonged drought, dry conditions, and poor irrigation and water supply management result in a significant reduction in water reserves and resources [1]. Irrigation water reductions are based on studies of percentage reductions in the required volume of irrigation water and crop responses to these reductions. The required irrigation volume for each crop is calculated based on the evapotranspiration (ET) of each crop, which has been empirically found (Epan) in the past [2]. However, the need for an accurate measurement of ET today remains imperative and is an integral element in the decision-making of Water Policy Bodies. The continuous decrease in rainfall in Cyprus [3,4] (Cyprus Meteorological Services, 2020/2021) in recent decades has contributed to an increase in irrigation and, therefore, the volumes of water consumed in agriculture. However, the reckless use of irrigation water combined with reduced rainfall caused a serious blow to the storage of water in Cyprus [5].

2. Methods

ETc forecasts based on remote sensing are almost universally used due to the following advantages of remote sensing: abundant raw data, low cost, and weather forecasting. It is worth noting that Landsat 7 ETM+ and 8 OLI images cover almost the entire island of

Cyprus, and their resolution is good for hydrological studies. The study area is in the village of Mandria in the Paphos Region (west of Cyprus), which is the main agricultural area of the Paphos Region and absorbs the largest amount of water (26%) in the water supply system of Cyprus. Remote sensing techniques have been applied to estimate the accurate irrigation volumes per crop and finally compared to the empirical values that farmers are already using.

3. Results and Discussions

To produce these results, evapotranspiration maps were created where users can see the needs of their crops on a daily or even monthly basis (Figure 1). These maps are essentially the satellite images of the area, which were converted into maps using the SEBAL algorithm [6] after its application in the ERDAS Imagine software (version 16).

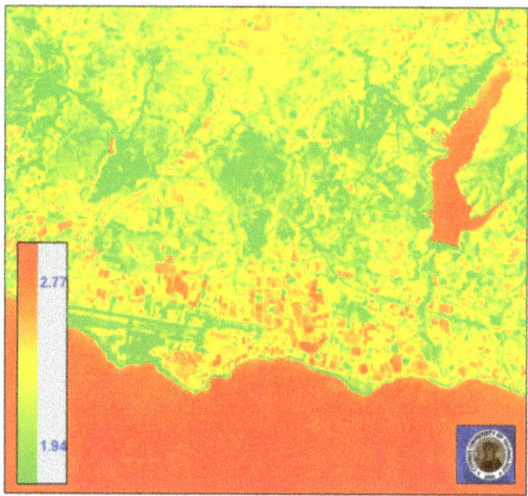

Figure 1. Example of a crop evapotranspiration map creation in Cyprus.

Table 1 shows the results of the project for the crops under investigation. The aim was to create an irrigation plan from which the producer would know the required irrigation volumes. It is noted that the results of this project, in terms of evapotranspiration, are given in mm/day, which, however, are easily converted to $m^3/ha/month$ for comparison purposes with previous research data referred to in the "Norm Input Output Data for the main crop and livestock enterprises of Cyprus" [7], which are listed in that format. The specific manual (Norm) published by the Agricultural Research Institute is the only technical and economic database in Cyprus for issues related to agricultural crops.

Table 1. Irrigation needs ($m^3/ha/month$) for specific crops in Cyprus.

Crop	J	F	M	A	M	J	J	A	S	O	N	D	Total
Potatoes	-	-	-	-	-	-	450	850	1200	1550	1300	-	5330
Groundnuts	-	-	-	-	-	-	620	1450	1650	300	-	-	4020
Beans	-	-	-	-	-	-	450	850	1200	990	-	-	3490
Peas	-	-	200	800	480	-	-	-	-	-	-	-	1480

The research data of this project were converted into irrigation costs after multiplying with the cost of irrigation water, which amounts to EUR 0.17/ton (Table 2). So, the project data are now comparable to the data mentioned in the Norm. This study focuses on the cost of irrigation, which appears to be a major expense for producers today. More specifically, the cost of irrigation for the crops amounts to 19%, 25%, 23%, and 5% of the variable costs, respectively, and to 12%, 19%, 17%, and 4% of the total costs, respectively. It is, therefore, indisputable that the cost of irrigation contributes to a reduction in the profit of the producer since it can contribute 20% to the formation of the cost of cultivation. A reduction in the cost of irrigation has positive inductive effects on the producer's profit, which is the final demand for producer–entrepreneurs.

Table 2. Percentage of marginal cost and profit in irrigation expenses.

Irrigation Cost as for	Potatoes	Beans	Groundnuts	Peas
Variable Cost (before)%	0.19	0.25	0.23	0.05
Total Cost (before)%	0.13	0.15	0.14	0.03
Variable Cost (after)%	0.12	0.19	0.17	0.04
Total Cost (after)%	0.08	0.11	0.10	0.02
Profit increase%	1.29	1.20	2.23	1.05

4. Conclusions

It is clear from the new data produced by this research project that the producer's profit increases. In addition to the benefit of the producer, however, at the microeconomic level, there are also social benefits. The use of satellite remote sensing in the field of irrigation and water resources management can contribute at a microeconomic level to the maximization of the producer's profit. The reduction in irrigation costs, using the optimal amounts of water for the needs of the crops, contributes dynamically to the profit margin of the producer while, at the same time, having a positive effect on the storage of surface water resources stored in the dams. All these positive research results fade out unless there is a proper advisory service that disseminates this knowledge. Although the Ministry of Agriculture, Rural Development and Environment in Cyprus has undertaken a great effort to establish an efficient system for agricultural knowledge and innovation (AKIS) to promote innovation in the agri-food sector, it should be improved by becoming even more integrated and powerful providing farmers better access to education and training programs.

Author Contributions: Conceptualization, G.P.; methodology, G.P.; software, G.K.; validation, G.P., D.H. and M.T.; formal analysis, G.P.; investigation, G.K.; resources, D.H.; data curation, G.P.; writing—original draft preparation, M.T.; writing—review and editing, G.P.; visualization, G.K.; supervision, G.P.; project administration, G.P. All authors have read and agreed to the published version of the manuscript.

Funding: This project was funded by the European Union. The views and opinions expressed are, however, those of the author(s) only and do not necessarily reflect those of the European Union or Research Executive Agency. Neither the European Union nor the granting authority can be held responsible for them.

Institutional Review Board Statement: Not applicable.

Informed Consent Statement: Not applicable.

Data Availability Statement: The data are contained within the article.

Acknowledgments: The authors would like to express their appreciation to the European Union under the CARBONICA project (HORIZON-WIDERA-2022).

Conflicts of Interest: The authors declare no conflict of interest.

References

1. Tsakiris, G.P.; Loucks, D.P. Adaptive Water Resources Management Under Climate Change: An Introduction. *Water Resour. Manag.* **2023**, *37*, 2221–2223. [CrossRef]
2. Kustas, W.P.; Jackson, R.D.; Asrar, G. Chapter 16: Estimating surface energy-balance components from remotely sensed data. In *Theory and Applications of Optical Remote Sensing*; Asrar, G., Ed.; John Wiley and Sons: New York, NY, USA, 1989; pp. 605–627.

3. Cyprus Meteorological Service. Cyprus Average Annual Precipitation 1901–2008. 2020. Available online: http://www.moa.gov.cy/moa/MS/MS.nsf/DMLclimet_reports_en/DMLclimet_reports (accessed on 15 July 2023).
4. Cyprus Meteorological Service. Monthly Rainfall in Cyprus during the Hydrometeorological Year 2008–2009 and 2009–2010. 2021. Available online: http://www.moa.gov.cy/moa/MS/MS.nsf/DMLmeteo_reports_en/DMLmeteo_reports (accessed on 20 July 2024).
5. Papadavid, G.; Hadjimitsis, D.G. Spectral signature measurements during the whole life cycle of annual crops and sustainable irrigation management over Cyprus using remote sensing and spectro-radiometric data: The cases of spring potatoes and peas. *Proc. SPIE Remote Sens. Agric. Ecosyst. Hydrol. XI* **2009**, *7472*, 747215. [CrossRef]
6. Bastiaanssen, W.G.M.; Menenti, M.; Feddes, R.A.; Holtslag, A.A.M. A remote sensing surface energy balance algorithm for land (SEBAL), part 1: Formulation. *J. Hydrol.* **1998**, *212–213*, 198–212. [CrossRef]
7. Markou, M.; Papadavid, G. *Norm Input-Output Data for the Main Crop and Livestock Enterprises of Cyprus*; Agricultural Research Institute: Nicosia, Cyprus, 2008.

Disclaimer/Publisher's Note: The statements, opinions and data contained in all publications are solely those of the individual author(s) and contributor(s) and not of MDPI and/or the editor(s). MDPI and/or the editor(s) disclaim responsibility for any injury to people or property resulting from any ideas, methods, instructions or products referred to in the content.

Proceeding Paper

Opinions and Perceptions on Sustainable Weed Management: A Comparison between Greek and Tunisian Farmers [†]

Efstratios Michalis [1,*], Ahmed Yangui [2], Athanasios Ragkos [1], Mohamed Kharrat [3] and Dimosthenis Chachalis [4]

1. Agricultural Economics Research Institute, Hellenic Agricultural Organization (ELGO-DIMITRA), Kourtidou 56-58, 111 45 Athens, Greece; ragkos@elgo.gr
2. Agricultural Economic Laboratory (LER), National Institute of Agronomic Research of Tunisia (INRAT), University of Carthage, Ariana 1004, Tunisia; yangui.ahmed@gmail.com
3. Field Crop Laboratory (LGC), National Institute of Agronomic Research of Tunisia (INRAT), University of Carthage, Ariana 1004, Tunisia; kharrat.mohamed@inrat.ucar.tn
4. Laboratory of Weed Science, Benaki Phytopathological Institute, Stefanou Delta 8, Kifisia, 145 61 Athens, Greece; d.chachalis@bpi.gr
* Correspondence: efstratiosmichalis@gmail.com
† Presented at the 17th International Conference of the Hellenic Association of Agricultural Economists, Thessaloniki, Greece, 2–3 November 2023.

Abstract: Societal awareness, demand for innovative food systems and increasing herbicide resistance have induced policy, regulatory and research actions towards the adoption of sustainable weed management, which is based on sustainable, integrated and ecological principles. The study investigates farmers' perceptions with regard to sustainable weed management, considering that the adoption of relevant practices depends on a set of farmer-specific and innovation-specific attributes. To achieve this purpose, an on-site survey was conducted in Greece and Tunisia based on a structured questionnaire, which was completed by 105 arable farmers in total. The questionnaire was designed to record farmers' opinions and preferences regarding aspects related to sustainable weed management, such as innovation and the decision making process. Using descriptive statistics methods, the study pinpointed significant differences between the responses of Greek and Tunisian farmers due to their particular needs and characteristics, suggesting thus the integration of targeted approaches towards the expansion of sustainable weed management.

Keywords: weeds; questionnaire survey; innovation; decision making

1. Introduction

Weeds constitute the most important biotic constraints to agricultural production in both developing and developed countries [1], as they compete with crops, leading to the overuse of natural resources and agricultural inputs. The reduction in crop productivity due to weeds is a major issue related to food security, taking into account the rapidly growing human population worldwide.

At the same time, herbicide resistance—due to misuse or overuse of chemical herbicides [2]—is considered one of the most serious challenges associated with weed management, as, by November 2018, resistance to herbicides had been reported in 255 weed species in 70 countries [3].

All these challenges have induced policy, regulatory and research actions towards sustainable weed management practices, which, however, are not always accepted and their adoption depends on a combination of farmer-specific and innovation-specific attributes.

The purpose of the study is to shed light on how farmers from two Mediterranean countries (Greece and Tunisia) perceive aspects related to sustainable weed management. Using descriptive statistics methods, the study showed that responses vary significantly between Greek and Tunisian farmers, especially in terms of their decision making.

2. Methods

2.1. Survey Profile

The study presents the results of an on-site survey of farmers—specialized in the cultivation of annual arable crops—in typical rural areas of Greece and Tunisia. For the purpose of the main analysis, a structured questionnaire comprising two parts was developed to record the perceptions, attitudes, motivations and aspirations of farmers in both countries regarding sustainable weed management. The first part of the questionnaire, as shown in Table 1, recorded the personal profile of the respondent (gender, age, education level). In the second part, participants were asked to respond to different sets of questions—using a 5-point Likert-scale ranging from 1 (=Totally disagree/Never) to 5 (=Totally agree/Very often)—aiming to evaluate their attitudes towards innovation and sustainable weed management practices. In total, 105 farmers were interviewed in Greece and Tunisia from June 2021 to August 2022.

Table 1. Respondents' profiles.

Variable	Frequency	Percentage (%)
Region		
Greece	61	58.1
Tunisia	44	41.9
Gender		
Male	99	94.3
Female	6	5.7
Age		
20–29	12	11.4
30–39	22	21.0
40–49	24	22.9
50–59	29	27.6
>60	18	17.1
Education		
Primary education	24	22.9
Secondary education	40	38.1
Technical graduate school	21	20.0
University education	20	19.0

2.2. Methodological Background

The methodological approach used to analyze the categorical (ordinal) data in this study involved a descriptive analysis of responses, aiming to acquire a general viewpoint of interviewees' opinions and attitudes. The Mann–Whitney non-parametric test was used to determine the level of significance of the differences between the responses provided by Greek and Tunisian farmers. This test can be applied for the ordinal data of two independent groups, without a normality assumption, to examine whether one variable has a higher value than the other [4]. The analysis in this study was performed with the statistical package SPSS, version 24.

3. Results and Discussion

Table 2 summarizes farmers' perceptions on innovations in agriculture. It seems that their beneficial role was acknowledged by respondents. In particular, the necessity of innovations received the highest attention, followed by their contribution to increasing farm productivity, their support to food security, their ability to improve standards of living as well as their conduciveness to the production of high-quality products. On the other hand, respondents were neutral regarding the contribution of innovations to environmental protection and about the guaranteed result of their application. Differences between Greek and Tunisian farmers were found with regard to the complicated use of innovations and the degree to which they are subsidized by the State. Especially for the latter, Greek farmers were significantly more negative.

Table 2. Respondents' perceptions regarding innovations in agriculture.

	Means		Medians	
	Greece	Tunisia	Greece	Tunisia
Innovations are necessary	4.29	4.45	4.00	5.00
Innovations are expensive	3.96	3.70	4.00	4.00
Innovations require training and specific knowledge	4.04	4.11	4.00	4.00
The use of innovations is complicated **	2.93	3.47	3.00	4.00
Innovations improve standards of living	3.86	3.93	4.00	4.00
Innovations increase farm productivity	4.11	3.88	4.00	4.00
The result of innovations is not guaranteed	3.27	2.88	3.00	3.00
Innovations are adequately subsidized by the State **	2.83	4.15	3.00	5.00
Young farmers tend to adopt innovations more easily	3.11	3.11	3.00	3.00
Innovations contribute to the production of high-quality products	3.88	3.86	4.00	4.00
Innovations contribute to environmental protection *	3.22	2.61	3.00	3.00
Innovations contribute to food security	3.98	3.75	4.00	4.00

* indicates significant difference between the medians of Greek and Tunisian farmers at the 5% level. ** indicates significant difference between the medians of Greek and Tunisian farmers at the 1% level.

Table 3 presents the means and medians of items describing sources of information based on which farmers make decisions regarding weed management. The vast majority of respondents make such decisions based on their own knowledge and expertise, which received by far the highest score. In contrast, training courses and seminars, public services, universities and research institutes, internet and other mass media were considered unreliable sources by farmers. Here, discrepancies between the responses of Greek and Tunisian farmers were notable in more items compared to their perceptions on innovations in agriculture. Greek farmers consulted other farmers, members of their family, internet as well as private advisors and agronomists more often than Tunisians. Finally, farmers from both countries ignored the role of public services, but Greeks did so to a higher degree.

Table 3. Respondents' decision making regarding weed management practices.

	Means		Medians	
	Greece	Tunisia	Greece	Tunisia
Based on my knowledge and expertise	4.55	4.65	5.00	5.00
I ask other farmers I trust **	2.75	1.84	3.00	2.00
I ask members of my family **	3.36	1.77	4.00	2.00
I attend training courses and seminars	1.70	1.65	2.00	2.00
Private advisors—Consultants/Agronomists **	3.85	1.47	4.00	1.00
Public services *	1.18	1.59	1.00	1.00
University/Research Institute *	1.60	1.31	1.00	1.00
Internet **	2.26	1.59	2.00	1.00
TV/Radio shows/Mass media/Books/Magazines *	1.60	1.25	1.00	1.00

* indicates significant difference between the medians of Greek and Tunisian farmers at the 5% level. ** indicates significant difference between the medians of Greek and Tunisian farmers at the 1% level.

4. Conclusions

The need to shift towards sustainable and environmentally friendly practices that can also ensure effective weed control is constantly gaining attention. Innovative practices, however, are not always accepted by farmers, as an outcome of their specific characteristics and requirements. The study detected significant differences between the responses of Greek and Tunisian farmers regarding innovations in agriculture and especially regarding their decision making about weed management.

These findings present orientations for strategic and policy design towards the expansion of sustainable weed management, taking into account that the diverse socio-economic profiles of farmers and their different attitudes towards innovation require targeted approaches.

Author Contributions: Conceptualization, E.M., A.Y. and A.R.; methodology, E.M.; validation, A.Y. and A.R.; formal analysis, E.M. and A.R.; investigation, E.M., A.Y. and M.K.; data curation, E.M.; writing—original draft preparation, E.M.; writing—review and editing, A.Y. and A.R.; supervision, M.K. and D.C.; funding acquisition, D.C. All authors have read and agreed to the published version of the manuscript.

Funding: This research is supported by the PRIMA program under the project "ZeroParasitic" (Section 2, 2018 Call). The PRIMA program is supported and funded under Horizon 2020, the Framework European Union's Program for Research and Innovation.

Institutional Review Board Statement: Not required according to Institutional Regulations.

Informed Consent Statement: Available by the authors.

Data Availability Statement: Data will be available to interested parties upon request from the corresponding author and "ZeroParasitic" project coordinator after the completion of the embargo period (two years after the completion of the project).

Acknowledgments: We would like to show our gratitude for the respondents contributing to the survey.

Conflicts of Interest: The authors declare no conflicts of interest. The funders had no role in the design of the study; in the collection, analyses, or interpretation of data; in the writing of the manuscript; or in the decision to publish the results.

References

1. Chauhan, B.S. Grant challenges in weed management. *Front. Agron.* **2020**, *1*, 3. [CrossRef]
2. Gage, K.L.; Krausz, R.F.; Walters, S.A. Emerging challenges for weed management in herbicide-resistant crops. *Agriculture* **2019**, *9*, 180. [CrossRef]
3. Moss, S. Integrated Weed Management (IWM): Why are farmers reluctant to adopt non-chemical alternatives to herbicides? *Pest Manag. Sci.* **2019**, *75*, 1205–1211. [CrossRef]
4. Hart, A. Mann-Whitney test is not just a test of medians: Differences in spread can be important. *Br. Med. J.* **2001**, *323*, 391–393. [CrossRef]

Disclaimer/Publisher's Note: The statements, opinions and data contained in all publications are solely those of the individual author(s) and contributor(s) and not of MDPI and/or the editor(s). MDPI and/or the editor(s) disclaim responsibility for any injury to people or property resulting from any ideas, methods, instructions or products referred to in the content.

Proceeding Paper

Fava Santorinis: Brining Added Value to a Protected Designation of Origin (PDO) Product through the Security of the Traditional Cultivar and Farmers Network [†]

Elissavet Ninou [1,*], Fokion Papathanasiou [2], Christos Alexandris [3], Elisavet Chatzivassiliou [4], Garyfallia Economou [5], Dimitrios Vlachostergios [6], Konstantinos Koutis [7], Anthoula Tsipi [8] and Ioannis Mylonas [8,*]

1. Department of Agriculture, International Hellenic University, 57400 Sindos, Greece
2. Department of Agriculture, University of Western Macedonia, 50100 Koila Kozanis, Greece; fpapathana-siou@uowm.gr
3. Cooperative Union of Thiraic Products, SantoWines Winery, Pyrgos Kalistis, 84700 Santorini, Greece; alexandris@santowines.gr
4. Plant Pathology Laboratory, Faculty of Crop Science, Agricultural University of Athens, Iera Odos 75, 11855 Athens, Greece; echatz@aua.gr
5. Laboratory of Agronomy, Faculty of Crop Science, Agricultural University of Athens, Iera Odos 75, 11855 Athens, Greece; economou@aua.gr
6. Institute of Industrial & Forage Crops, ELGO-DIMITRA, Kourtidou 56–58 & Nirvana, 11145 Athens, Greece; vlachostergios@elgo.gr
7. AEGILOPS—Network for Biodiversity and Ecology in Greece Ano Lechonia, 37300 Agria Volos, Greece; koutisresfarm@gmail.com
8. Institute of Plant Breeding and Genetic Resources, ELGO-DIMITRA, Kourtidou 56–58 & Nirvana, 11145 Athens, Greece; anthitsipi@gmail.com

* Correspondence: lisaninou@gmail.com or lisaninou@agr.teithe.gr (E.N.); ioanmylonas@yahoo.com (I.M.)
† Presented at the 17th International Conference of the Hellenic Association of Agricultural Economists, Thessaloniki, Greece, 2–3 November 2023.

Abstract: The characterization of "Fava Santorinis" as a PDO product does not protect the cultivated genetic material that produces this product, since this is not registered as a traditional cultivar in the National Common Catalogue. The failure to include this information presents a significant hazard to the genetic diversity of these cultivars, potentially resulting in the loss of their distinct traits, reduced crop yields, and quality. Furthermore, it seeks to comply with established procedures for characterizing and subsequently register this traditional cultivar in the National List of Varieties. The "Santorini Fava" (Lathyrus sp.) is a renowned agricultural product that is unique to Santorini, and it has played a pivotal role in upholding the island's traditional agriculture. Today, the local agricultural cooperation continues the cultivation of this crop, preserving it as an indispensable facet of the island's cultural heritage. The objective of the project M16SYN2-00135 is to guarantee and secure this indigenous variety, from which the PDO product in question originates, by applying official description protocols and making use of the existing know-how for the description of the genetic material, the definition of the landrace, and its description for registration in the National Catalog of Varieties. At the same time, the sustainable management of viral diseases and the rational management of its seed production will lead to an increase in productivity, its stabilization, and ultimately, its shielding. The product will be utilized by the cooperative contributing to the sustainability of the holdings and the prevention of commercial exploitation of the traditional variety beyond the area of origin based on the best practices for the preservation of the varieties.

Keywords: Fava Santorinis; *Lathyrus clymenum* L.; biodiversity protection; local varieties; added value

1. Introduction

"Fava Santorinis" is a PDO product [1] that is produced from the seeds of the botanical species *Lathyrus clymenum* L., a leguminous crop cultivated on the Cyclades islands in the South Aegean Sea; it is confirmed that the seeds of *Lathyrus clymenum* L. were found in archaeological residues dated back to the 16th century B.C. [2]. However, the PDO characterization does not protect the genetic material from which "Fava Santorinis" is produced, since this is not registered in the National Common Catalogue. This work is in alignment with the application of the EC 2008/62/EK (official Greek Gazette) FEK 165/30-01-2014 that provides the necessary regulations to protect the traditional cultivar. The legume Fava Santorinis (Lathyrus sp.) has all the attributes that have been qualified by researchers, such as Zeven [3], Camacho Villa et al. [4], and Newton et al. [5], for landraces. Observing the beginning of agriculture at the end of the 8th millennium BC., the archaeobotanical traces support the local origin and the continuous route of the lathyrus in the Aegean until the present day [6,7]. Presently, the local agricultural cooperative persists in cultivating this crop, safeguarding it as a vital element of the island's cultural legacy. It is imperative to highlight that without protective interventions, there is an inevitable risk of genetic erosion, endangering the ongoing cultivation of "Fava Santorinis" on the island. Currently, "Fava Santorinis" is cultivated on approximately 150 hectares, and its cultivation plays a pivotal role in the local agricultural community of Santorini, contributing significantly to the economy, which is estimated at EUR one million.

The aim of project M16SYN2-00135 is to ensure the protection and preservation of this native variety, from which the PDO product in question is derived. This will be achieved through the implementation of official description protocols and by leveraging existing expertise to characterize the genetic material.

2. Materials and Methods

The pilot study for the 1st year of the project focused on assessing the landraces and identifying plants that show improved productivity and quality and maintaining all the characteristics of the plants producing the products. The genetic material used was seed of the "Santorini Fava" variety, which came from the Association of Cooperatives of Theraic Products SANTO. The experiment was carried out on the farm of the Institute of Genetic Improvement and Plant Genetic Resources in Thessaloniki therme in the 2022–2023 growing season. In total, seeds were sown in 500 plant positions. The positions were 50 cm apart. All observations were taken at the individual plant level and related to a range of agronomic and descriptive traits (Table 1).

Table 1. Mean value and coefficient of variation (CV %).

Characteristic	Mean Value	CV (%)
Number of Pods/Plants	12.3	48.0
Pod Length (cm)	5.0	11.0
Pod Width (cm)	0.9	10.2
Pod Thickness (cm)	0.6	12.0
Number of Seeds per Pod	4.1	26.0
Seed Yield (g/plant)	6.1	78.0

Implementation methodology:

1. Study of genetic variability, description of the variety, and removal of deviating genotypes: In the pilot fields of min 4 ha, assessment of genetic variability will be conducted, and an improvement program will be implemented with a mild selection scheme to remove low-yielding plants that deviate and carry viral diseases and are not resistant in drought.

2. Compilation of file: At the final stage, a full description of the variety will be made as required by the protocol, and the file will be compiled for registration by the Cooperative. Even for a complete profile, molecular techniques and qualitative analyses of seeds will be conducted.

3. Training for proper seed production and production: This involves actions to train the cooperative's staff and farmers to produce high-quality healthy seed material and improve agricultural practices to produce the product.

3. Results and Discussion

Utilizing landraces for breeding purposes is a strategy that is employed to enhance both the yield and yield consistency within agricultural systems that are characterized by limited inputs [8]. The stagnation of yields in specific regions can largely be attributed to the restricted genetic diversity that is found in recently developed high-yielding varieties [9]. Consequently, the introduction of well-adapted germplasm from the primary centers of diversity for the crop can prove to be advantageous.

An essential prerequisite for enhancing a landrace is the identification of the existing genetic variability within that landrace. This step is crucial to establishing an effective breeding program aimed at improving landraces. The statistical measures for the performance components of "Fava Santorini" are given in Table 1. The greatest variability was observed in the characteristic of seed yield per plant, indicating the possibility of selection within the population for this characteristic, which would increase and stabilize the performance of the landrace. From the description, it was revealed that the "Fava Santorini" variety is an annual plant, reaching a height of 25–53 cm. The middle and upper leaves form two to six leaflets with a spiral arrangement. The green pods end in a curved tip. The dry pods have moderate constrictions, and the shape of the cross-section is elliptical. The seeds are brown or green in color, with a smooth surface and a spherical shape.

4. Conclusions

This multifaceted approach of the project M16SYN2-00321, funded in the context of the Agricultural Development Program 2014–2020 (Measure 16), and in particular Sub-Measure 16.1–16.2, will enhance the safeguarding and subsequent utilization of this valuable resource through:

- The registration and identification of the landrace that presents the PDO product based on the new legislation and EU directives by applying the description protocol.
- The definition of the mentioned protected variety and application for registration in the National List of Varieties.
- The establishment and implementation of an innovative framework/process for the dissemination of good conservation/seed production practices of the landrace in the region of origin to ensure certification and adequate purity of the seed. It will be implemented by the SANTO cooperation with the support of the researchers of different research institutes and agricultural universities.
- Authentication with morphological and qualitative characteristics and DNA techniques.
- Consulting services to improve farming techniques for farmers: field schools, e-learning, online applications, and networking through an online platform.
- Documentation of reduced product inputs.

Author Contributions: Conceptualization, E.N. and I.M.; methodology, E.N., I.M., C.A., E.C., G.E., F.P., D.V., K.K. and A.T.; validation, I.M., F.P., E.N. and A.T.; formal analysis, investigation, E.N., I.M., C.A., E.C., G.E., F.P., D.V., K.K. and A.T.; resources I.M., F.P., E.N. and A.T.; data curation, E.N., I.M., C.A., E.C., G.E., F.P., D.V., K.K. and A.T.; writing—original draft preparation, E.N., I.M., C.A., E.C., G.E., F.P., D.V., K.K. and A.T.; writing—review and editing, E.N., I.M., C.A., E.C., G.E., F.P., D.V., K.K. and A.T.; visualization, I.M., F.P. and E.N.; supervision E.N., I.M. and A.T.; project administration, E.N., I.M. and A.T.; funding acquisition, E.N., I.M., C.A., E.C., G.E., F.P., D.V., K.K. and A.T. All authors have read and agreed to the published version of the manuscript.

Funding: This work was funded in the context of the Agricultural Development Program 2014–2020 (Measure 16), and in particular Sub-Measure 16.1–16.2. Establishment and operation of Operational Groups of the European Innovation Partnership for the productivity and sustainability of agriculture (Project No M16SYN2-00135).

Institutional Review Board Statement: Not applicable.

Informed Consent Statement: Not applicable.

Data Availability Statement: Data are only available on request due to privacy restrictions.

Conflicts of Interest: The authors declare no conflict of interest.

References

1. Available online: https://eur-lex.europa.eu/legal-content/EN/TXT/?uri=CELEX:32010R0901&qid=1706524114607 (accessed on 1 September 2023).
2. Korpetis, E.; Ninou, E.; Mylonas, I.; Ouzounidou, G.; Xynias, I.N.; Mavromatis, A.G. Bread Wheat Landraces Adaptability to Low-Input Agriculture. *Plants* **2023**, *12*, 2561. [CrossRef] [PubMed]
3. Villa, T.C.C.; Maxted, N.; Scholten, M.; Ford-Lloyd, B. Defining and Identifying Crop Landraces. *Plant Genet. Resour.* **2005**, *3*, 373–384. [CrossRef]
4. Newton, A.C.; Akar, T.; Baresel, J.P.; Bebeli, P.J.; Bettencourt, E.; Bladenopoulos, K.V.; Czembor, J.H.; Fasoula, D.A.; Katsiotis, A.; Koutis, K.; et al. Cereal Landraces for Sustainable Agriculture. A Review. *Agron. Sustain. Dev.* **2010**, *30*, 237–269. [CrossRef]
5. Valamoti, S.M. Grain versus Chaff: Identifying a Contrast between Grain-Rich and Chaff-Rich Sites in the Neolithic of Northern Greece. *Veg. Hist. Archaeobotany* **2005**, *14*, 259–267. [CrossRef]
6. Kotsakis, S.-M.V. Kostas Transitions to Agriculture in the Aegean: The Archaeobotanical Evidence. In *The Origins and Spread of Domestic Plants in Southwest Asia and Europe*; Routledge: London, UK, 2007; ISBN 978-1-315-41761-5.
7. Available online: https://ec.europa.eu/agriculture/eambrosia/geographical-indications-register/details/EUGI00000013927 (accessed on 1 September 2023).
8. Pecetti, L.; Boggini, G.; Gorham, J. Performance of Durum Wheat Landraces in a Mediterranean Environment (Eastern Sicily). *Euphytica* **1994**, *80*, 191–199. [CrossRef]
9. Annicchiarico, P.; Pecetti, L. Yield vs. Morphophysiological Trait-Based Criteria for Selection of Durum Wheat in a Semi-Arid Mediterranean Region (Northern Syria). *Field Crops Res.* **1998**, *59*, 163–173. [CrossRef]

Disclaimer/Publisher's Note: The statements, opinions and data contained in all publications are solely those of the individual author(s) and contributor(s) and not of MDPI and/or the editor(s). MDPI and/or the editor(s) disclaim responsibility for any injury to people or property resulting from any ideas, methods, instructions or products referred to in the content.

Proceeding Paper

An Empirical Investigation of Ethical Food Choices: A Qualitative Research Approach [†]

Georgios Roumeliotis [1,*], Elena Raptou [1], Konstantinos Polymeros [2] and Konstantinos Galanopoulos [1]

[1] Department of Agricultural Development, Democritus University of Thrace, 68200 Orestiada, Greece; elenra@agro.duth.gr (E.R.); kgalanop@agro.duth.gr (K.G.)
[2] Department of Food Science and Nutrition, University of Thessaly, 43100 Karditsa, Greece; polikos@uth.gr
* Correspondence: geroumel@agro.duth.gr
[†] Presented at the 17th International Conference of the Hellenic Association of Agricultural Economists, Thessaloniki, Greece, 2–3 November 2023.

Abstract: Why do customers incorporate concerns about social and environmental issues into the decision-making process? How ethical are food choices in the modern world? Answers to these questions have often revolved around how informed consumers might be and whether they have the appropriate skills to act on concerns they might have. Today, ethical food consumption is a growing market where consumers' behavior shifts from the rational manner focusing on the products price and attributes to the food ethics associated with environment, social welfare, public health, and morality. Using data selected from a purposive sample of 20 consumers, this study employed a qualitative research procedure to explore the main dimensions that influence the decision-making process and eating preferences in the post-COVID 19 era and within an economically turbulent environment. The main results showed that health protection, sustainability, and social wefare constitute the main axes of ethical food consumption. Participants were found to be more individualists than altruists since the "personal health" dimension was the most prevalent. Future research should extend these findings and explore variations in the ethical consumption factors among various consumer segments.

Keywords: ethical food choice; personal in-depth interviews; health protection; environmental protection; social welfare

1. Introduction

Climate change impacts are now visible all over the world through numerous physical and biological changes in agriculture, environment, and society [1]. These have contributed to the urgent need for enhancing sustainable development and the necessity for efficient coordination between governance and international societal systems [2]. Since modern food systems are estimated to produce about a third of global greenhouse gas emissions [3], there is an ever-increasing interest in ethical issues in agricultural production and the food industry with consumers becoming more environmentally conscious and susceptible to the social impacts of agro-food production [4,5]. In addition, the rapid technological development of recent decades and the modern consumer lifestyle in the western world have brought to the forefront new ethical dilemmas and concerns [6,7]. Therefore, the interlinkage of agriculture with environment, food seasonality and locality, farmers and employees' rights, animal welfare, and the protection of local businesses seem to have a crucial role in establishing ethical consumer's food choices and purchase behaviors [5,8,9].

The present study delved into the various dimensions of ethical consumption, namely the health, the environmental, and the socioeconomic dimension [10], to explore consumer ethics and values that influence the decision-making process and eating preferences. Furthermore, it shed light onto the main aspects of food choice motives in Greece in the post-COVID 19 era and within an economically turbulent environment.

2. Materials and Methods

A qualitative research procedure was adopted, and primary data were selected through personal in-depth interviews based on a semi-structured questionnaire with open-ended questions [11]. To achieve the objectives of the present research, a non-probabilistic purposive sample of 20 adult consumers from Northern Greece (Eastern Macedonia and Thrace area) was employed in order to reach data saturation and achieve the objectives of the present research [12,13]. All personal in-depth interviews were tape-recorded, and the interview duration ranged from 50 to 90 min. Health, environment, economy, and society were the main axes for the construction of the informal questionnaire according to the recent body of literature [14–17]. Data selection started in June 2021 and lasted for approximately three months.

3. Results

Our sample covered a wide range of sociodemographic characteristics. In particular, respondents were 18–83 years old, and the great majority were women over 35 years old (14 women out of 20 participants).

The main findings showed that consumers identify "ethical consumption" with "consumer health", "environment (sustainability)", and "social development" (benefits for the society and economy). Apparently, most participants (19 out of 20 consumers) interrelated ethical foods with pesticide- and chemical fertilizer-free products because "they seem to be healthier", whereas five respondents linked such foods with environmental protection and sustainability. Seasonality in food consumption and locality (locally produced foods) were also underlined by a significant proportion of respondents that seemed to associate these perceptions with better quality in food products and more nutritious food choice, whereas the most prominent selection criteria was health protection. Less promoted food brands were also considered more ethical choices for consumers, because, as respondents explained, they provide more "natural products" to the food supply chain, and food items are produced with less environmentally invasive methods.

Although the main factors underlying ethical consumption are "health protection", "environment (sustainability)", and "benefits for the society/economy", consumers seem to be more individualists than altruists with "personal health protection" selection criteria outweighing the "environmental protection" and the "social welfare" criteria. Given that the qualitative research procedure was conducted after the COVID-19 isolation measures, consumers' increased awareness toward personal health protection at the expense of the other two criteria could be justified and explained.

Our results further support the previous literature on the values and motives driving ethical food choice, stating individual's health [18], environmental protection and sustainability [5,9,19,20], and social welfare [19,21] as the key axes of ethical consumption.

4. Discussion

Ethical consumption of agro-food products is a multi-dimensional process. Our findings further support recent research indicating that the main motives for ethical food consumption include individual/personal factors, environmental preservation, and social well-being. However, it seems that Greek consumers are less informed on food ethics since they were found to mostly identify ethical consumption motives with self-advancement, and more specifically personal health protection. Ethical consumption motives associated with environmental protection and society were less prevalent and reported by a smaller proportion of consumers, highlighting the necessity for educating customers on the different aspects/dimensions of food ethics and their implications for action in the modern food systems. Of particular interest is the fact that none of the participants correlate their food choices with animal welfare or social welfare.

These findings will be elaborated in a subsequent quantitative research design to provide a thorough picture of the values, the motivation process, and perceptions toward ethical food choices and build the profile of "ethical consumer". Future research should also take into consideration these findings and explore variations in ethical perceptions and food consumption patterns among various consumer segments.

Author Contributions: Conceptualization, G.R. and E.R.; methodology, G.R. and E.R.; investigation, G.R.; data curation, G.R.; writing—original draft preparation, G.R. and E.R.; writing—review and editing, K.P. and K.G. All authors have read and agreed to the published version of the manuscript.

Funding: This research received no external funding.

Institutional Review Board Statement: Not applicable.

Informed Consent Statement: Informed consent was obtained from all subjects involved in the study.

Data Availability Statement: The data that support the findings of this study are available from the corresponding author upon reasonable request.

Conflicts of Interest: The authors declare no conflict of interest.

References

1. Okon, E.M.; Falana, B.M.; Solaja, S.O.; Yakubu, S.O.; Alabi, O.O.; Okikiola, B.T.; Awe, T.E.; Adesina, B.T.; Tokula, B.E.; Kipchumba, A.K.; et al. Systematic review of climate change impact research in Nigeria: Implication for sustainable development. *Heliyon* **2021**, *7*, e07941. [CrossRef] [PubMed]
2. Fuso Nerini, F.; Sovacool, B.; Hughes, N.; Cozzi, L.; Cosgrave, E.; Howells, M.; Tavoni, M.; Tomei, J.; Zerriffi, H.; Milligan, B. Connecting climate action with other Sustainable Development Goals. *Nat. Sustain.* **2019**, *2*, 674–680. [CrossRef]
3. Jameel, S. Climate change, food systems and the Islamic perspective on alternative proteins. *Trends Food Sci. Technol.* **2023**, *138*, 480–490. [CrossRef]
4. Hoek, A.C.; Malekpour, S.; Raven, R.; Court, E.; Byrne, E. Towards environmentally sustainable food systems: Decision-making factors in sustainable food production and consumption. *Sustain. Prod. Consum.* **2021**, *26*, 610–626. [CrossRef]
5. Vasco, C.; Salazar, D.; Cepeda, D.; Sevillano, G.; Pazmiño, J.; Huerta, S. The Socioeconomic Drivers of Ethical Food Consumption in Ecuador: A Quantitative Analysis. *Sustainability* **2022**, *14*, 13644. [CrossRef]
6. Caruana, R.; Gloser, S.; Eckhardt, G.M. 'Alternative Hedonism': Exploring the Role of Pleasure in Moral Markets. *J. Bus. Ethics* **2020**, *166*, 143–158. [CrossRef]
7. Yilmaz, L. A quantum cognition model for simulating ethical dilemmas among multi-perspective agents. *J. Simul.* **2020**, *14*, 98–106. [CrossRef]
8. Paredes, M.; Cole, D.C.; Muñoz, F.; April-Lalonde, G.; Valero, Y.; Prado Beltrán, P.; Boada, L.; Berti, P.R. Assessing Responsible Food Consumption in Three Ecuadorian City Regions. In *Sustainable Food System Assessment*; Blay-Palmer, A., Conaré, D., Meter, K., Battista, A.D., Johnston, C., Eds.; Routledge: London, UK, 2019; pp. 195–215.
9. Tomșa, M.M.; Romonți-Maniu, A.I.; Scridon, M.A. Is sustainable consumption translated into ethical consumer behavior? *Sustainability* **2021**, *13*, 3466. [CrossRef]
10. Cecchini, L.; Torquati, B.; Chiorri, M. Sustainable agri-food products: A review of consumer. *Agric. Econ.* **2018**, *64*, 554–565.
11. Strauss, A.; Corbin, J. *Basics of Qualitative Research: Techniques and Procedures for Developing Grounded Theory*; Sage Publications: London, UK, 1998.
12. Guest, G.; Namey, E.; Chen, M. A simple method to assess and report thematic saturation in qualitative research. *PLoS ONE* **2020**, *15*, e0232076. [CrossRef] [PubMed]
13. Patton, M. *Qualitative Research & Evaluation Methods: Integrating Theory and Practice*, 4th ed.; Sage Publications: Thousand Oaks, CA, USA, 2015.
14. Djafarova, E.; Foots, S. Exploring ethical consumption of generation Z: Theory of planned behavior. *Young Consumers* **2022**, *23*, 413–431. [CrossRef]
15. Kushwah, S.; Dhir, A.; Sagar, M. Understanding consumer resistance to the consumption of organic food. A study of ethical consumption, purchasing, and choice behavior. *Food Qual. Prefer.* **2019**, *77*, 1–14. [CrossRef]
16. López-Fernández, A.M. Price sensitivity versus ethical consumption: A study of Millennial utilitarian consumer behavior. *J. Mark. Anal.* **2020**, *8*, 57–68. [CrossRef]
17. Pinna, M. Do gender identities of femininity and masculinity affect the intention to buy ethical products? *Psychol. Mark.* **2020**, *37*, 384–397. [CrossRef]
18. Apaolaza, V.; Hartmann, P.; D'Souza, C.; López, C.M. Eat organic–Feel good? The relationship between organic food consumption, health concern and subjective wellbeing. *Food Qual. Prefer.* **2018**, *63*, 51–62. [CrossRef]
19. Papoutsis, G.; Noulas, P.; Tsatoura, K. Animals or Humans: What Do Greek Consumers Care More about When Buying Feta Cheese? *Sustainability* **2023**, *15*, 316. [CrossRef]

20. Thi Khanh Chi, N. Ethical consumption behavior towards eco-friendly plastic products: Implication for cleaner production. *Clean. Responsible Consum.* **2022**, *5*, 100055.
21. Christiaensen, L.; Rutledge, Z.; Taylor, E. Viewpoint: The future of work in agri-food. *Food Policy* **2021**, *99*, 101963. [CrossRef] [PubMed]

Disclaimer/Publisher's Note: The statements, opinions and data contained in all publications are solely those of the individual author(s) and contributor(s) and not of MDPI and/or the editor(s). MDPI and/or the editor(s) disclaim responsibility for any injury to people or property resulting from any ideas, methods, instructions or products referred to in the content.

Proceeding Paper

Unraveling the Research Landscape of Happiness through Agro, Agri, and Rural Tourism for Future Directions [†]

Sofia Karampela *, Aigli Koliotasi and Konstantinos Kostalis

Department of Tourism, Ionian University, 49100 Corfu, Greece; aspkoliotasi@ionio.gr (A.K.); k.kostalis@gmail.com (K.K.)
* Correspondence: skarampela@ionio.gr
[†] Presented at the 17th International Conference of the Hellenic Association of Agricultural Economists, Thessaloniki, Greece, 2–3 November 2023.

Abstract: "Agro", "agri", and "rural" tourism have gained significant attention as emerging forms of tourism that provide unique experiences rooted in agricultural and rural settings. Beyond their economic and cultural contributions, these forms of tourism have been found to have a profound impact on individual happiness and well-being. This piece of work delves into the mechanisms underlying the relationship between "agro", "agri", or "rural" tourism and happiness, drawing from research in sociology and environmental science. By understanding the science behind this connection, we can further promote the development and implementation of "agro", "agri", and "rural" tourism initiatives that foster happiness and well-being. This study aims to examine existing research on "agro", "agri", or "rural" tourism and happiness, assess the implications of relevant scientific articles, and identify potential areas for future research. A systematic process was employed to identify articles related to terms such as "agrotourism", "agro tourism", "agro-tourism", "agritourism", "agri tourism", "agri-tourism", or "rural tourism" and happiness in the Scopus database. The selection criteria focused on articles that explored the above terms in their titles, abstracts, and keywords. The findings equally rely on qualitative and quantitative assessments, predominantly from the demand side, followed by the supply side and residents' views.

Keywords: agrotourism; agritourism; rural tourism; happiness; literature

Citation: Karampela, S.; Koliotasi, A.; Kostalis, K. Unraveling the Research Landscape of Happiness through Agro, Agri, and Rural Tourism for Future Directions. *Proceedings* **2024**, *94*, 51. https://doi.org/10.3390/proceedings2024094051

Academic Editor: Eleni Theodoropoulou

Published: 20 February 2024

Copyright: © 2024 by the authors. Licensee MDPI, Basel, Switzerland. This article is an open access article distributed under the terms and conditions of the Creative Commons Attribution (CC BY) license (https://creativecommons.org/licenses/by/4.0/).

1. Introduction

"Agro", "agri", and "rural" tourism offer tourists an opportunity to engage with agricultural activities, explore natural landscapes, and experience the rural way of life. From the perspective of supply, it is the result of urbanization development, the improvement of residents' income, and the optimization of tourism product structure; from the perspective of demand, it is the psychological demand of citizens to escape from urban pollution and fast-paced lifestyles and return to the countryside [1]. While previous studies have highlighted the economic benefits and cultural significance of these forms of tourism, the exploration of their impact on happiness is relatively novel. Happiness is increasingly used by social scientists as a synonym for a subjective enjoyment of life, while psychologists formally refer to this construct as subjective well-being [2]. Understanding the mechanisms that link agro, agri, and rural tourism with happiness can help to guide policymakers, tourism planners, and stakeholders in creating environments that maximize well-being outcomes for tourists and local communities. Engaging in agricultural activities as part of "agro", "agri", and "rural" tourism experiences has been linked to various benefits. Participating in farming activities such as planting, harvesting, and interacting with animals can promote a sense of accomplishment, self-efficacy, and mindfulness, and can stimulate positive emotions and evoke nostalgic memories, further contributing to happiness.

2. Materials and Methods

In order to explore the various viewpoints in scientific research pertaining to "agro", "agri", or "rural" tourism and happiness, a systematic approach was employed. Firstly, specific title and keyword criteria were established to search for relevant documents on these forms of tourism. Terms such as "agrotourism", "agro tourism", "agro-tourism", "agritourism", "agri tourism", "agri-tourism", or "rural tourism" and happiness were utilized to search the title, abstract, and keywords of scientific articles in the Scopus database. The search was conducted in mid-2023 and provided just 13 documents. Subsequently, the resulting most-referred keywords from these articles were presented (based on the idea of [3,4], enriched by the authors). Additionally, the journals in which these articles were published and the countries of origin of the authors were identified. During the next phase, the full papers were thoroughly examined and classified under the following themes: various factors influencing the (a) supply, (b) demand, and (c) residents within the tourism sector; the (d) countries in which the case studies were conducted; and the research methods employed, including both (e) qualitative and (f) quantitative approaches. It is important to note that these themes and approaches are not mutually exclusive, as each paper may be classified under multiple themes and/or approaches.

3. Results and Discussion

Based on the analysis of the keywords extracted from scientific papers (see summarized Table 1), it can be observed that a significant proportion of keywords include geographic information, such as specific countries, regions, and characteristics of the selected case study areas. Additionally, various forms of tourism are discussed and the terminology is used interchangeably, including terms like agritourism and rural and farm tourism. Furthermore, there appeared keywords related to emerging trends, such as value co-creation, mindfulness and memory, experience and experiential satisfaction, and quality, with a particular emphasis on the economic dimensions. Surprisingly, certain topics like sustainability, environment/ecology, and planning are under-represented in the scientific literature.

Table 1. Keywords from "Agro-", "agri-", or "rural" tourism and happiness articles in the Scopus database (May 2023).

Keywords in Categories	Number of Results (N)
Countries/regions/geographic position characteristics	10
Form(s) of tourism	8
Happiness	8
Methods	8
"New trend" keywords	7
Experience and experiential satisfaction and quality	7
Economics	7
Agriculture	6
Rural/regional development	6
Tourism management and development	5
Community	5
Rural Development and Experience Economy	4
Rural Areas and Environments	4
Sustainability	3
Farm	3
Environment/ecology	3
Planning	2
Gender	2
Behavior	2
Total	100

Source: https://www.scopus.com (accessed on 2 May 2023), processed by the authors.

The majority of research papers focusing on "agro-," "agri-", or "rural" tourism and happiness are predictably published in tourism journals. In Table 2, the authors' country affiliations are presented, revealing China at the first position with five authors, followed by Portugal, Spain, and the USA with two authors. At the next phase of our analysis, an important finding emerged. The articles use qualitative and quantitative methods at the same rate, particularly focusing on the demand side of these special interest forms of tourism, followed by the supply side and resident views, with China, Spain, the USA, Turkey, Croatia, Hungary, Romania, and Bhutan constituting the case study countries. Moreover, the papers predominantly examine single case studies, with a notable absence of comparisons between different cases (with the exception of [5]).

Table 2. "Agro-", "agri-", or "rural" tourism and happiness papers by country of authors' affiliation.

Country of Authors' Affiliation	Number of Results (N)
China	5
Portugal	2
Spain	2
United States	2
Bhutan	1
Finland	1
Hungary	1
Malaysia	1
Romania	1
Taiwan	1
Turkey	1
United Kingdom	1
Total	19

Source: https://www.scopus.com (accessed on 2 May 2023), processed by the authors.

4. Conclusions

This study explored the scientific articles from the Scopus database that examined the relationship between "agro", "agri", or "rural" tourism and happiness. The number of publications has increased in the last five years. The results suggest that these dimensions of tourist experience are positively and significantly associated with happiness [6], and also underscore the role of education in farmers' happiness, with higher levels of education being associated with higher levels of happiness [7].

Author Contributions: Conceptualization, S.K.; methodology, S.K.; validation, S.K., A.K. and K.K.; formal analysis, S.K.; investigation, S.K.; data curation, S.K., A.K. and K.K.; writing—original draft preparation, S.K.; writing—review and editing, S.K., A.K. and K.K.; supervision, S.K. All authors have read and agreed to the published version of the manuscript.

Funding: This research received no external funding.

Informed Consent Statement: Not applicable.

Conflicts of Interest: The authors declare no conflicts of interest.

References

1. Zhou, G.; Chen, W. Agritourism experience value cocreation impact on the brand equity of rural tourism destinations in China. *Tour. Rev.* **2023**, *78*, 1315–1335. [CrossRef]
2. Nawijn, J.; Mitas, O. Resident attitudes to tourism and their effect on subjective well-being: The case of Palma de Mallorca. *J. Travel Res.* **2012**, *51*, 531–541. [CrossRef]
3. Lane, B.; Kastenholz, E. Rural tourism: The evolution of practice and research approaches–Towards a new generation concept? *J. Sustain. Tour.* **2015**, *23*, 1133–1156. [CrossRef]
4. Karampela, S.; Andreopoulos, A.; Koutsouris, A. "Agro", "Agri", or "Rural": The different viewpoints of tourism research combined with sustainability and sustainable development. *Sustainability* **2021**, *13*, 9550. [CrossRef]

5. Smith, M.; Slusariuc, G.C. A cross-cultural comparison of generation Y attitudes to nature, wellbeing and rural tourism. In *Strategic Tools and Methods for Promoting Hospitality and Tourism Services*; Nedelea, A.-M., Korstanje, M., George, B., Eds.; IGI Global: Hershey, PA, USA, 2016; pp. 159–178. [CrossRef]
6. Loureiro, S.M.; Breazeale, M.; Radic, A. Happiness with rural experience: Exploring the role of tourist mindfulness as a moderator. *J. Vacat. Mark.* **2019**, *25*, 279–300. [CrossRef]
7. Aydoğdu, M.H.; Cançelik, M.; Sevinç, M.R.; Çullu, M.A.; Yenigün, K.; Küçük, N.; Karlı, B.; Ökten, Ş.; Beyazgül, U.; Doğan, H.P.; et al. Are you happy to be a farmer? Understanding indicators related to agricultural production and influencing factors: GAP-Sanliurfa, Turkey. *Sustainability* **2021**, *13*, 12663. [CrossRef]

Disclaimer/Publisher's Note: The statements, opinions and data contained in all publications are solely those of the individual author(s) and contributor(s) and not of MDPI and/or the editor(s). MDPI and/or the editor(s) disclaim responsibility for any injury to people or property resulting from any ideas, methods, instructions or products referred to in the content.

 proceedings

Proceeding Paper

The Impact of the Improved Genetic Material to the Economic Value of Plake Fasoli Prespon PGI Product †

Elissavet Ninou [1,*], Fokion Papathanasiou [2], Iosif Sistanis [2], Anastasia Kargiotidou [3], Sonia Michailidou [4], Konstantinos Koutis [5], Anthoula Tsipi [6] and Ioannis Mylonas [6,*]

1. Department of Agriculture, International Hellenic University, 57400 Sindos, Greece
2. Department of Agriculture, University of Western Macedonia, 50100 Florina, Greece; fpapathana-siou@uowm.gr (F.P.); aff00018@uowm.gr (I.S.)
3. Institute of Industrial & Forage Crops-ELGO DIMITRA, Kourtidou 56-58 & Nirvana, 11145 Athens, Greece; kargiotidou@elgo.gr
4. Agricultural Cooperative of Bean Producers of the Prespa National Forest "PELEKANOS", 53007 Laimos Florinas, Greece; s.mixailidou.1@gmail.com
5. AEGILOPS—Network for Biodiversity and Ecology in Greece Ano Lechonia, 37300 Agria Volos, Greece; koutisresfarm@gmail.com
6. Institute of Plant Breeding and Genetic Resources-ELGO DIMITRA, Kourtidou 56-58 & Nirvana, 11145 Athens, Greece; anthitsipi@gmail.com
* Correspondence: lisaninou@gmail.com or lisaninou@agr.teithe.gr (E.N.); ioanmylonas@yahoo.com (I.M.)
† Presented at the 17th International Conference of the Hellenic Association of Agricultural Economists, Thessaloniki, Greece, 2–3 November 2023.

Abstract: The project aims to safeguard this local variety by comprehensively studying its genetic variability. Furthermore, it seeks to follow official protocols for the description and subsequent registration of the variety in the National List of Varieties, increasing the product's value and securing its identity. Experimentation targets evaluation of the landrace to select plants with improved productivity and quality. The profit from implementing the program will come from a combination of higher productivity due to the use of improved genetic material, improved consulting services related to agricultural techniques, and increased values due to higher prices due to authenticating the product. This initiative aspires to provide benefits for the Agricultural Cooperative of Bean producers of the Prespes area. At the same time, the farmers will be trained for good seed reproduction and production of the landrace.

Keywords: Plake Fasoli Prespon; PGI; local landrace; added value

1. Introduction

In Greece, beans assume a central role among pulse crops, forming the cornerstone of the traditional Mediterranean diet. Over recent years, the cultivated area dedicated to beans has seen a noticeable expansion [1]. The common bean enjoys a prestigious status as a globally esteemed legume, offering a substantial source of high-quality proteins, carbohydrates, vitamins, minerals, dietary fiber, phytonutrients, and antioxidants, all vital for human nutrition. These components have been recognized for their significant positive impacts on human health. Consequently, the common bean holds promise as a valuable functional food [2,3]. These beans are typically grown for their dried seeds during the spring and summer seasons.

The primary region for their cultivation is located in northern Greece, particularly in areas characterized by relatively high altitudes and cool climatic conditions like Prespa area. Belonging to the Fabaceae family, common beans and lathyrus both play essential roles in promoting sustainable agriculture due to their ability to fix atmospheric nitrogen. This capability reduces the need for excessive fertilizer applications, contributing to environmentally friendly farming practices.

The designation of Plake Fasoli Prespon as a Protected Geographical Indication (PGI) product does not provide protection for the cultivated genetic material because it is not registered as a traditional cultivar in the National Catalog of Varieties [EC 2008/62/EK (official Greek Gazette) FEK 165/30-01-2014]. This oversight contributes to the erosion of the variety's genetic diversity, with the associated risks of losing its distinct identity, reduced yields, and compromised quality. The project's (M16SYN2-00181) primary objective is to safeguard this local variety by comprehensively studying its genetic variability. Furthermore, it aims to follow official protocols for the description and subsequent registration of the variety in the National List of Varieties, which will increase the value of the product and secure its identity.

2. Material and Methods

Three research Institutions (ELGO-Dimitra, University of Western Macedonia, and International Hellenic University), the local Agricultural Community represented by the Agricultural Cooperative of Florina "Pelekanos", an NGO Aegilops, and an Advisor (Tsipi Anthoula) are cooperating under the PAA M16.1–16.2 project (M16SYN2-00181) to achieve this goal. he project's first-year inception phase revolved around the examination of genetic diversity within landraces. The primary objective was to identify plants that exhibited enhanced productivity and quality while retaining all the essential plant characteristics required for product production. To facilitate this research, the genetic materials utilized were seeds from "Fasoli Prespon", sourced from the "Pelekanos" Agriculture cooperative. The experiment was conducted during the 2022–2023 growing season and was carried out at the University of Western Macedonia-Department of Agriculture's farm, as illustrated in Figure 1. A total of 600 plant positions were established for each landrace in low plant density. All observations were made at the individual plant level and pertained to a range of agronomic and physiological traits.

Figure 1. Field experiment of common bean in low plant density.

3. Results and Discussion

The designation of Plake Fasoli Prespon as a Protected Geographical Indication (PGI) product does not provide protection for the cultivated genetic material because it is not registered as a traditional cultivar in the National Catalog of Varieties [EC 2008/62/EK (official Greek Gazette) FEK 165/30-01-2014]. This oversight contributes to the erosion of the variety's genetic diversity, with the associated risks of losing its distinct identity, reduced yields, and compromised quality. Following the primary objective of this project, e.g., the study of genetic variability, it was found that yield ranged from 50 to 500 g per plant, the number of pods from 30 to 400 per plant, flowering began from 48 to 63 days from sowing and continued to the end of growing season. Similar variability was found for physiological characteristics of chlorophyll and photosynthesis measurements.

The cultivation in question is of great importance to the local agricultural community in Prespes area, making a substantial contribution to the local economy. It is estimated that Plake Fasoli Prespon is cultivated on ~300 ha, with a yield of 3000 Kg/ha. and price of ~2.8 €/Kg, respectively, with an essential economic contribution (2.52 million) to the local Agricultural community of Western Macedonia. The expected advantages of implementing this program are estimated to result in a crucial annual profit. This boost in profit will be the outcome of several factors, including enhanced productivity resulting from the use of improved genetic materials, more effective consulting services, and an increased market value due to the verified origin and quality of the product.

4. Conclusions

This work will present the parameters connected with the description of the unique identity of this product, its origin, traceability, local agricultural practices, and specific product characteristics that will contribute to this. The product will be utilized by the Agricultural Cooperative of Florina, and at the same time, the farmers will be trained for the good seed reproduction and production of the product. This initiative promises several benefits for the Agricultural cooperative and producers of the Prespes area.

The comprehensive strategy of project M16SYN2-00181, funded within the framework of the Agricultural Development Program 2014–2020 (under Measure 16), with a specific focus on Sub-Measure 16.1–16.2, is designed to significantly enhance the preservation and effective utilization of a valuable agricultural resource. This endeavor involves a multi-faceted approach that encompasses the following key actions:

1. Registration and Identification: The project will diligently describe and identify the landrace, ensuring compliance with the latest legislative guidelines and EU directives. A meticulous description protocol will be employed for this purpose.
2. Protected Variety Designation: A crucial step involves officially designating the landrace as a protected variety and initiating the process of its registration in the National List of Varieties.
3. Innovative Conservation and Seed Production: The project will pioneer an innovative framework and process for disseminating best practices in conservation and seed production within the landrace's region of origin. This initiative will ensure the preservation of the landrace and guarantee certification and seed purity. Collaboration with "Pelekanos" Agricultural cooperation and the valuable support of researchers from esteemed research institutes and Agricultural Universities will be instrumental in this effort.
4. Authentication: The project will employ a rigorous approach to authenticate the landrace. This will involve a detailed examination of morphological and qualitative characteristics and advanced DNA techniques to establish and confirm the landrace's unique identity.
5. Consulting Services for Farmers: The project is dedicated to providing consulting services to improve farming techniques among local farmers. This will encompass a range of educational methods, including field schools, e-learning opportunities, user-friendly online applications, and the creation of a supportive online networking platform.
6. Reduced Input Documentation: The project will diligently document and demonstrate the reduction in product inputs as a result of the implemented measures, showcasing the economic and environmental benefits of the program.

Author Contributions: Conceptualization, I.M., F.P. and E.N.; methodology, F.P., E.N., I.M., I.S., A.K., S.M., K.K. and A.T; validation, I.M., F.P., E.N. and A.T., formal analysis, F.P., E.N., I.M., I.S., A.K., S.M., K.K. and A.T; investigation, I F.P., E.N., I.M., I.S., A.K., S.M., K.K. and A.T; resources I.M., F.P. and E.N.; data curation, F.P., E.N., I.M., I.S., A.K., S.M., K.K. and A.T.; writing—original draft preparation, F.P., E.N., I.M., I.S., A.K., S.M., K.K. and A.T; writing—review and editing, I.M., F.P. and E.N.; visualization, I.M., F.P. and E.N.; supervision F.P., E.N. and I.M.; project administration, F.P., E.N. and I.M.; funding acquisition F.P., E.N., I.M., I.S., A.K., S.M., K.K. and A.T. All authors have read and agreed to the published version of the manuscript.

Funding: This work was funded in the context of the Agricultural Development Program 2014–2020 (Measure 16), and in particular Sub-Measure 16.1–16.2. Establishment and operation of Operational Groups of the European Innovation Partnership for the productivity and sustainability of agriculture (Project No M16SYN2-00181).

Institutional Review Board Statement: Not applicable.

Informed Consent Statement: Not applicable.

Data Availability Statement: Data are only available on request due to privacy restrictions.

Conflicts of Interest: The authors declare no conflict of interest.

References

1. Graham, P.H.; Ranalli, P. Common Bean (*Phaseolus vulgaris* L.). *Field Crops Res.* **1997**, *53*, 131–146. [CrossRef]
2. Cardador-Martínez, A.; Loarca-Piña, G.; Oomah, B.D. Antioxidant Activity in Common Beans (*Phaseolus vulgaris* L.). *J. Agric. Food Chem.* **2002**, *50*, 6975–6980. [CrossRef] [PubMed]
3. Chávez-Mendoza, C.; Sánchez, E. Bioactive Compounds from Mexican Varieties of the Common Bean (*Phaseolus vulgaris*): Implications for Health. *Molecules* **2017**, *22*, 1360. [CrossRef] [PubMed]

Disclaimer/Publisher's Note: The statements, opinions and data contained in all publications are solely those of the individual author(s) and contributor(s) and not of MDPI and/or the editor(s). MDPI and/or the editor(s) disclaim responsibility for any injury to people or property resulting from any ideas, methods, instructions or products referred to in the content.

Proceeding Paper

Rural Infrastructure Using Dry-Stone Walling, an Asset for Sustainable Development in a Regional and Local Context [†]

Evangelia Stathopoulou [1,*], Eleni Theodoropoulou [1,*], Antony Rezitis [2] and George Vlahos [2]

1   Department of Economics and Sustainable Development, Harokopio University of Athens, 17676 Athens, Greece
2   Department of Agricultural Economics and Rural Development, Agricultural University of Athens, 11855 Athens, Greece; arezitis@aua.gr (A.R.); gvlahos@aua.gr (G.V.)
*   Correspondence: estathopoulou@hua.gr (E.S.); etheodo@hua.gr (E.T.)
†   Presented at the 17th International Conference of the Hellenic Association of Agricultural Economists, Thessaloniki, Greece, 2–3 November 2023.

Abstract: In this study, dry-stone walling was assessed by the public to map perceptions on the recognition, durability, appeal, food production aspects, biodiversity advocacy, and other characteristics and functions of dry-stone walling. The survey's goal was to define how informed the public is about the functions performed by dry-stone walling. The answers were expected to reveal whether the returns of dry-stone walling are widely acknowledged by the public, what the key factors are for the dissemination of these profits, and if there is solid ground for the reintroduction of dry-stone walling as a cutting-edge choice for new projects.

Keywords: sustainability; dry stone; rural; agri-food; landscape; natural resources; environmental protection; local economy

---

Citation: Stathopoulou, E.; Theodoropoulou, E.; Rezitis, A.; Vlahos, G. Rural Infrastructure Using Dry-Stone Walling, an Asset for Sustainable Development in a Regional and Local Context. *Proceedings* **2024**, *94*, 53. https://doi.org/10.3390/proceedings2024094053

Academic Editors: Elias Giannakis and Stavriani Koutsou

Published: 20 February 2024

Copyright: © 2024 by the authors. Licensee MDPI, Basel, Switzerland. This article is an open access article distributed under the terms and conditions of the Creative Commons Attribution (CC BY) license (https://creativecommons.org/licenses/by/4.0/).

1. Introduction

In view of the growing interest in sustainability and the related challenges that need to be addressed, rural areas play multiple roles with interweaving and often conflicting functions (Guštin et Slavič [1]); they are food providers, keepers of biodiversity and natural resources, economic components, social cohesion buffers, and cultural treasuries. To motivate synergies that can promote sustainability, it is important to make the most of opportunities, reconcile conflicts, and address challenges in a positive way. We need to be innovative, resourceful, and willing to examine new solutions, but we should also note the possibilities and context of reintroducing longstanding and effective solutions.

As such, dry-stone (DS) building (building without the aid of mortar) is an asset, mainly in rural areas, that conveys the identity of a place, as in the past, it supported a range of agricultural and community infrastructure (Allport, [2]). The technique has produced elements that have shaped landscapes and facilitated the interrelation of humans with the natural environment through the development of culture and socioeconomic organization. It represents the anonymous rural builders who combined their efforts, intelligence, and expertise to create amazing infrastructure that was fully integrated in the natural environment.

Mediterranean landscapes are characterized by DS walling, where the construction of terraces allowed the transformation of steep terrains into highly productive agricultural land (Druguet, A. [3]). The use of DS walling also supported a whole range of agricultural and community infrastructure, such as field boundaries, livestock amenities, supportive huts for storage and shelter, threshing floors, mills, irrigation canals, retaining walls, bridges, housing, and more. Cultural changes resulted in the abandoned and rapidly forgotten DS walling technique since the knowhow was transmitted orally and through observation and experience according to the special characteristics of each area. (Pagratiou, [4])

Nowadays, dry-stone walling is recognized as an important cultural element of rural communities (UNESCO, [5]) and an excellent example of optimizing natural and human resources (Picuno, [6]); however, it is not confined to that. Due to the technique, the produced infrastructure offers a wide range of advantages for the environment, the biodiversity (Manenti, [7]), the development of socioeconomic capital through culture (Rose, [8]), local jobs (HISTORIC ENGLAND, [9]), and rural tourism (Greffe, [10]).

The scientific community acknowledges the environmental, sociocultural, aesthetic, and economic value of DS walling, which therefore integrates all three aspects of sustainability. Consequently, DS walling—although marginalized in the past—is becoming more relevant in accordance with the 2015 United Nations action plan "Transforming our World: the 2030 Agenda for Sustainable Development" (COUNCIL OF EUROPE, [11]).

It is of interest to systematically examine the sustainability aspect of DS building and the advantages it can produce for local and regional sustainable development. The identification of the whys and wherefores of DS walling will provide a thorough understanding on the benefits that can be produced and the potentials that emerge, aiming to provide insights on opportunities and restrictions as well.

On this basis, in this paper, dry-stone walling capital was assessed by the public to map perceptions on the recognition, durability, appeal, food production aspects, biodiversity advocacy, and other characteristics and functions of DS walling. The objective of this work was the pilot-phase implementation of an upcoming questionnaire survey, and we intended to check the questionnaire for structural failures and weaknesses in the utilization of the information it produced. The pilot survey was implemented in a small scale; however, some interesting findings were noticed regarding the public perceptions of DS walling. The answers to the future survey are expected to reveal whether the returns of the DS capital are widely acknowledged by the public, what the key factors are for the dissemination of these profits, and if there is solid ground for the reintroduction of DS walling as a cutting-edge choice for new projects.

2. Methods

The survey was quantitative; the data were collected through an Internet questionnaire survey (Google Forms) and analyzed using SPSS 22.0. The target group consisted of individuals older than 16 years old that lived in Greece. In total, 132 participants self-administrated the closed-format questionnaire between 3 March 2023 and 27 March 2023. Due to the pilot character of the survey at this stage, focus was not set on the composition of the sample, which resulted in imbalances that were expected. For the measure of rank correlations, Spearman's correlation coefficient was used. To investigate whether there was a statistically significant difference between the means of two independent groups, the non-parametric Mann–Whitney U rank test was used.

3. Results and Discussion

The research analyzed data collected from 132 participants whose main characteristics are summarized in Table 1. Most of the participants regarded most features of rurality, such as nutrition and health, culture, recreation, and environmental protection, as moderately important. Regarding DS walling recognition, 49.2% of the respondents claimed to be aware of DS walling; however, up to 62.1% of them chose the correct definition of DS walling when asked.

The main fields in which the importance of DS walling is highly acknowledged are culture, natural resource depletion, and resilience/durability. It is important to note that significant features of dry-stone walling did not appear to be highly accredited. Its role in landscape, biodiversity, tourism, and local development was most often stated of to be of fair importance.

Table 1. Sample characteristics.

Characteristic	Percent
Women	63.6
Age 40–49	48.5
University, Polytechnic, master's degree	48.4
Employed in the private sector	35.6
Scientific and technical professional experience	32.6
Residents of urban areas (over 2000 residents)	94.7
Apartment-dwellers	54.5
Rare visitors of rural areas (3–6 times per year)	37.1

As for attitudes towards the adoption of DS walling in new private constructions, 71.8% of the respondents were in favor mainly due to the aesthetic value of the construction, the use of ecological material, and its resilience/durability over time, while biodiversity advocacy was a key factor for only 28.4% of them. Interestingly, as far as public constructions are concerned, approval reached 78.3% among the individuals, while case resilience/durability was not so highly considered, and the biodiversity factor seemed to gain impact when the public space was considered (42.3%).

The main types of construction preferred by individuals mainly regarded enclosures and elements such as benches, water features, etc.; however, for public constructions, it is interesting that bigger-scale interventions were highly preferred. These include the incorporation of DS walling in the design and development of open public spaces and building surroundings, as well as the layering of surfaces.

Concerning the willingness to pay for the assignment of dry-stone walling, 75.8% of the respondents preferred DS over another type of construction, if it would not cost more, with only a small amount of the respondents being willing to pay up to a maximum of 20% more for this.

Furthermore, age is a significant factor that affected the participants' opinions regarding the importance of DS walling over various criteria. Older individuals expressed more positive opinions regarding the criterion "In favor of the landscape", "Biodiversity advocacy", and "Use and reuse of natural resources and the avoidance of environmental pollution" ($p < 0.05$).

Additionally, the type of area of the respondents' main residence is a factor that differentiated their opinions regarding the importance of DS over various criteria; in particular, citizens whose main residence was in a rural area expressed more positive opinions regarding the criterion "Durability and endurance over time" than citizens whose main residence was in an urban area ($p < 0.05$).

4. Conclusions

In conclusion, our main results indicate that DS walling branding is not especially strong, although the particularity of the technique (the absence of mortar) is well identified, maybe because of the name itself. As a practice of the past, it plays an important role in terms of tradition, which can be beneficial but also confining. Perceptions over the significance of DS walling do not highly acknowledge important components and synergies that are involved, such as the impact on the landscape and the connection with tourism, the protection of biodiversity, and the promotion of local masonry jobs, although the Mediterranean landscape is characterized by agricultural terraces and other DS constructions.

This points out the necessity of promoting DS branding to increase recognition and raise awareness of the profits that DS walling can provide locally, at an environmental as well as a socioeconomic level. Additionally, older participants were more aware of the beneficial role that DS walling plays in various criteria for sustainability, while residents of rural areas had a better understanding of the fact that DS walling is a competent, long-lasting choice for construction. It seems that there is a gap in information and awareness, possibly due to alterations in lifestyle and urbanization. Nowadays, more people, and especially younger individuals, are more detached from rurality, resulting in the loss of valuable tools that can be important assets for development.

Nevertheless, DS walling is generally highly accepted as a choice for contemporary works, especially in public spaces, which perhaps also pinpoints a need to differentiate construction methods because of the uprising concerns over the environment and the consequences of climate change. In this respect, rebranding DS walling as a modern choice would empower recognition and the diffusion of the rewards to a much wider audience. Supposing that DS walling costs more than the alternatives, cost may be an issue. In this case, subsidies for the extra cost could be necessary to support the maintenance of existing DS structures before they become ruins and to further develop dry-stone walling capital with the implementation of new works. This fact seems to be recognized by policy makers, since specific Rural Development Policy measures are supporting such endeavours.

Overall, DS walling is a multidimensional asset that provides benefits to all aspects of sustainability. Although it is a traditional practice, DS walling still could address present and future concerns; thus, it is crucial to ensure the survival of the traditional technique. While environmental and socioeconomic crises are becoming more common globally, there is considerable skepticism about the choices that have been made. In this setting, DS walling seems to have the dynamics for a comeback that would produce opportunities for local and regional development.

Author Contributions: All the authors, E.S., E.T., A.R. and G.V., have contributed equally to all stages of the paper. All authors have read and agreed to the published version of the manuscript.

Funding: This research received no external funding.

Informed Consent Statement: Informed consent was obtained from all subjects involved in the study.

Data Availability Statement: Data are contained within the article.

Conflicts of Interest: The authors declare no conflict of interest.

References

1. Guštin, Š.; Slavič, I.P. Conflicts as catalysts for change in rural areas. *J. Rural. Stud.* **2020**, *78*, 211–222. Available online: https://www.sciencedirect.com/science/article/pii/S0743016719308174 (accessed on 1 June 2023). [CrossRef]
2. Allport, S. *Sermons in Stone; The Stone Walls of New England and New York*; W. W. Norton: New York, NY, USA, 1990.
3. Druguet, A. Concilier agriculture et conservation d'un paysage de terrasses à la périphérie du Parc national des Cévennes. *Géocarrefour [Online]* **2007**, *82/4*, 199–207. [CrossRef]
4. Pagratiou, E.; Ioannina, A.N.E.Z. *Based on: Murs de Pierres Seches: Manuel pour la Construction et la Refection*; Fondation, Actions en Faveur de l'Environnement: Bern, Switzerland, 1996.
5. UNESCO. Decision of the Intergovernmental Committee: 13.COM 10.b.10, Culture-Intangible Heritage, UNESCO. 20 December 2018. Available online: https://ich.unesco.org/en/Decisions/13.COM/10.b.10 (accessed on 1 June 2023).
6. Picuno, P. Use of traditional material in farm buildings for a sustainable rural development. *Int. J. Sustain. Built Environ.* **2016**, *5*, 451–460. Available online: https://www.sciencedirect.com/science/article/pii/S2212609016300048?via=ihub (accessed on 1 June 2023). [CrossRef]
7. Manenti, R. Dry stone walls favour biodiversity: A case-study from the Appennines. *Biodivers. Conserv.* **2014**, *23*, 1879. [CrossRef]
8. Rose, M. Dwelling as Marking and Claiming. *Environ. Plan. D Soc. Space* **2012**, *30*, 757–771. [CrossRef]
9. Historic England. Building Value: Public Benefits of Historic Farm Building and Drystone Wall Repairs in the Yorkshire Dales National Park. 2 April 2007. Available online: https://historicengland.org.uk/images-books/publications/building-value-yorkshire-dales-national-park/building-value-3/ (accessed on 1 June 2023).
10. Greffe, X. Is rural tourism a lever for economic and social development? *J. Sustain. Tour.* **1994**, *2*, 22–44. [CrossRef]
11. Council of Europe. Document of the Secretariat General of the Council of Europe, 10th Council of Europe Conference on the European Landscape Convention. 23 July 2019. Available online: https://rm.coe.int/council-of-europe-european-landscape-convention-10th-council-of-europe/168096844b (accessed on 1 June 2023).

Disclaimer/Publisher's Note: The statements, opinions and data contained in all publications are solely those of the individual author(s) and contributor(s) and not of MDPI and/or the editor(s). MDPI and/or the editor(s) disclaim responsibility for any injury to people or property resulting from any ideas, methods, instructions or products referred to in the content.

Proceeding Paper

Social Innovation and Women's Agricultural Cooperatives: Applying Social Change Theory [†]

Ioannis Sotiriadis, George Sidiropoulos and Maria Partalidou *

Department of Agricultural Economics, School of Agriculture Forestry and Natural Environment, Aristotle University of Thessaloniki, 541 24 Thessaloniki, Greece; giannis-sotiriadis@hotmail.com (I.S.); georgesidiro3@gmail.com (G.S.)
* Correspondence: parmar@agro.auth.gr
[†] Presented at the 17th International Conference of the Hellenic Association of Agricultural Economists, Thessaloniki, Greece, 2–3 November 2023.

Abstract: In most investments, businesses, or even organizations, results and their value are calculated in terms of profit and economic terms. But what if you have to calculate the value and work of a social enterprise? What is that thin line that separates one business from another? The way to evaluate the efficiency of a business includes the social contribution and the social footprint of the business. Is it possible for a successful farmer cooperative that wants to increase its activity to remain as a social enterprise, or must it change its legal form? In an agricultural cooperative that shows remarkable success, how aligned are the opinions of the members with the vision of the cooperative and to what extent do the cooperative's vision and its reason to exist change? The above questions were the reasons behind why this study was carried out and the realization of the primary research presented in this article. The research presented herein is based on qualitative research tools, and this study involved carrying out a case study of a women's agricultural cooperative in Agios Antonios, a village in the prefecture of Thessaloniki, Northern Greece.

Keywords: women's cooperatives; social effect; economic benefit

---

Citation: Sotiriadis, I.; Sidiropoulos, G.; Partalidou, M. Social Innovation and Women's Agricultural Cooperatives: Applying Social Change Theory. *Proceedings* **2024**, *94*, 54. https://doi.org/10.3390/proceedings2024094054

Academic Editor: Eleni Theodoropoulou

Published: 28 February 2024

Copyright: © 2024 by the authors. Licensee MDPI, Basel, Switzerland. This article is an open access article distributed under the terms and conditions of the Creative Commons Attribution (CC BY) license (https://creativecommons.org/licenses/by/4.0/).

1. Introduction

When we talk about business, a single word automatically comes to mind: profit. Rarely when evaluating the course of a conventional business (individual company, joint stock company, etc.) do we consider its social footprint. However, even in the case of a social enterprise, things do not seem to differ sharply in practice, contrary to their definition. The value of a social business depends on and is measured in financial terms, and in some cases, without profits, a business can cease operating. Despite the minimal importance given to the social results, however, this is precisely what practically separates social businesses from other forms of businesses in modern economic and fully competitive environments. By definition, social enterprises are based on the concept of the social economy: "The totality of economic, business, productive and social activities, which are undertaken by legal persons or associations of persons, whose statutory purpose is the pursuit of the collective benefit and the service of general social interests" (Vairami, 2015) [1]. Agricultural enterprises are also included among social enterprise cooperatives. Their viability is judged almost exclusively by their financial benefits. Social contributions are not recorded as profit. When any agricultural cooperative attempts to find financial support, such as in the form of a loan from a financial institution (e.g., banks), the only evaluation criteria are financial and accounting situations, and the social footprint the cooperative might offer is not considered. In fact, evaluating and calculating the social impact of an enterprise is inversely proportional to the typical method used to evaluate and calculate the success of an enterprise or cooperative. The success of a cooperative and its economic growth create some conditions that affect the cooperative's vision, goals, and reason for existing

Agricultural cooperatives, which have managed to create strong scale economies, have expanded significantly and exponentially in terms of their memberships, and in some cases, this has resulted in each member becoming disconnected from the decisions of the cooperative and its overall course. Previously, the benefits obtained by each member were only financial (improved selling price). Interpersonal relationships between members, due to the increase in the sizes of cooperatives, stopped existing, and the ultimate aim was world market dominance. The result was ultimately a change in law which led agricultural cooperatives to become public limited companies. Through our research in this particular field and carrying out a case study of a cooperative that promotes social innovation through the integration and empowerment of rural women, we set out to achieve some very specific goals, namely to capture the social significance of an agricultural cooperative, to emphasize the differences in social businesses from the trivial and speculative perspectives, and to evaluate, with the use of a research tool, whether the visions of both the cooperative as a whole and its individual members change over time.

2. Materials and Methods

Primary research was conducted on a women's agricultural cooperative in the nearby area of Thessaloniki, Northern Greece. Our research focused on Agios Antonios Women's Agricultural Cooperative of Traditional Products in the village of Agios Antonios. This particular choice it is not accidental. On the contrary, we specifically targeted and studied this cooperative. This specific women's cooperative is a fairly well known and successful one, despite the difficulties it has faced from time to time. It has specific terms of registration for new members (female farmers exclusively from their local community), is located at a very close distance from Thessaloniki (33 km), and its economic activity is constantly increasing. This women's cooperative has been in operation since 1999. It is active in the agri-food sector (produces locally traditional products). At the headquarters of the cooperative there, is an organized dining area that operates daily. It consists of 16 members, all of which are exclusively from the local community of Agios Antonios (which has a population of 647); all members are female farmers, and all ages are represented. The research that we carried out is purely qualitative and based on data obtained from holding in-depth discussions with a total of 7 women from the cooperative (with each member being interviewed individually), as well as through a collective discussion (a focus group discussion). Furthermore the qualitative research tool "Journey of Change" was utilized, through which we sought to measure the effects of the changes resulting from each activity the cooperative carried out to benefit interested parties and not for profit (Baker & Courtney, 2018) [2], and these activities contributed to the organization of our discussions and facilitated our qualitative research. At the same time, this tool helped to concentrate our research and organize the thoughts of the participants, and it is a primary tool for quantifying qualitative data. Our application of Journey of Change was structured in three time phases (short term, medium term, and long term) which were defined by the members of the cooperative themselves. At the same time, our research raises concerns about the cooperative's auxiliary accelerators and the obstacles that it currently faces or will face in the future. Our research process was organized into two stages. In the first, the participants were asked to complete the Journey of Change form, which concerned the individual characteristics of each participant and her individual opinion about her participation in the cooperative. They were asked to define the benefits they have obtained from participating in the cooperative as well as the goals they have set individually for the cooperative in the future. In the second phase, the participants collectively filled out the Journey of Change form. In both phases, a total of 7 women (of which 3 belonged to the board of the cooperative) participated in the research process. The implementation of the first phase was preceded by the research team providing information and guidance to the cooperative members.

3. Results

Based on the qualitative processing of the data collected during the first stage (individual) of the research process, important elements emerged. In all three-time phases, almost all of the participants (five out of seven) noted exclusively the social benefits instead of the economic ones as the primary reason behind their participation in the cooperative. As a collective, the members reported networking and cooperation as the main reason for participation in the cooperative, and this aspect was also listed as a benefit derived from their participation. Our results also included answers related to self-improvement, and some members sought employment in the cooperative as way for them to escape from everyday life. Answers related to financial benefits were more or less absent, and financial benefits were mentioned the least out of all the benefits (mentioned by two out of seven participants, and the two who did mention financial benefits held a relatively high position in the cooperative). Individual future financial benefits were also scarcely mentioned by the respondents. Regarding the second stage, no answers regarding the economic development of the cooperative and increasing its income were recorded. On the contrary, many of the members' answers pertained to social future desires and goals. Specifically, at the stage wherein the group collectively completed the Journey of Change form (Figure 1), the participants stated that their future and main goal was to increase cooperation with the municipality to promote their area. Alongside responses pertaining to the fear of a reduced number of membership renewals from women in the cooperative, the respondents mentioned the need to add new members to the cooperative and boost its development in the field of agro-tourism and the home industry to increase the amount of job opportunities for the local community.

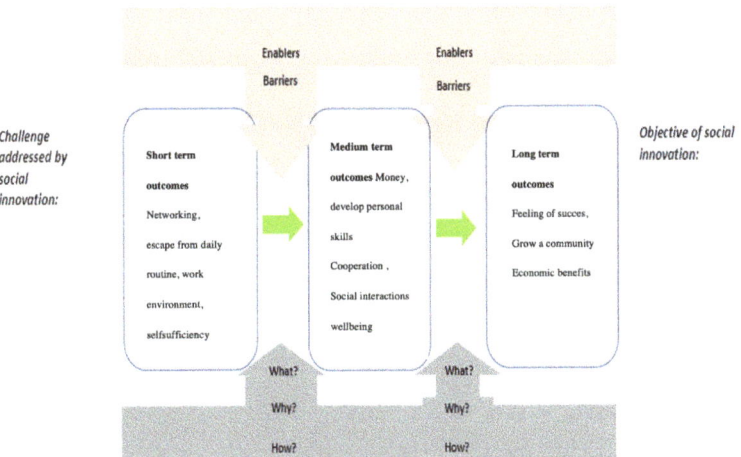

Figure 1. Journey of change for social innovation solutions. **Source: field research, 2023.** Adapted with the permission from "Social Return on Investment (SROI) Guide", 2023, P. Courtney, J. Powell, K. Kubinakova and C. Baker.

4. Discussion

Based on the above results, it appears that the participants have enjoyed personal benefits from their participation in the cooperative which have mainly affected the social aspects of their lives. After a comparison between the individual and group responses, it became clear that topics such as the economic reasons for participation and increasing the financial capacity of the cooperative in the future, which were recorded in the individual forms, were not mentioned at all in the Journey of Change form that was completed collectively. This fact reflects that due to the majority of the participants' emphasis on an open and collective discussion and dialogue in the second stage and the social impacts that participation in the cooperative have had on their daily lives, economic-centered answers were not considered important and were limited and/or not provided at all by any participant. At the same time, despite the success of the cooperative, none of the participants stated that there was a need for any significant changes (operational, tax or legal). They all said that the future plans of the cooperative will simply revolve around ensuring that it still exists and remains as it is.

5. Conclusions

In summary, this specific cooperative (as a case study example of social innovation and integration among female farmers) demonstrates the social value that a cooperative has. It was shown that the social benefits derived from participation in the cooperative are quite important, as they were mentioned in, on average, about 80% of the individual answers and prioritized in most of the answers given (85% exclusively social responses, 75% hierarchically higher social responses). The success of an agricultural business, as it turns out, is not only related to profit. Measuring the value and importance of a cooperative should not be carried out using only monetary criteria and values; on the contrary, it should be carried out by considering other benefits and social factors. The above research process proved the social importance of a cooperative, highlighted the fact that social benefits can be of greater value to social enterprises than financial ones, and captured the challenges between personal views and collective ones and how these influence discussions within the cooperative. However, our research focused on a case study and the perspectives of the cooperative's members. Future research efforts should concern measuring the perceptions of local communities, consumers, and visitors on the social impact of a cooperative in rural areas.

Author Contributions: Conceptualization, M.P. and I.S.; methodology, I.S., I.S. and G.S.; writing—original draft preparation, I.S., G.S. and M.P.; writing—review and editing, all; visualization, I.S.; supervision, M.P.; All authors have read and agreed to the published version of the manuscript.

Funding: This research received no external funding.

Institutional Review Board Statement: Not applicable.

Informed Consent Statement: Informed consent was obtained from all subjects involved in the study.

Data Availability Statement: Data is unavailable due to privacy.

Conflicts of Interest: The authors declare no conflict of interest.

References

1. Vairami, B.N. Social Enterprises in Greece: Strategic Direction and Performance. Doctoral Dissertation, Aristotle University of Thessaloniki, Thessaloniki, Greece, 2015.
2. Baker, C.; Courtney, P. Conceptualizing the wider societal outcomes of a community health programme and developing indicators for their measurement. *Res. All* **2018**, *2*, 93–105. [CrossRef]

Disclaimer/Publisher's Note: The statements, opinions and data contained in all publications are solely those of the individual author(s) and contributor(s) and not of MDPI and/or the editor(s). MDPI and/or the editor(s) disclaim responsibility for any injury to people or property resulting from any ideas, methods, instructions or products referred to in the content.

Proceeding Paper

The Greek Perspective on Foreign Farm Workers and Agricultural Labor [†]

Lykourgos Chatziioannidis * and Maria Partalidou

Department of Agricultural Economics, School of Agriculture, Aristotle University of Thessaloniki Greece, 541 24 Thessaloniki, Greece; parmar@agro.auth.gr
* Correspondence: lyk.chatz@gmail.com
[†] Presented at the 17th International Conference of the Hellenic Association of Agricultural Economists, Thessaloniki, Greece, 2–3 November 2023.

Abstract: Apart from immigrants in Greece who have papers, and perhaps can enjoy greater stability in their lives, there is a very large number of informal immigrants who are faced with the fear of deportation from the country daily. With this in mind, qualitative research was carried out by conducting in-depth interviews with farmers (head of the farm) and quantitative, online research was undertaken using students studying agronomy and/or people who lived in rural areas; the research material was distributed through agricultural/agronomic forums in order to better understand perceptions of agricultural work and find out the main reasons as to why the integration of immigrants and farm workers in Greece is considered to be so difficult.

Keywords: agricultural labor; field research; countryside

1. Introduction

A major obstacle for the integration of immigrants into Greek society is socially constructed perceptions defined by xenophobia and racism; varied actions need to be taken and strategies need to be designed and put into practice (Maroukis, 2012) [1].

The main objective of this research was to identify what the perception in Greece is regarding rural life, but also the perspective that people have of farm work, its requirements, and its negatives and positives. At the end of the survey, it was considered important to examine what the sample of respondents thought about rural life and living conditions and to make an effort to establish how realistic a depiction of the countryside the sample had. Through the questionnaires that were drawn up and distributed to 365 people, it was possible to draw very important conclusions and, after their processing, to adequately answer the aforementioned questions.

In the end, the results of the research were basically quite in line with the questions that were formulated at the beginning, as many of our initial assumptions were verified; however, as it will be seen below, there were some results that were beyond what was expected.

2. Materials and Methods

The main goal of the survey was to better understand the Greek perspective on foreign farm workers and agricultural labor. To do this, two types of questionnaires were created which, due to the COVID-19 pandemic, were not only compiled online but also answered via the internet, since it was considered too risky to conduct the questionnaires in person. The first questionnaire was aimed exclusively at people who were farm heads and was drawn up with the main purpose of elaborating on the perspective that farmers themselves have on farm work. The second questionnaire was essentially aimed at urbanites and dealt with the perspective on working in the countryside. Here, it was deemed necessary that the sample should be big enough for its results to have significance, and so it was

answered by a larger number of people who varied in age, social status, work, whether they were employed in the agricultural sector or not, and several other characteristics. The large participation in this questionnaire was considered to be positive as the large sample of 365 respondents, who were found after the distribution of the questionnaire to social networks, groups of agricultural students, and agricultural forums, enabled us to better understand the opinion held by a part of society on the present issue. The questionnaires were compiled through Google Forms and the results were processed in autumn 2022; the most important questions will be presented through tables and diagrams in Word to make them easier to read and understand.

3. Results and Discussion

Working in the countryside was considered to be very demanding and have a higher degree of difficulty than most occupations (61.1% and 57.5%, respectively), but most of the respondents did not believe that rural work offers a higher income than conventional occupations, nor that it has more free hours. An important element of the survey, which was characterized as being unexpected, was the answers given to the question of whether working in the countryside is mainly a "male" occupation, since 56.1% of the sample disagreed with this wording and only 20.2% agreed, with the remaining 23.8% not taking a clear position. When the questionnaire was drawn up, it was assessed that there was a significant probability that most respondents would agree with this wording because of the very common manual nature of working in the countryside, but this was not verified, possibly (also) because of the fairly high educational level of the sample. Moreover, a majority agreed that working in the countryside offers a sense of freedom and independence (46.5%) and is very demanding (74.6%), while few believed that experience in other similar occupations is required (just 17.8%). Finally, most of the respondents seemed to believe that the effect of foreign workers living in rural areas on the quality of life there either depends on the amount of workers living in each area (33.2%) or is negative (31.5%) (see Figures 1–4).

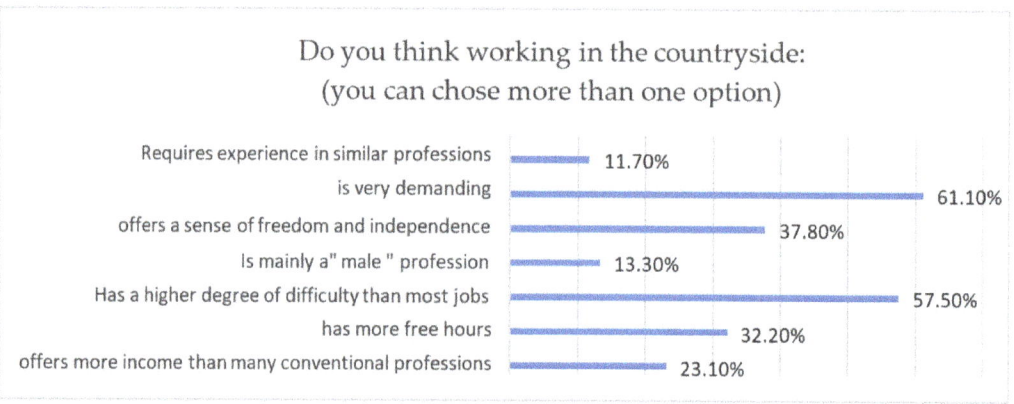

Figure 1. Views on working in the countryside. Source: Field Research 2022.

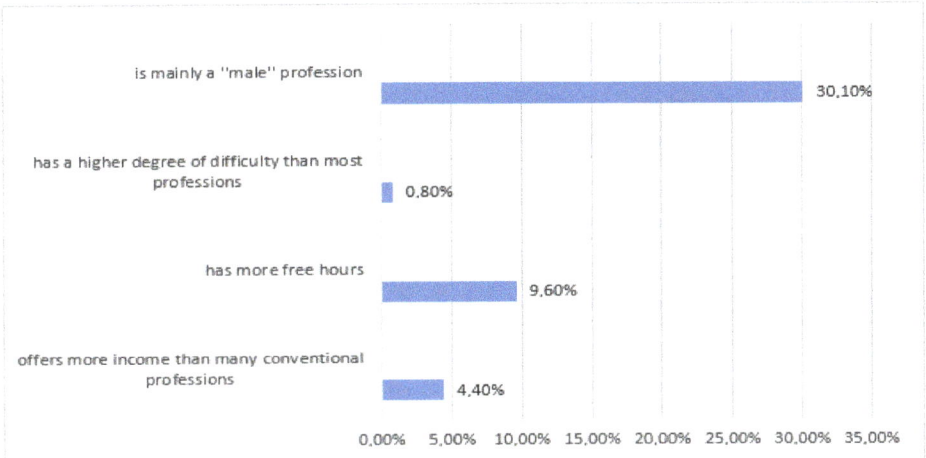

Figure 2. Views on agricultural work in Greece. Source: own edit, 2022.

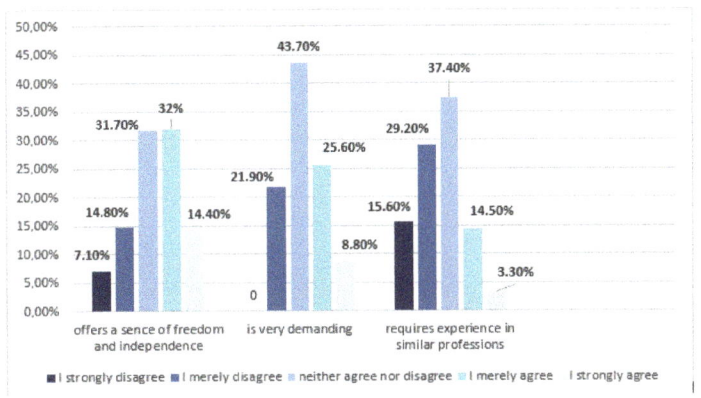

Figure 3. Perceptions on agricultural work in Greece (2). Source: own edit, 2022.

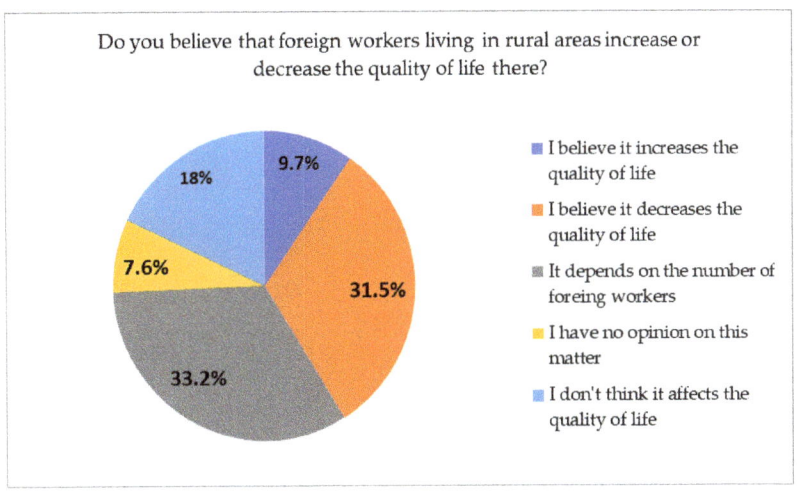

Figure 4. Perspective on foreign farm workers. Source: own edit, 2022.

4. Conclusions

Thanks to the results of the research, we managed to draw some very useful conclusions in relation to the questions that we asked at the beginning. Perceptions on agricultural labor show us that there are still mostly antiquated notions on the issue, and it turns out that there is still a lot of work to be done so that conservative perceptions change and the urbanite can better understand what agricultural labor really is and evaluate it more objectively.

Author Contributions: Conceptualization, M.P. and L.C.; methodology, M.P. and L.C.; software, L.C.; validation, L.C.; formal analysis, L.C.; investigation, L.C.; resources, M.P. and L.C.; writing—original draft preparation, M.P. and L.C.; writing—review and editing, M.P. and L.C.; visualization, L.C.; supervision, M.P. All authors have read and agreed to the published version of the manuscript.

Funding: This research received no external funding.

Institutional Review Board Statement: Ethical review and approval were waived for this study, due to the fact that he study was conducted in accordance with the Declaration of Helsinki and the EU General Data Protection Regulation.

Informed Consent Statement: Informed consent was obtained from all of the subjects involved in the study.

Data Availability Statement: Data are unavailable due to privacy concerns.

Conflicts of Interest: The authors declare no conflicts of interest.

Reference

1. Venetis, E.; Tzogopoulos, G. Briefing Note: "US Military Withdrawal from Iraq". 2012, pp. 5–8. Available online: https://www.eliamep.gr/en/publication/%CE%B5%CE%BD%CE%B7%CE%BC%CE%B5%CF%81%CF%89%CF%84%CE%B9%CE%BA%CF%8C-%CF%83%CE%B7%CE%BC%CE%B5%CE%AF%CF%89%CE%BC%CE%B1-%C2%AB%CE%BC%CE%B5%CF%84%CE%B1%CE%BD%CE%AC%CF%83%CF%84%CE%B5%CF%85%CF%83%CE%B7/ (accessed on 27 February 2024).

Disclaimer/Publisher's Note: The statements, opinions and data contained in all publications are solely those of the individual author(s) and contributor(s) and not of MDPI and/or the editor(s). MDPI and/or the editor(s) disclaim responsibility for any injury to people or property resulting from any ideas, methods, instructions or products referred to in the content.

Proceeding Paper

Agricultural Cooperatives as a Vehicle for Small-Scale Farmer's Viability and Sustainable Practices [†]

Myrto Paraschou * and Panagiota Sergaki

Department of Agricultural Economics, School of Agriculture, Aristotle University of Thessaloniki, 541 24 Thessaloniki, Greece; gsergaki@auth.gr
* Correspondence: mirtoparaschou1@gmail.com
[†] Presented at the 17th International Conference of the Hellenic Association of Agricultural Economists, Thessaloniki, Greece, 2–3 November 2023.

Abstract: Nowadays, the agricultural sector is poised to undergo significant transformations towards sustainability. Small-scale farmers' restricted accessibility to resources hinders their ability to effectively adapt to such advancements. This research paper investigates the potential role of agricultural cooperatives as deus ex machina, offering an idea for solving the challenges faced by small-scale farmers. Additionally, it examines the potential benefits agricultural cooperatives could provide to large-scale farmers while simultaneously advocating sustainable agricultural practices. To gather accurate data, individuals who were members of cooperatives in the Larissa region of Greece were interviewed using questionnaires. The sixty qualitative interviews conducted shed light on the fact that cooperatives play a significant role in promoting sustainable agriculture and offer numerous benefits to their members, particularly small-scale farmers.

Keywords: agricultural cooperatives; sustainability; small-scale farmers

Citation: Paraschou, M.; Sergaki, P. Agricultural Cooperatives as a Vehicle for Small-Scale Farmer's Viability and Sustainable Practices. *Proceedings* **2024**, *94*, 56. https://doi.org/10.3390/proceedings2024094056

Academic Editor: Eleni Theodoropoulou

Published: 28 February 2024

Copyright: © 2024 by the authors. Licensee MDPI, Basel, Switzerland. This article is an open access article distributed under the terms and conditions of the Creative Commons Attribution (CC BY) license (https://creativecommons.org/licenses/by/4.0/).

1. Introduction

In recent years, agriculture has adopted more environmentally friendly and sustainable methods. Most environmentally friendly agriculture operations need expensive equipment, funding, and expertise. Thus, small-scale producers may struggle to adapt to these strategies. Small-scale producers may need to unite and consolidate to remain relevant in the developing environmentally conscious agricultural movement.

Agricultural cooperatives, when operating well, act as a means of obtaining a wide range of valuable resources, including information, technology, marketing, credit, purchasing power, and equipment (van Dijk et al., 2019) [1] (p. 176). According to Papageorgiou, individuals engaged in farming operations may enjoy numerous advantages by participating in cooperatives, such as reduced production expenses and enhanced managerial practices (Papageorgiou, 2015) [2] (pp. 75–76).

To examine the impact of agricultural cooperatives on the viability of small-scale producers and the adoption of sustainable practices, a qualitative research study was conducted. In this study, we used a sample of 60 individuals, consisting of small-scale producers who are affiliated with cooperatives. Based on the findings of this research, it was reported that a significant majority of the small-scale producers who were surveyed and who are currently members of agricultural cooperatives expressed that these organizations have played a crucial role in ensuring their economic sustainability. Furthermore, it appears that in certain instances, these cooperatives have even assisted with the expansion of their operations. Additionally, it is worth noting that a significant majority of the participants reported that cooperatives have played an important role in furnishing them with valuable knowledge, guidance, and resources that are required for enhancing their output by using novel agricultural techniques. When specifically questioned regarding the utilization of techniques such as composting waste or precision farming, an overwhelming majority expressed the belief that these approaches are unfeasible to implement without involvement in cooperatives due to their substantial financial requirements.

Within each agricultural cooperative, there exists a heterogeneous composition of individuals, encompassing both small-scale and large-scale producers. However, similar to other industries, individuals with limited means encounter significant hurdles within the agricultural sector (Oleg Nivievskyi et al., 2023) [3] (p. 19). This study examines cooperatives' role in sustainability, their benefits to members, and their protection of small-scale farmers in the context of the current environmentally conscious agricultural paradigm. This study examines cooperative membership as a solution to small-scale producers' complex problems.

2. Methods

The data presented in this research paper were derived from qualitative interviews conducted with a sample of 60 small-scale farmers who were members of various agricultural cooperatives in the Larissa region of Thessaly, Greece, during the year 2023. Larissa was once home to a prominent union of agricultural cooperatives, which, regrettably, faced bankruptcy in the year 2012. Subsequently, numerous producers within the region have experienced an erosion in their trust in cooperatives. Conversely, within the Larissa region, a multitude of producers operating on a small-scale can be found. The choice of location was undertaken with the intention of highlighting the potential for cooperatives to support and revive small-scale producers. Furthermore, it was intended to demonstrate that the unfortunate and sad incident experienced by the union is now, and has been for a long time, a distant memory. Cooperatives have emerged as a promising trajectory in the agricultural sector, holding significant potential for the future.

In this research, we employed a sampling technique. Purposive sampling, often known as a judgmental procedure, was the sampling technique employed. This technique depends on the researcher's discretion when choosing the individuals and instances (small-scale producers who are members of cooperatives) that can offer the most useful data to meet the study's goals. Initially, we initiated communication with individuals who were affiliated with several agricultural cooperatives in Larissa. The participants were then divided into small-scale and large-scale farmers according to agricultural holding size based on the questionnaire. Sergaki and Michailidis assert that there exists a multitude of classifications pertaining to small-scale producers (Sergaki & Michailidis, 2020) [4]. According to the FAO's definition, small-scale food producers are individuals engaged in farming or entrepreneurial activities who have limited opportunities and operate under structural limitations, including inadequate access to resources, technology, and markets (FAO, 2017) [5]. Following the segregation of the two cohorts, we proceeded to administer questionnaires to the cohort of small-scale producers, constituting our sample size of 60 individuals.

It is essential to bear in mind that the findings derived from this research were obtained from a limited sample size of 60 individual cases, all of which originate from a single geographic region. Hence, it is probable that the obtained results may not accurately represent all situations in their entirety.

3. Results

3.1. Cooperatives and Their Role in the Promotion of Sustainable Practices

The majority of the cooperative members reported that they engaged in sustainable practices and programs and possessed knowledge of initiatives. Educative webinars and expert consultations helped to disseminate this information. Moreover, most respondents reported that their cooperatives actively promote investments in heavy machinery, equipment, and technology that enable sustainable agricultural approaches like precision farming and waste composting (Table 1).

Table 1. Key questions and results regarding the role of cooperatives in the promotion of sustainability.

Question	Positive
Access to information pertaining to sustainable practices, programs, and initiatives available through the cooperative	88%
Access to educational seminars about more eco-friendly practices available through the cooperative	81%
Access to information about important agricultural issues and recent developments available through the cooperative	76%
Provision of consultations on sustainable practices with esteemed specialists and advisers by the cooperative	65%
Promoting investments through means that facilitate the implementation of sustainable methods by the cooperative	61%

3.2. Cooperatives and Their Role in the Viability of Small-Scale Producers

The results of our study demonstrate that small-scale producers can gain significant benefits in terms of enhanced purchasing power through their participation in cooperatives. Additionally, small-scale farmers with limited resources might increase their household income by working in a cooperative. Another significant advantage that small-scale producers derive from cooperatives is their ability to become members with minimal financial investment. In addition to this, individuals gain access to a type of organization that has experienced significant growth over the course of several years, attracting substantial investments, all through the payment of a nominal charge. Another salient aspect to consider is that cooperatives engage in purchasing products from all of their members at an agreed-upon rate price. When a large-scale producer with more bargaining power demands a higher price from the cooperative, the cooperative buys from the small-scale producer at the same price (Table 2).

Table 2. Key questions and results regarding the role of cooperatives in the viability of small-scale producers.

Question	Answer
Does the cooperative buy products at the same price from all members?	100% stated cooperatives buy products from members at the same price
Is farming the only thing you do for a living?	79% had a distinct primary occupation
Cooperatives contributing to the augmentation of household income	37% positive answers
In the absence of cooperative membership, would you opt to divest your land?	21% stated they would consider it

4. Discussion and Conclusions

The examination of the efficacy of cooperatives in the Larissa region holds significance, as these entities have faced long-standing stigmatization in the aftermath of a regrettable incident.

The present manuscript focuses on a comprehensive analysis of the advantages that small-scale producers derive from their participation in cooperatives. Additionally, this study investigates the techniques employed by cooperatives to foster sustainability, thereby enabling their members to stay abreast of contemporary environmentally conscious agricultural practices.

Members regard access to information and facilitation through the provision of technology, equipment, and heavy gear as the most significant advantages cooperatives provide. Hence, the Greek government's prioritization of reinforcing the cooperative education regulation holds significant importance. Additionally, the convening of member meetings holds significance in facilitating the exchange of perspectives and plays a crucial role in the decision-making process.

Author Contributions: Conceptualization, M.P.; methodology, M.P.; validation, M.P.; formal analysis, M.P.; investigation, M.P.; writing—original draft preparation, M.P.; writing—review and editing, P.S.; supervision, P.S. All authors have read and agreed to the published version of the manuscript.

Funding: This research received no external funding.

Informed Consent Statement: Not applicable.

Data Availability Statement: Data sharing is not applicable due to privacy.

Conflicts of Interest: The authors declare no conflicts of interest.

References

1. van Dijk, G.; Sergaki, P.; Baourakis, G. *The Cooperative Enterprise: Practical Evidence for a Theory of Cooperative Entrepreneurship (Cooperative Management)*, 1st ed.; Zopounidis, C., Baourakis, G., Eds.; Springer: Cham, Switzerland, 2019; Volume 6, p. 176.
2. Papageorgiou, K. *Sustainable Cooperative Economy Theory and Practice*, 3rd ed.; Stamoulis: Athens, Greece, 2015; pp. 75–76.
3. Nivievsky, O.; Iavorsky, P.; Donchenko, O. Assessing the role of small farmers and households in agriculture and the rural economy and measures to support their sustainable development. *arXiv* **2023**, arXiv:2307.11683.
4. Sergaki, P.; Michailidis, A. Small-Scale Food Producers: Challenges and Implications for SDG2. *Zero Hunger* **2020**, 787–799. [CrossRef]
5. FAO. *Defining Small Scale Food Producers to Monitor Target 2.3 of the 2030 Agenda for Sustainable Development*; FAO Statistics Division: Rome, Italy, 2017.

Disclaimer/Publisher's Note: The statements, opinions and data contained in all publications are solely those of the individual author(s) and contributor(s) and not of MDPI and/or the editor(s). MDPI and/or the editor(s) disclaim responsibility for any injury to people or property resulting from any ideas, methods, instructions or products referred to in the content.

Proceeding Paper

The Effect of the Regulatory Role of Collective Organizations in Relation to the Consumption of Fruits and Vegetables from Cooperatives †

Aristotelis Batzios [1,*] **and Maria Tsiouni** [2,*]

1. Agricultural Economy, School of Agriculture, Forestry and Natural Environment, Aristotle University of Thessaloniki, 54124 Thessaloniki, Greece
2. School of Agriculture, International Hellenic University, 57400 Thessaloniki, Greece
* Correspondence: a1990batzios@gmail.com (A.B.); mtsiouni@agr.teithe.gr (M.T.)
† Presented at the 17th International Conference of the Hellenic Association of Agricultural Economists, Thessaloniki, Greece, 2–3 November 2023.

Abstract: European agri-food-chains are characterized by strong interconnections among all partners, their complexity, their resilience in a period of uncertainty, and their shared commitment to continue to strive for food safety and quality. The regulatory role of Greek collective organizations thus empowers their members and enables small farmers to achieve the above agri-food-chain goals. A large number of academic articles on collective organizations focus on economic analysis of their performance, but there is little research on the impact of regulation on consumer behavior. The objective of this study is (a) to analyze the Greek market of fruit and vegetable cooperatives, (b) to identify consumers' opinions with regard to the regulatory role of Greek collective organizations in the fruit and vegetable supply chain, and (c) to assess whether consumers and producers benefit from the cooperative movement.

Keywords: regulatory role; collective organizations; agri-food-chain

1. Introduction

According to [1], one in three European citizens (33%) reported that they do not consume any fruits or vegetables daily, while only 12% of the European Union population consumes five portions of fruits and vegetables or even more than five per day. In Greece, the average monthly expenditure per household for fresh fruit and vegetables was found to be EUR 41.87, of which EUR 18.69 concerned fresh fruits and EUR 23.18 was for fresh vegetables [2].

Alternative food networks in Greece are primarily based on social entrepreneurship and solidarity economies (such as women's agro-tourism cooperatives, social cooperative enterprises, and community-supported agriculture).

In order to study consumers, we utilized shopping center sampling, which is primarily used for marketing research [3]. A questionnaire survey was used to collect data and the sample size was determined to be 400 consumers in the urban complex of Thessaloniki [4]. In terms of statistical analysis, PCA (Principal Component Analysis) is used to identify the common factors that contribute to variation [5]. According to a study, two common factors explain why consumers buy fruits and vegetables from cooperatives. This research fills the gap regarding the importance of a regulatory role for consumers.

2. Methods

The urban complex of Thessaloniki was the study area. The city's urban complex is the second largest in terms of population and has a mix of urban, industrial, and working citizens. It also has a sufficient number of government officials. Furthermore, Thessaloniki is a representative city since it is the administrative, cultural, and spiritual center of Northern Greece. In the regional unit of Central Macedonia, there are 8 agricultural cooperatives that operate as a joint venture. As a matter of fact, this joint venture produces fifty percent of Greek dessert peaches. Collective organizations establish networks to reduce transaction costs and facilitate the exchange of information and resources, increasing their economic efficiency and competitiveness [6].

This survey took place in 2020. The structured questionnaire had questions that were classified into three sections. In the literature, groupings and typologies are usually based on Principal Component Analysis (PCA). PCA is a common method in the social sciences that can describe, to a significant extent, the framework of a set of data [7]. PCA was performed using SPSS v26.0. The significance level of all statistical analyses was predetermined at $p < 0.05$. After the extraction of factors, Orthogonal Rotation Maximum Variation was used because it aims for minimization of the number of variables that appear as high weightings in each factor. Then, reliability analysis was implemented using Cronbach's alpha, which is based on the average of correlations between variables (items). The Kaiser–Meyer–Olkin (KMO) test was used to test sampling adequacy [8]. Also, Barlett's Sphericity Test was utilized to generate the correlation matrix as well as the identity matrix. PCA was limited to indicators with a variation coefficient (VC) over 50% [9].

3. Results and Discussion

Based on descriptive statistics (Table 1), 85.7% of consumers consider agri-food cooperatives to play an important regulatory role. A significant percentage of consumers' agreement, according to the regulatory role of collective organizations, began to prevail after 2016, along with the fact that the first legislative act regarding the concept of the "Solidarity Economy" was introduced in Greece [10,11].

Table 1. Do you consider the regulatory role of collective organizations to be important in the food supply chain?

	Frequency	Percentage %
No	57	14.3
Yes	342	85.2
Total	399	99.8
Missing Value	1	0.3
Total	400	100.0

The results of the Principal Component Analysis (PCA) demonstrated that the first two factors explained 54.97% of the total variance. The selection of the factors was based on the percentage of their variation and their eigenvalue (>1). The Kaiser–Meyer–Olkin index was calculated with a value of 0.856 > 0.8 and Barlett's Sphericity Test was statistically significant ($p < 0.01$). These statistical tests determine the suitability of PCA for a sample [12].

The new factors with their content are given below according to corresponding variables:
- E11_1, E11_2, E11_3, E11_4, E11_8, E11_9, E11_10 (Factor 1: Trust)
- E11_5, E11_6, E11_7 (Factor 2: Corroboration)

  ❖ E11_1: I trust fruit and vegetable cooperatives;
  ❖ E11_2: I believe in cooperatives and cooperative movement;
  ❖ E11_3: I believe that agricultural cooperatives form a reliable marketing channel for producers;
  ❖ E11_4: I consider it important to reinforce the incomes of cooperative producers;
  ❖ E11_5: Producers ensure fair prices for their products through cooperatives;
  ❖ E11_6: Consumers ensure fair prices for the products that they buy through cooperatives;

- E11_7: Agricultural cooperatives have the possibility to reduce the cost of distribution in products from cooperatives;
- E11_8: I trust the quality of products from cooperatives;
- E11_9: I find them to be of better quality compared to fruits and vegetables that are not from cooperatives;
- E11_10: The market of fruits and vegetables from cooperatives creates more job positions in the local community.

Cronbach's alpha coefficient was calculated for new factors. Cronbach's alpha amounted to 0.839 for Factor 1 (Trust) and Factor 2 (Corroboration) amounted to 0.682.

4. Conclusions

This study fills the gap in the analysis of collective organization regulation in Greece; evaluation of the regulatory role is difficult, especially when it refers to all stakeholders in the agri-food-chain. A first step towards improving the regulatory function of collective organizations is to identify consumer perceptions about products from cooperatives. Most consumers buy their food through alternative food networks, supporting the incomes of producers. Consumers emphasize trust and corroboration as key factors in purchasing products from cooperatives. This tendency is based on consumers' belief in the cooperative movement and the regulatory role of collective organizations.

Moreover, consumers consider that producers of collective organizations secure fair prices and a reduction in the distribution cost of their agricultural products, improving the relation between price and quality. Consumers' satisfaction with regard to better qualitative categorization of fruits and vegetables from cooperatives will lead to a rise in sales.

In conclusion, executive officers of collective organizations should meet consumers' demands for quality and reasonable food prices. For this reason, collective organizations have to develop innovative sales techniques to meet the special preferences of consumers.

Author Contributions: Conceptualization, A.B. and M.T.; methodology, A.B.; software, M.T.; validation, A.B., M.T.; formal analysis, A.B.; investigation, A.B.; resources, M.T.; data curation, A.B.; writing—original draft preparation, M.T.; writing—review and editing, A.B.; visualization, M.T.; supervision, A.B.; project administration, A.B.; funding acquisition, M.T. All authors have read and agreed to the published version of the manuscript.

Funding: This research received no external funding.

Informed Consent Statement: Informed consent was obtained from all subjects involved in the study.

Data Availability Statement: Data are contained within the article.

Conflicts of Interest: The authors declare no conflict of interest.

References

1. Eurostat. How Much Fruits and Vegetables Do You Eat Daily? Available online: https://ec.europa.eu/eurostat/web/products-eurostat-news/-/ddn-20220104-1 (accessed on 18 September 2022).
2. Elstat. The Greek Economy. Available online: https:www.statistics.gr (accessed on 26 March 2021).
3. Aaker, D.; Kumer, V.; Day, G. *Marketing Research*, 5th ed.; John Willey & Sons: New York, NY, USA, 1995; p. 150.
4. Farmakis, N. *Introduction to Sampling*, 1st ed.; Christodoulidi: Thessaloniki, Greece, 2000; p. 56.
5. Dafermos, B. *Factor Analysis with SPSS, LISREL, AMOS, EQS, and STATA*, 1st ed.; Ziti: Thessaloniki, Greece, 2013; p. 36.
6. Karantininis, K. The network form of the cooperative organization. In *Vertical Markets and Cooperative Hierarchies: The Role of Cooperatives in the Agri-Food Industry*, 1st ed.; Karantininis, K., Nilsson, J., Eds.; Springer: Dordrecht, The Netherlands, 2007; Volume 2, pp. 19–34.
7. Stevens, J.P. *Applied Multivariate Statistics for the Social Sciences*, 4th ed.; Lawrence Erlbaum Associates: Mahwah, NJ, USA, 2002; p. 46.
8. Odongo, W.; Dora, M.; Molnar, A.; Ongeng, D.; Gellynck, X. Performance perceptions among food supply chain members. A triadic assessment of the influence of supply chain relationship quality on supply chain performance. *Br. Food J.* 2016, *118*, 1783–1799. [CrossRef]
9. Ruiz, F.; Mena, Y.; Castel, J.; Guinamard, C.; Bossis, N.; Caramelle-Holz, E.; Contu, M.; Sitzia, M.; Fois, N. Dairy goat grazing systems in Mediterranean regions: A comparative analysis in Spain, France and Italy. *Small Rumin. Res.* 2009, *85*, 42–49. [CrossRef]

10. Bekridaki, G. The New Greek Law on SSE: Progressing in the Right Direction. Available online: http://www.ripess.eu/the-new-greek-law-on-sse-progressing-in-the-right-direction/#more-2478 (accessed on 20 August 2022).
11. Petropoulou, E.A. Social and solidarity economy. The case of an Urban Consumption Co-operative in Greece. *Paco* **2018**, *11*, 70–94. [CrossRef]
12. Sharma, S. *Applied Multivariate Techniques*, 1st ed.; John Wiley & Sons: New York, NY, USA; University of North Carolina: Chapel Hill, NC, USA, 1996; p. 90.

Disclaimer/Publisher's Note: The statements, opinions and data contained in all publications are solely those of the individual author(s) and contributor(s) and not of MDPI and/or the editor(s). MDPI and/or the editor(s) disclaim responsibility for any injury to people or property resulting from any ideas, methods, instructions or products referred to in the content.

Proceeding Paper

Automated Sorting System for Skeletal Deformities in Cultured Fishes †

George Bellis [1,*], Paris Papaggelos [1], Evangeli Vlachogianni [1], Ilias Laleas [1], Stefanos Moustos [1], Thanos Patas [2], Sokratis Poulios [2], Nikos Tzioumakis [2], Giannis Giakas [3], Dimitris Bokas [4], Christos Kokkotis [5] and Dimitris Tsaopoulos [6]

1. Biomechanical Solutions (BME), 43100 Karditsa, Greece; ppapangel@gmail.com (P.P.); gb@bme.gr (E.V.); laleasilias@yahoo.gr (I.L.); stefmoy74@gmail.com (S.M.)
2. Polytech S.A., 41222 Larissa, Greece; gaspar611@gmail.com (T.P.); spoulios@polytech.com.gr (S.P.); polytech@lar.forthnet.gr (N.T.)
3. Department of Physical Education and Sport Science, University of Thessaly, 38221 Trikala, Greece; ggiakas@uth.gr
4. PLAGTON S.A., 30100 Agrinio, Greece; bokas@otenet.gr
5. Department of Physical Education and Sport Science, Democritus University of Thrace, 69100 Komotini, Greece; ckokkoti@affil.duth.gr
6. Center for Research and Technology Hellas, Institute for Bio-Economy & Agri-Technology (IBO-CERTH), 60361 Volos, Greece; d.tsaopoulos@certh.gr
* Correspondence: ge.mpellis@gmail.com; Tel.: +30-2441024575
† Presented at the 17th International Conference of the Hellenic Association of Agricultural Economists, Thessaloniki, Greece, 2–3 November 2023.

Abstract: Anomaly occurrence is a constant worldwide problem in aquaculture and it raises economic and animal welfare issues. The early-stage removal of abnormal fish from the stocks is necessary, and the sorting process remains manual worldwide, causing a significant increase in personnel cost and delays in the production cycle. The purpose of this project is to develop an integrated automated system for the valid sorting of farmed fishes by removing these with shape or colour anomalies or skeletal deformities. The sorting will be based on vision analysis and shape pattern recognition techniques.

Keywords: fry; fish; anomalies; deformities; sorting

1. Introduction

The appearance of anomalies is common in the juvenile stages of the fish biological cycle and may include the abnormal shape of a body part, miscolouring, deficient fin growth, a problematic swim bladder, skeletal deformities, etc. Some examples are presented in Figure 1. Although fish farms aim to the retain of increased standards of fish wellbeing and disease prevention, anomalies still exist. There are multiple generative factors with a synergistic effect for anomaly occurrence in fry fish. Some of these are heredity, total fish treatment, high stress levels [1–4], malnutrition, disease [5–9], and an improper farming environment [10].

Fry fish with obvious anomalies may later present a slower growth rate, or die, or ultimately have a reduced commercial value or be discarded. In each case, anomalies financially impact fish farms negatively, so relevant actions are necessary. Early-stage fish sorting is considered a necessity for fish farms [11] and it also aligns with the concept of animal wellbeing [12]. The sorting procedure is proven to consume far fewer resources than using resources to let all the fish grow.

Sorting fish with anomalies remains a manual procedure worldwide and has many disadvantages. It is a labour-intensive procedure, as it demands workers that will observe and sort fish manually. Also, this type of fish-by-fish observing occurs under intense lighting conditions and is an arduous procedure to perform. Furthermore, the process needs experienced workers for effective sorting. The threshold between commercially severe and non-severe deformities is empirically and subjectively defined by workers, due to the lack of a precise quality scale to allow skeletal development to be connected to the external morphology of the fish in various growing phases. This clearly means that manual sorting can be subjective and depend on the experience and condition of the worker that performs it, and obviously this time-consuming procedure causes a bottleneck in the production cycle of a fish hatchery.

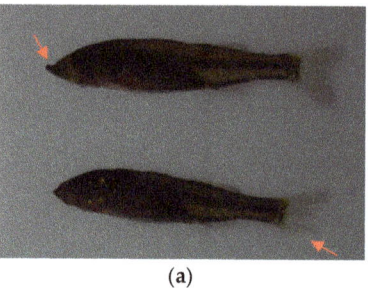
(a)

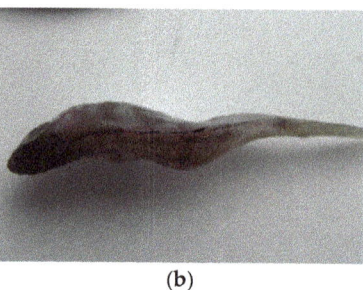
(b)

Figure 1. Fish with anomalies. (**a**) Arrows indicating an abnormal jaw above, and a deficient tail below. (**b**) Example of a fish with a skeletal deformity (scoliosis).

The proper time phase for a single sorting to proceed in a hatchery is when the critical body parts of fry fish have mostly developed and their size is adequate for workers to observe potential anomalies. A multi-stage sorting strategy in various fry fish growth-cycle phases could lead to earlier anomaly detection and reduce the required nutritional resources for hatcheries. As fish sorting is manual, a multi-stage procedure would significantly increase labour costs and is infeasible.

The purpose of this project is the development of an integrated automated system for the valid sorting of farmed fish in the early stages of the growth cycle. The sorting will be based on vision analysis and shape pattern recognition techniques. The practical implication of the system is an improvement in sorting procedure efficiency and a significant reduction in the required time, the labour intensity, and the cost of the process.

2. Methods

The system is initially designed to function as a collaborative robot (co-bot) that combines a configuration with a fish feeding system and a closed cabin with controlled lighting conditions and components that feed visual content to a classification process running on a computational system, using advanced intelligent machine learning algorithms. Under this configuration, the co-bot device relies on a user to perform tasks that are easily performed by human hands under the instruction of the user interface, which is presented on a monitor. The user accesses the fish handling plate through an opening hatch on the cabin surface. The complete process is presented in Figure 2. The practical implication of such a co-botic system that performs the sorting is a significant reduction in labour intensity due to less required time for fish observance and the elimination of human subjectivity.

The algorithms running on the computational system perform multiple tasks, such as the recognition of fish spots over the handling plate, the recognition of probable cases of fish that overlap so that the user is notified to move them apart, and the recognition of visual anomalies and skeletal deformities. The feeding system provides batches of fish and, after a processing procedure, the user is given instructions over the monitor, in order to perform various tasks.

The training of the specific machine learning algorithms, started by manually creating a dataset of multiple pictures of fish, under the instruction of fish-sorting workers, is presented in Figure 3. Experienced sorting workers played a significant role in helping our team to tag the acceptable fish and sort those with anomalies, which were also tagged, as rejectable.

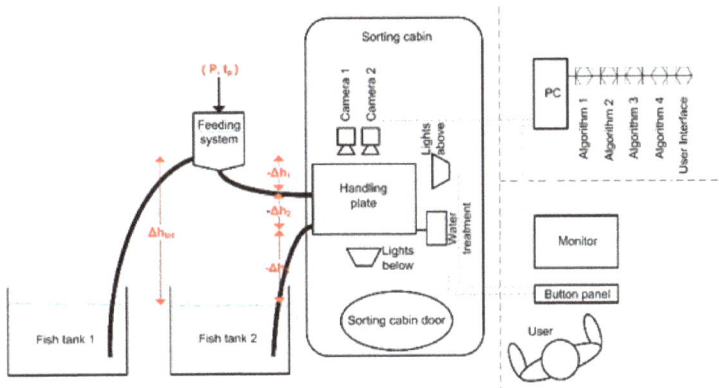

Figure 2. Flowchart of the fish–sorting co–bot device.

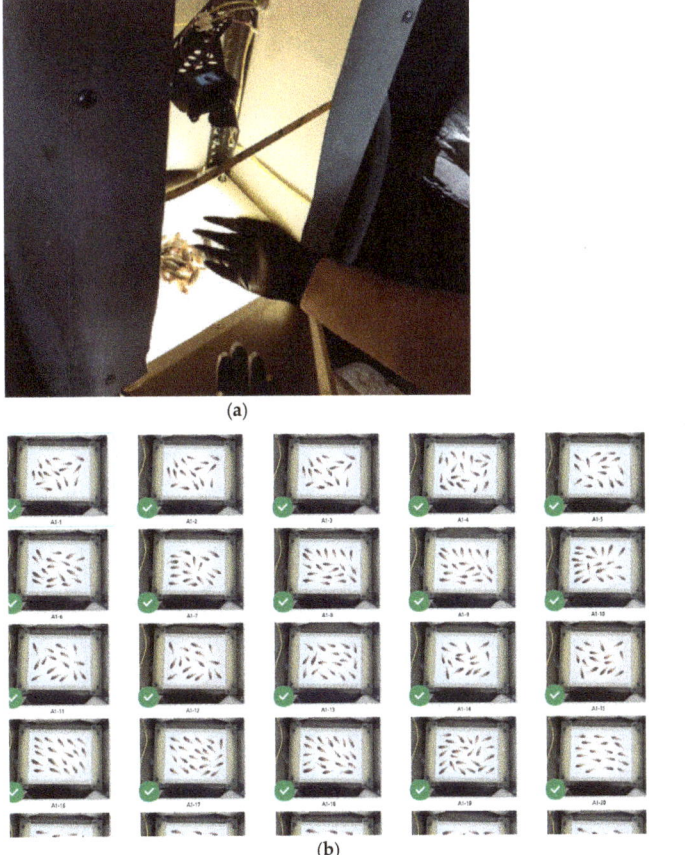

Figure 3. Dataset creation for the training of a machine learning algorithm. (**a**) Purpose-built device for the dataset collection. (**b**) Thousands of photographed fish, tagged in categories.

Significant geometric and color differences were studied in selected areas of interest between acceptable and discarded fish, as presented in Figure 4, so that an efficient percentage of recognition was achieved.

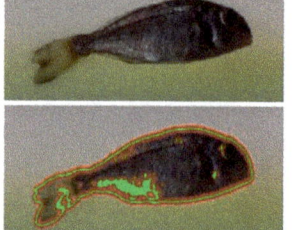 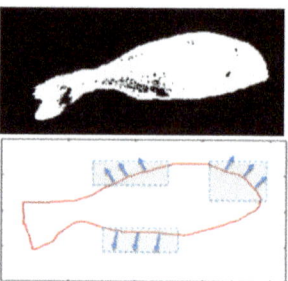

Figure 4. Geometric and colour differentiation recognition.

3. Results

For one of the trials for the validation of the proper function of the prototype device, the following tests were conducted:

- Step 1: A batch of a random number of fish was acquired, and the fish were manually counted to achieve 1394 fish.
- Step 2: The batch was fed to the system, and 124 fish were classified as "discard". Also, the total number of the fish was automatically counted to 1394, which was valid.
- Step 3: The fish that were classified as "discard" were manually separated and examined by an experienced worker, and two (2) were sorted as "acceptable".
- Step 4: The initial batch was formed again by mixing fish and then was manually sorted by an experienced worker. The number of fish that were found to be non-acceptable was 138. This led to the conclusion that the device had an 88.4% recognition success rate.
- Step 5: The manually pre-sorted batch that contained 1256 acceptable fish was fed to the device, and three were classified as "discard".
- Step 6: The manually pre-sorted batch that contained 138 non-acceptable fish was fed to the device, and 123 of them were classified as "discard", which led to the conclusion of an 89.1% recognition success rate, which is slightly higher than that in the previous test.

4. Discussion and Conclusions

The main feature of the prototype device is the non-acceptable fish recognition. In the conducted function tests, the lower percentage of recognition rate was 88.4%, which is considered to be acceptable for a prototype device. The slight difference in the recognition percentage between two tests could occur for various reasons, which could be due to random and uncontrollable facts, such as a small difference in the lighting reflections on the fish. The future goals of the project team include the improvement of the system classification rate.

As a future work, it would be interesting to calculate a break-even point for the device classification success rate, over which the cost of farming non-acceptable fish that were misclassified overcomes the labor costs of manual fish sorting.

Author Contributions: Conceptualization, G.G. and D.T.; methodology, G.B. and C.K.; software, C.K. and I.L.; practical implementation, P.P., E.V., S.M., N.T., D.B., T.P. and S.P.; writing—original draft preparation, G.B.; writing—review and editing, G.B. All authors have read and agreed to the published version of the manuscript.

Funding: This project was funded by Hellenic General Secretariat of Research and Innovation. Project code: T2EDK-00981.

Institutional Review Board Statement: Not applicable.

Informed Consent Statement: Not applicable.

Data Availability Statement: Data sharing is not applicable to this article.

Conflicts of Interest: The authors George Bellis, Paris Papaggelos, Evangeli Vlachogianni, Ilias Laleas and Stefanos Moustos are employed by Biomechanical Solutions (BME). The author Thanos Patas, Sokratis Poulios and Nikos Tzioumakis are employed by Polytech S.A. Dimitris Bokas is employed by PLAGTON S.A. The remaining authors declare that the research was conducted in the absence of any commercial or financial relationships that could be construed as potential conflicts of interest.

References

1. Katavic, I. Diet Involvement in Mass Mortality of Sea Bass (*Dicentrarchus labrax*) Larvae. *Aquaculture* **1986**, *58*, 45–54. [CrossRef]
2. Foscarini, R. A review: Intensive farming procedure for red sea bream (*Pagrus major*) in Japan. *Aquaculture* **1988**, *72*, 191–246. [CrossRef]
3. Carmichael, G.J.; Tomasso, J.R. Swim bladder stress syndrome in largemouth bass. *Tex. J. Sci.* **1984**, *35*, 315321.
4. Divanach, P.; Papandroulakis, N.; Anastasiadis, P.; Koumoundouros, G.; Kentouri, M. Effect of water currents during postlarval and nursery phase on the development of skeletal deformities in sea bass (*Dicentrarchus labrax* L.) with functional swim-bladder. *Aquaculture* **1997**, *156*, 145–155. [CrossRef]
5. Browder, A.J.; McClellan, D.B.; Harper, D.E.; Michael, G. Kandrashoff, M.G.; Walter Kandrashoff, W. A major developmental defect observed in several Biscayne Bay, Florida, fish species. *Environ. Biol. Fishes* **1993**, *37*, 181–188. [CrossRef]
6. Koumoundouros, G.; Gagliardi, F.; Divanach, P.; Boglione, C.; Cataudella, S.; Kentouri, M. Normal and abnormal osteological development of caudal fin in *Sparus aurata* L. fry. *Aquaculture* **1997**, *149*, 215–226. [CrossRef]
7. Koumoundouros, G.; Oran, G.; Divanach, P.; Stefanakis, S.; Kentouri, M. The opercular complex deformity in intensive gilthead sea bream (*Sparus aurata* L.) larviculture. Moment of apparition and description. *Aquaculture* **1997**, *156*, 165–177. [CrossRef]
8. Koumoundouros, G.; Divanach, P.; Savaki, A.; Kentouri, M. Effects of three preservation methods on the evolution of swimbladder radiographic appearance in sea bass and sea bream juveniles. *Aquaculture* **2000**, *182*, 17–25. [CrossRef]
9. Koumoundouros, G.; Sfakianakis, S.; Maingot, E.; Divanach, P.; Kentouri, M. Osteological development of the vertebral column and of the fins in *Diplodus sargus* (Teleostei: Perciformes: Sparidae). *Mar. Biol.* **2001**, *139*, 853–862.
10. Corti, M.; Loy, A.; Cataudella, S. Form changes in the sea bass, *Dicentrarchus labrax* (Moronidae: Teleostei), after acclimation to freshwater: An analysis using shape coordinates. *Environ. Biol. Fishes* **1996**, *47*, 165–175. [CrossRef]
11. Korwin-Kossakowski, M.; Myszkowski, L.; Kaminski, R. A simple method to detect body morphological abnormalities in juvenile cyprinid fish—A case study on ide *Leuciscus idus*. *Aquaculture* **2017**, *25*, 915–925. [CrossRef]
12. Boglione, C.; Gisbert, E.; Gavaia, P.; Witten, P.E.; Moren, M.; Fontagné, S.; Koumoundouros, G. Skeletal anomalies in reared European fish larvae and juveniles. Part 2: Main typologies, occurrences and causative factors. *Rev. Aquac.* **2013**, *5*, 121–167. [CrossRef]

Disclaimer/Publisher's Note: The statements, opinions and data contained in all publications are solely those of the individual author(s) and contributor(s) and not of MDPI and/or the editor(s). MDPI and/or the editor(s) disclaim responsibility for any injury to people or property resulting from any ideas, methods, instructions or products referred to in the content.

Proceeding Paper

The Role of Cooperatives in the Interconnection of the Agri-Food and Tourism Sectors, Kyllini, 14/09/2023 †

Panagiota Pantazi

Agricultural Cooperative Kampos Andravidas, PC 27068 Kyllini, Greece; pl.pantazi@gmail.com
† Presented at the 17th International Conference of the Hellenic Association of Agricultural Economists, Thessaloniki, Greece, 2–3 November 2023.

Abstract: The interconnection of cooperatives with tourist products is the assumption of the quality of their products as an incentive for tourists to visit the country and propose it further through their gastronomic experience. The high quality of this gastronomic experience is the proposal to link tourism (tertiary sector) with agri-food (primary sector). The research questions that arise through the analysis of the agri-food and tourism industry, concern the finding of the reasons that hinder the interconnection of agri-food and tourism through gastronomy as well as the advantages arising from the interconnection between them. A combination of qualitative and quantitative research has been chosen as a research methodology.

Keywords: social and solidarity economy; cooperative enterprises; agri-food; gastronomy; tourism

1. Introduction

Tourism, from 2019 onwards, due to the COVID-19 crisis, suffered a severe blow [1]. This volatile situation that has emerged due to the ongoing economic, geopolitical, and health crises raises questions about the channeling of both tertiary sector services and the distribution of primary production, creating a vicious circle of crises of economic inadequacy. The existing tourism model of development of over-tourism, appears saturated, with its consequences visible in the ecological destruction of entire regions and the cultural degradation of local communities. The UN, as stated in its "2030 Agenda for Sustainable Development", aims (goal 8) "by 2030, to design and implement policies to promote sustainable tourism that creates jobs and promotes local culture and products".

As an alternative approach to the problems presented above, a new trend has emerged in recent years, that of "Gastronomic Tourism", based on food, which becomes an important incentive to travel [2]. Traditional products and dishes become new tourist attractions shaping the choice of a destination or even the type of agricultural production [3]. This new relationship between agriculture and tourism is being studied because of the positive effects that can result from their effective cooperation. This cooperation requires differentiation of the final product produced and offered as the consumer-tourist is interested in quality. This qualitative difference is possible within agricultural cooperatives through their quality systems [4], ensuring fresh products with a low environmental footprint and strengthening the social fabric through the branching and interconnection of several different economic sectors.

The different facets of gastronomic tourism, including food ethics, social bonding, hospitality, local development, and sustainability make it the subject of further research.

Thus, the challenge of (a) gastronomic tourism, which does not expose the local environment to its limits, but is integrated into local communities aiming at gastronomic satisfaction and the economic regeneration of the local community, and (b) the development and strengthening of agricultural cooperatives that satisfy the demand for fresh products of high nutritional quality from tourists, promoting the strengthening of the primary sector,

Citation: Pantazi, P. The Role of Cooperatives in the Interconnection of the Agri-Food and Tourism Sectors, Kyllini, 14/09/2023. *Proceedings* **2024**, *94*, 59. https://doi.org/proceedings2024094059

Academic Editor: Eleni Theodoropoulou

Published: 7 March 2024

Copyright: © 2024 by the author. Licensee MDPI, Basel, Switzerland. This article is an open access article distributed under the terms and conditions of the Creative Commons Attribution (CC BY) license (https://creativecommons.org/licenses/by/4.0/).

which in the long run, through appropriate actions and networking, can lead the primary and tertiary sector to new cross-sectoral cooperation.

This research examines precisely this interconnection of the agri-food and tourism sectors and the role that cooperatives can play between them. The research is limited to the region of Western Greece (P.D.E.) and more specifically to the Regional Unit of Ilia (P.E. Ilia) due to the particular characteristics of the primary and tertiary sectors there [5]. The purpose of this analysis, in P.D.E., is to understand the reasons why cooperation between the agri-food and tourism sectors has not yet flourished to the extent needed and to highlight the advantages for both sectors.

2. Materials and Methods

For the scope of this analysis, qualitative and quantitative research was used also primary and secondary sources such as statistics, scientific books, journal articles, research presentations, and websites. Good practices were also sought in Greece and abroad. Particularly important was the use of quantitative data by ELSTAT, DAOK ILIA, and the Chamber of ILIA.

The sample included:
1. the total population—number (87) of hotel accommodation in the Prefecture of Ilia and 9 of the type "Rented accommodation"
2. the total population—number (27) of active cooperatives in the Prefecture of Ilia and
3. 39 production units (in the sector of fresh fruit and vegetables, beverages, standardized products such as dairy, yeasts, pastries, juices, jams, sweets).

The total sample is 162 respondents.

As instruments for measuring the research process, 3 different questionnaires were designed and compiled. The design of the three different questionnaires aimed to find detailed results for each sector. The first concerned tourist units. This was followed by the questionnaire for the cooperatives and finally the questionnaire for the production units. Each of them consists of sections and each section of questions in the form of short completion, multiple choice, and Likert scale. The conduct of the interviews and completion of the questionnaires took place from 15 June 2022 to 31 December 2022.

3. Results

Summarizing the key features of the 3 sections, we find that:

- in the 87 tourist units, food, and drink revenues, in all-star categories, constitute 1/3 of their total revenue. Food and beverage purchases exceed 50% while domestic food and beverage supplies exceed 60%, especially in 5-star and 4-star hotels. The largest influx of supplies is served by intermediate wholesalers.
- Of the 27 cooperatives in the region of Ilia, only 2 cooperate with tourist accommodation as they all have exclusively export activity (over 90%).
- On the contrary, out of the 39 production bodies, 61.50% cooperate with tourist units, especially in the category of local wines and fresh fruits and vegetables.

As concern «the difficulty of cooperation between all three sectors» the main reasons are:

- Limited or unreliable distribution network,
- Lack of certification (quality assurance systems),
- Incomplete/ineffective cooperation,
- Unfair behaviour, ineffective cooperation,
- The low prices offered by hotels,
- The inability to cover the range of products that tourist units want.

The research has also shown that solving the above difficulties is feasible and as solutions are proposed:

- the existence of a certification program for the locality of products at the regional level,

- the existence of a thematic tourism programme within which a local quality pact operates,
- easy access/updating/ordering of products through an online platform,
- the use of a networking program of different partners at the regional level,
- the existence of advisory support from a body of the Region like an Agri-Food Partnership.

Finally, regarding the development of gastronomic tourism for all three categories of respondents, despite the different degrees of priority that has been given, common points are distinguished such as:

- the absence of networks and synergies between agencies,
- the absence of a coordinating body at the regional level,
- the insufficient staffing of tourist units,
- the ineffective promotion of gastronomic wealth,
- the difficulties with quality labels (although the tourist units as a whole do not have quality labels, except 5* and 4*, there is a Greek breakfast offer. On the other hand, cooperatives, due to their export orientation, have great quality production systems as a whole).

4. Discussion

The research highlighted the important role that P.D.E acquires in coordinating a series of actions both for the development of cooperation between all sectors and actions for the development of gastronomic tourism. It appears as a conscious choice and can be transformed into a tool for creating added value in the distribution cycle to tourist accommodation, forming a first gastronomic identity.

In the above challenges, P.D.E., understanding its role, is already undertaking actions towards the connection of agri-food with gastronomic tourism with the participation of all 27 cooperatives, the 87 tourist units, and the other productive companies in the region of Ilia.

5. Conclusions

The research in P.E. Ilia revealed many encouraging elements of conditional cooperation. The acquisition of a new sustainable tourism consciousness, in cooperation with all involved companies, seems feasible. Any barriers and knowledge deficits can be overcome through educational activities related to sustainability. In addition to the above proposals, this paper was an occasion, for those who participated in the research, for new reflections and further thoughts on what kind of tourism we are looking for, how this event affects our future prosperity, and the role of everyone in its realization. The promise by the majority of respondents for an annual repetition of the survey and their participation in it, with the development of proposals, to monitor the degree of maturation of gastronomic tourism as an alternative and sustainable way of development, is for the region of Ilia an element of hope and future development with a social orientation.

Funding: This research received no external funding.

Institutional Review Board Statement: Ethical review and approval were waived for this study, due to the research presents no more than minimal risk of harm to participants and involves no procedures for which written consent is normally required outside the research context.

Informed Consent Statement: Informed consent was obtained from all subjects involved in the study. All the participants and i give our permission to MDPI for publication of this article.

Data Availability Statement: The data used in this study are available on request from the corresponding author.

Conflicts of Interest: The authors declare no conflict of interest.

References

1. Report for the Year 2020, Bank of Greece. Available online: https://www.bankofgreece.gr (accessed on 15 December 2022).
2. SETE. Association of Greek Tourism Businesses, Gastronomy in Greek Tourism Marketing. 2009. Available online: https://sete.gr/_fileuploads/gastro_files/100222gastronomy_f.pdf (accessed on 8 December 2022).
3. Toposophy. Gastronomy as a Factor Enriching the Travel Experience and Upgrading the Tourist Product. Study by INSETE. 2019. Available online: https://insete.gr/wp-content/uploads/2020/04/Gastronomy_full.pdf (accessed on 3 November 2022).
4. Kontogeorgos, A.; Sergaki, P. *Management Principles of Agricultural Cooperatives*; Kallipos, Open Academic Publications: Zografou, Greece, 2015. Available online: https://hdl.handle.net/11419/3684 (accessed on 15 December 2022).
5. Dianeosis. Reconstruction of Ilia: An Overview of the Current Situation. Available online: https://www.dianeosis.org/2022/12/anasygrotisi-tis-ilias-mia-episkopisi-tis-simerinis-katastasis/ (accessed on 10 January 2023).

Disclaimer/Publisher's Note: The statements, opinions and data contained in all publications are solely those of the individual author(s) and contributor(s) and not of MDPI and/or the editor(s). MDPI and/or the editor(s) disclaim responsibility for any injury to people or property resulting from any ideas, methods, instructions or products referred to in the content.

Proceeding Paper

Consumers' Behavior toward Plant-Based Milk Alternatives †

Dimitris Alexandridis *, Christina Kleisiari and George Vlontzos

Agricultural Economics and Consumer Behavior Lab, University of Thessaly, 38446 Volos, Greece; chkleisiari@uth.gr (C.K.); gvlontzos@uth.gr (G.V.)
* Correspondence: dalexandridis@uth.gr; Tel.: +30-6984799546
† Presented at the 17th International Conference of the Hellenic Association of Agricultural Economists, Thessaloniki, Greece, 2–3 November 2023.

Abstract: In recent years, more people have expressed interest in Plant-Based Milk Alternatives (PBMAs). Our research focused on Greek consumers to examine consumer behavior regarding PBMAs. Using relevant literature, a questionnaire was designed and distributed both online and through personal interviews. The sample was random and concerned 576 consumers from the Greek mainland, of which 53.5% were women and 46.5% were men, aged 18 to 80. The Health Belief (HBM) and Stimulus Organism Response (SOR) models were used to design the questionnaire, while Principal Component Analysis (PCA) was applied for the interpretation of the survey results. PCA showed that consumers' perception of PBMAs, and their willingness to consume them or influence others to do so, are the most significant variables. Furthermore, Linear Regression Analysis revealed that PBMAs are primarily purchased by younger and more highly educated consumers. The results of the research can contribute to the improvement of PBMA retail marketing strategies in Greece.

Keywords: PBMAs; consumer behavior; principal component analysis (PCA)

Citation: Alexandridis, D.; Kleisiari, C.; Vlontzos, G. Consumers' Behavior toward Plant-Based Milk Alternatives. *Proceedings* **2024**, *94*, 60. https://doi.org/10.3390/proceedings2024094060

Academic Editor: Eleni Theodoropoulou

Published: 11 March 2024

Copyright: © 2024 by the authors. Licensee MDPI, Basel, Switzerland. This article is an open access article distributed under the terms and conditions of the Creative Commons Attribution (CC BY) license (https://creativecommons.org/licenses/by/4.0/).

1. Introduction

In recent years, the consumption of plant-based beverages has shown an upward trend. According to Nielsen IQ [1], for the year 2021–2022, milk and dairy alternatives' total dollar growth increased by 6.8% compared to the previous year and 34.6% over the previous 3 years. The same research states that even though alternative milk consumption has increased significantly, the plant-based industry accounts for only 15% of total milk sales. Markets & Markets [2] projects that the global market for dairy alternatives will grow by USD 17.5 billion until 2027, reaching USD 44.8 billion. Despite the growth tendency, many consumers still find it difficult to switch from traditional milk to milk alternatives. More specifically, a recent European study, including German, French and Polish consumers, indicated that people are not ready to give up on dairy products entirely, whilst the reason for adding dairy substitutes to their diets is primarily out of curiosity, the need to explore new things and to expand their diet [3]. According to Schiano et al. [4], many parents still link dairy milk with positive characteristics, making them more persuasive than plant-based alternatives. On the other hand, Moss et al.'s [5] survey states that consumers have correlated milk alternatives with health benefits, sustainability and sensory attributes. Research from Denmark by Martinez-Padilla et al. [6] indicates that taste, followed by health and naturalness, are the main motivators for the consumption of PBMAs, while negative indicators include the belief that PBMAs are artificial and heavily processed.

Research opinions on the consumption of plant-based beverages are divided. Considering that plant-based milk alternatives are an ongoing trend, this study aimed to examine Greek consumers' attitudes regarding those products. The goal is, therefore, to be able to define the main factors leading consumers to either choose these beverages or not, in comparison to conventional milk, and their awareness of those products, their concerns and their future intentions to consume them, using a sufficient and representative sample of respondents. Furthermore, PCA was performed. The results showed that consumers' perceptions of PBMAs and their willingness to consume and promote them are the most important variables. Additionally, the Linear Regression Analysis indicated that PBMAs are predominantly bought by younger consumers with a higher education level.

2. Methods and Materials

2.1. Questionnaire Construction

Initially, a review of the literature on consumer behavior theories was conducted to identify the most essential aspects influencing consumer perceptions and preferences. The HBM was used to examine the relationship between health-related factors and consumer behavior. In addition, the SOR model was used to clarify how direct and indirect social environment stimuli contribute to making these decisions, and how the consumer evaluates the product through consumption experience. The questionnaire is divided into two main categories. The first category included demographic questions, while the second one concerned consumers' knowledge of plant-based milk alternatives, awareness of their consumption, consumer behavior, health-related questions, and willingness to consume. The questionnaire was distributed to consumers from the Greek mainland, aged 18 to 80, over the course of five months, starting from November 2022, through both online and in-person interviews. In-person interviews were used to encourage more people, especially older people and those with little or no technological background, to participate in the survey.

2.2. Statistical Analysis

PCA and Linear Regression Analysis was used to analyze the results. PCA allows a considerable amount of information to be condensed into a small number of variables, which indicate very distinct and clear characteristics of the phenomenon under assessment. The order in which the components appear reflects their relative weight, resulting in a systematic ranking of the factors. Thus, the first component, in order of appearance, holds more importance than the second, and the second more than the third, and so on. Variables were expressed on a Likert scale from 1 to 5, with 1 indicating total disagreement and 5 total agreement with each statement. KMO was calculated to ensure that PCA results were credible. Linear Regression Analysis was used to correlate the factors emerging from PCA with the socioeconomic characteristics of the sample.

3. Results

3.1. Demographic Information

Demographic data showed that 53.5% of the sample corresponded to female participants, while male participants accounted for the remaining 46.5%. Furthermore, the data indicated that 21.53% of the participants fell within the 18–24 age group, 20.49% in the 25–34 age group, 22.40% in the 35–44 age group, 20.66% in the 45–59 age group and 14.93% aged 60 and above.

3.2. Consumer Behavior Analysis

For the consumer behavior analysis, PCA was used to examine how the variables are interconnected, and if there are any relations among them. PCA was performed with variables from the second part of the questionnaire. The KMO indicator was 0.847, close to 1 and of high significance, meaning that the sample was adequate for PCA analysis and that the study was valid.

PCA produced seven components, of which the two most significant will be further discussed. Factor 1 indicates that consumers' perceptions of plant-based beverages and traditional milk are related to their interest in drinking plant-based milk alternatives and their willingness to influence others (Figure 1). All the variables in the second component are grouped under the same heading: health-related questions. Consumers are concerned about a variety of health issues that can be induced by the consumption of dairy milk, including diabetes, cardiovascular disease, osteoporosis, and chronic fatigue.

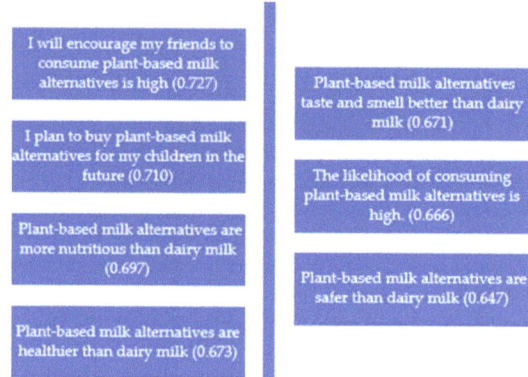

Figure 1. Questions related to the first component (factor loadings), authors' elaboration.

Linear Regression Analysis indicated that participants' age and educational status affect both the first and second component. The results for the first factor showed that most younger consumers believe that PBMAs are healthier than regular milk, whereas older consumers strongly disagree with this statement. In addition, even though younger people are more interested in buying PBMAs for their kids in the future, older people do not share the same interest. Another significant result shows that only highly educated consumers are willing to buy PBMAs for their children. The results of the second component reveal that older customers were more anxious about facing cardiovascular disease or osteoporosis than younger people. Furthermore, educated consumers are less concerned regarding these health issues, as was expected.

4. Discussion

It is important to mention that in this study consumers with higher educational status were more interested in purchasing PBMAs. This is verifying Kriwy and Mecking's survey [7], which connects a high level of education with organic food consumption. Differences in food perceptions are known to be influenced by traditions and cultures, as is the case for consumer surveys in Poland, Germany and France [3]. In most relevant research papers only a small group of young consumers participated, clarifying the parent perspective on PBMA consumption. For this reason, the findings of this study are significant to fulfill this need.

5. Conclusions

According to the research findings, there is a growing interest in plant-based milk alternatives among Greek consumers. This tendency, however, is more popular among younger and more educated consumers. Overall, the study can help inform retailers and marketers in Greece about this upward trend and the alternative ways to effectively promote and commercialize plant-based milk alternatives.

Author Contributions: Conceptualization, G.V.; methodology, G.V.; software, D.A. and C.K.; validation, C.K. and D.A.; formal analysis, D.A. and C.K.; investigation, C.K.; resources, C.K.; data curation, C.K.; writing—original draft preparation, D.A.; writing—review and editing, D.A. and C.K.; visualization, D.A.; supervision, G.V.; project administration, C.K.; funding acquisition, G.V. All authors have read and agreed to the published version of the manuscript.

Funding: This project has received funding from the European Union's HE research and innovation program under grant agreement No 101084647 and official acronym NATAE. Views and opinions expressed are, however, those of the authors only and do not necessarily reflect those of the European Union or the European Research Executive Agency (REA). Neither the European Union nor the granting authority can be held responsible for any use of the information contained in the document.

Institutional Review Board Statement: Not applicable.

Informed Consent Statement: Informed consent was obtained from all subjects involved in the study.

Data Availability Statement: All data used for this study are available upon request to the corresponding author.

Conflicts of Interest: Authors declare no conflicts of interest.

References

1. Nielsen IQ. Available online: https://nielseniq.com/wp-content/uploads/sites/4/2022/09/high-priority-revisions-pulse-on-plant-based-ebook-d01.pdf (accessed on 11 May 2023).
2. Markets & Markets. Available online: https://www.marketsandmarkets.com/Market-Reports/dairy-alternative-plant-milk-beverages-market-677.html (accessed on 11 May 2023).
3. Adamczyk, D.; Jaworska, D.; Affeltowicz, D.; Maison, D. Plant-Based Dairy Alternatives: Consumers' Perceptions, Motivations, and Barriers—Results from a Qualitative Study in Poland, Germany, and France. *Nutrients* **2022**, *14*, 2171. [CrossRef] [PubMed]
4. Schiano, A.N.; Nishku, S.; Racette, C.M.; Drake, M.A. Parents' implicit perceptions of dairy milk and plant-based milk alternatives. *J. Dairy Sci.* **2022**, *105*, 4946–4960. [CrossRef]
5. Moss, R.; Barker, S.; Falkeisen, A.; Gorman, M.; Knowles, S.; McSweeney, M.B. An investigation into consumer perception and attitudes towards plant-based alternatives to milk. *Food Res. Int.* **2022**, *159*, 111648. [CrossRef] [PubMed]
6. Faber, I.; Petersen, I.L. Perceptions toward Plant-Based Milk Alternatives among Young Adult Consumers and Non-Consumers in Denmark: An Exploratory Study. *Foods* **2022**, *12*, 385. [CrossRef]
7. Kriwy, P.; Mecking, A. Health and environmental consciousness, costs of behaviour and the purchase of organic food. *Int. J. Consum. Stud.* **2022**, *36*, 30–37. [CrossRef]

Disclaimer/Publisher's Note: The statements, opinions and data contained in all publications are solely those of the individual author(s) and contributor(s) and not of MDPI and/or the editor(s). MDPI and/or the editor(s) disclaim responsibility for any injury to people or property resulting from any ideas, methods, instructions or products referred to in the content.

Proceeding Paper

How Do Agricultural Education, Advisory, and Financial Factors Affect the Adoption of Precision Farming in Greece? †

Maria Tsiouni *, Georgios Kountios and Alexandra Pavloudi

Department of Agricultural Economics and Entrepreneurship, School of Agriculture, International Hellenic University, Sindos, 57400 Thessaloniki, Greece; gkountios@gmail.com (G.K.); apavlousi@gmail.com (A.P.)
* Correspondence: mtsiouni@agr.teithe.gr
† Presented at the 17th International Conference of the Hellenic Association of Agricultural Economists, Thessaloniki, Greece, 2–3 November 2023.

Abstract: The purpose of this paper is to conduct an empirical investigation of the theoretical and literature-based constructs related to the adoption of precision agriculture (PA) practices by young farmers. For this research, primary and secondary data are used. The sample includes 220 young farmers. Among the results of this research, farmers are aware of the positive effects of technology systems in agriculture. Also, young farmers seem to be familiar with precision agriculture and have already adopted some of its methods, but the high cost of investment prevents farmers from adopting such technology. Innovative technologies and production methods can help young farmers to be competitive in the worldwide market.

Keywords: young farmers; precision agriculture; agricultural education; agricultural advisory; financial factors

1. Introduction

Today, the increased use of chemicals and fertilizers and agricultural mechanization have created imbalances in natural resources. Increasing farm income and optimizing yields with a minimum of resources and financial inputs are major challenges for sustainable agriculture. Technology and data-driven decision making play important roles in the management of farms, along with the application of knowledge, skills, and experiences. Utilizing production resources efficiently and adopting advanced technologies are key to maximizing production. Maximizing profits while operating within the constraints of accessible resources is a fundamental priority for businesses. These resources encompass financial and credit assets, material support essential for production, and the requisite skills needed for the workforce to carry out their tasks effectively [1].

Precision agriculture (PA) can address this challenge. Precision agriculture, as a tool enabling farmers to enhance land management efficiency, exerts a significant and diverse influence on farm management practices. Global trends indicate a projected surge in the adoption of precision agriculture over the next four years, resulting in a doubling in the market value from USD 17.41 billion in 2022 to USD 34.1 billion by 2026 [2].

Information technology is used in precision agriculture to improve the accuracy of quantity, quality, timing, and location information when applying and utilizing inputs in agricultural production, thereby reducing seed, fertilizer, water, and pesticide costs; increasing yields; and increasing profitability [3]. Precision agriculture is also used to increase agricultural production in several ways [4]. Tools based on GPS technologies, information technology, farm management and economic knowledge, and sensor and application technologies are available [5].

The European Union, following the latest revision of the Common Agricultural Policy, encourages farmers to produce high-quality agricultural products using environmentally friendly farming practices [6]. To achieve these goals, it is necessary to import technology

---

Citation: Tsiouni, M.; Kountios, G.; Pavloudi, A. How Do Agricultural Education, Advisory, and Financial Factors Affect the Adoption of Precision Farming in Greece? *Proceedings* **2024**, *94*, 61. https://doi.org/10.3390/proceedings2024094061

Academic Editor: Eleni Theodoropoulou

Published: 11 March 2024

Copyright: © 2024 by the authors. Licensee MDPI, Basel, Switzerland. This article is an open access article distributed under the terms and conditions of the Creative Commons Attribution (CC BY) license (https://creativecommons.org/licenses/by/4.0/).

into agriculture. A one-way road to increase farming efficiency and minimize environmental impact seems to be the challenge of adopting precision agriculture technologies.

The purpose of this study is to conduct an empirical investigation of the theoretical and literature-based constructs related to the adoption of precision agriculture (PA) practices by young farmers, as young farmers are better equipped to interpret new information and to search for suitable tools to support production [7,8].

2. Methods

The study area was the Central Macedonia region. The sampling frame of the survey included the years 2020–2021. In the first stage, stratified random sampling was applied with a proportional distribution of the sample between the Regional Units of the Region. Each Regional Unit of the Region of Central Macedonia was considered to correspond to a layer. In the second stage of sampling, simple random sampling was applied with systematic selection according to the lists of beneficiaries of the grant programs. The study population was defined as young farmers (under 40 years old) who live in the Region of Central Macedonia. A total of 220 questionnaires were collected, which were filled out through personal interviews. Given the population of 1732 young farmers, this constitutes a satisfactory sample size for a margin of error of $\pm 5\%$ and a confidence level of 1% (z = 2.58). For the purposes of research, primary and secondary data were used.

This study focuses on young farmers (under 40 years old) because they constitute a dynamic group of individuals willing to adopt innovation—an integral part of the Greek rural community with a vital role to play in improving the competitiveness of the Greek agriculture sector. The questionnaire included questions related to views and attitudes on innovation related to PA, information and communication technologies (ICTs), agriculture education, information about the environment, and the cost of adopting the PA technology.

3. Results and Discussion

The results analysis shows that young farmers have a remarkable level of information regarding new technologies and innovations in general while presenting positives in the adoption of innovations. The main findings are presented below:

Attitudes towards innovation in agriculture

A proportion of 95.7% of the sample are familiar with innovation in agriculture, which is of particular importance because they are the ones who are expected to adopt innovations. It turns out that young farmers are more innovative.

Attitudes towards PA

A proportion of 56.9% of farmers have some information about PA. However, many of the farmers apply systems that fall within the concept of PA without knowing it. As for the benefits of PA, farmers seem to recognize most of them.

Attitudes towards ICTs

According to the results, most young farmers in the sample (94.5%) know what ICTs are.

Attitudes towards agricultural education

A proportion of 60.4% of young farmers believe that education has an important role. During recent years, respondents attended training programs. The rest of the young farmers attended compulsory training before 2017 when they joined the program. In addition to this mandatory program, "Young Farmers" was attended by 9 out of 10 participants. A proportion of 89.7% of young farmers attended some training in the last five years. A proportion of 25.2% joined seminars for computer learning, while a small percentage were trained in youth entrepreneurship. The majority of young farmers (77.2%) state that knowledge serves daily needs.

Attitudes towards information/advisory

A proportion of 75.6% of the farmers believe in the importance of information/advisory services. The most important sources of information for young farmers in the sample are the specialized information; they trust mainly agronomists of the local Directorate of Rural Development (31.7%) and less private agronomist-researchers (21.3%) and agronomist-trades of agricultural supplies (15.9%).

Attitudes towards the environment

Almost all farmers (93.3%) say that they are concerned about the environment. Most farmers are recognizing the negative effects of conventional agriculture practiced today.

Attitudes towards the cost of adopting the PA technology

A proportion of 82.3% of the farmers believe that financial factors prevent young farmers from adopting PA. The high investment cost and the high maintenance costs are barriers regarding PA.

4. Conclusions

A significant challenge in sustainable agriculture involves achieving maximum crop yields and boosting farm income while minimizing the use of resources and financial investments, as well as ensuring the protection of the environment. Precision Farming (PF) technologies can play a crucial role in tackling this challenge. The implementation of Precision Farming (PF) has become feasible due to advancements in various technologies, including geographic information systems, global navigation satellite systems (GNSSs), remote sensing (RS), satellite imagery, ground sensors, and components of mobile computing and telecommunication [8]. Despite the benefits of PF, these technologies are currently not widely adopted by farmers and, especially, by the elderly. Younger farmers are generally more receptive to innovative ideas and are more inclined to incorporate new technologies into their farming practices [9].

Having realized the everyday changes in the methods of production, processing, and marketing of agricultural products, government agencies should draw up strategic directions for the development of the agricultural sector in a timely manner. Also, they need to develop agricultural research on a modern basis, with emphasis on the fields of ICTs, marketing and management, and other scientific fields. At the same time, they should develop the technological infrastructure and prepare the farmers, train them in modern information and communication technologies, and investigate production methods such as precision agriculture. Thus, the following is highly recommended: (a) the creation and provision of specific incentives for the acceptance and use of information and communication technologies; (b) the development of appropriate infrastructure to support the use of ICTs in agriculture, with appropriate networking equipment and know-how; (c) the configuration of the integrated educational program; (d) the improvement and development of advisory services; (e) and the subsidy of the new technologies regarding PA.

This study is not without limits. In this paper, Precision Agriculture (PA) is treated as a unified concept, yet there exists a considerable body of information indicating that adoption rates differ significantly for various types of Precision Farming (PF) technologies. Nevertheless, it is important to note that the current study should be seen as an initial assessment of PF adoption within Greek farms. In this context, it serves as a foundational point for future research endeavors, which can delve into more specific Precision Agriculture Technologies (PATs) and their adoption patterns.

Author Contributions: Conceptualization, M.T. and G.K.; methodology, G.K.; software, A.P.; validation, M.T., G.K., and A.P.; formal analysis, M.T.; investigation, G.K.; resources, A.P.; data curation, G.K.; writing—original draft preparation, M.T.; writing—review and editing, M.T.; visualization, G.K; supervision, A.P.; project administration, G.K. All authors have read and agreed to the published version of the manuscript.

Funding: This research received no external funding.

Institutional Review Board Statement: Not applicable.

Informed Consent Statement: Informed consent was obtained from all subjects involved in the study.

Data Availability Statement: Data are contained within the article.

Conflicts of Interest: The authors declare no conflicts of interest.

References

1. Shmatkovska, T.O.; Dziamulych, M.; Vavdiiuk, N.; Petrukha, S.; Koretska, N.; Bilochenko, A. Trends and Conditions for the Formation of Profitability of Agricultural Enterprises: A Case Study of Lviv Region, Ukraine. *Univ. J. Agric. Res.* **2022**, *10*, 88–98. [CrossRef]
2. Karydas, C.; Chatziantoniou, M.; Tremma, O.; Milios, A.; Stamkopoulos, K.; Vassiliadis, V.; Mourelatos, S. Profitability Assessment of Precision Agriculture Applications—A Step Forward in Farm Management. *Appl. Sci.* **2023**, *13*, 9640. [CrossRef]
3. Say, S.M.; Keskin, M.; Sehri, M.; Sekerli, Y.E. Adoption of precision agriculture technologies in developed and developing countries. *Online J. Sci. Technol.* **2018**, *8*, 7–15.
4. Fraser, A. 'You can't eat data'?: Moving beyond the misconfigured innovations of Smart Farming. *J. Rural. Stud.* **2022**, *91*, 200–207. [CrossRef]
5. Reichardt, M.; Jurgens, C.; Kloble, U.; Huter, J.; Moser, K. Dissemination of precision farming in Germany: Acceptance, adoption, obstacles, knowledge transfer and training activities. *Precis. Agric.* **2009**, *10*, 525–545. [CrossRef]
6. Tsiouni, M.; Aggelopoulos, S.; Pavloudi, A.; Siggia, D. Economic and Financial Sustainability Dependency on Subsidies: The Case of Goat Farms in Greece. *Sustainability* **2021**, *13*, 7441. [CrossRef]
7. Barnes, A.P.; Soto, I.; Eory, V.; Beck, B.; Balafoutis, A.; Sanchez, B.; Vangeyte, J.; Fountas, S.; van der Wal, T.; Gomez-Barbero, M. Exploring the adoption of precision agricultural technologies: A cross regional study of EU farmers. *Land Use Policy* **2019**, *80*, 163–174. [CrossRef]
8. Paxton, K.; Mishra, A.; Chintawar, S.; Roberts, R.; Larson, J.A.; English, B.; Dayton, M.L.; Marra, M.C.; Larkin, S.L.; Reeves, J.M.; et al. Intensity of Precision Agriculture Technology Adoption by Cotton Producers. *Agric. Resour. Econ. Rev.* **2011**, *40*, 133–144. [CrossRef]
9. Paustian, M.; Theuvsen, L. Adoption of precision agriculture technologies by German crop farmers. *Precis. Agric.* **2017**, *18*, 701–716. [CrossRef]

Disclaimer/Publisher's Note: The statements, opinions and data contained in all publications are solely those of the individual author(s) and contributor(s) and not of MDPI and/or the editor(s). MDPI and/or the editor(s) disclaim responsibility for any injury to people or property resulting from any ideas, methods, instructions or products referred to in the content.

Proceeding Paper

Discovering Innovation, Social Capital and Farm Viability in the Framework of the United Winemaking Agricultural Cooperative of Samos [†]

Sofia Karampela [1,2,3,*], Thanasis Kizos [2] and Alex Koutsouris [3]

1. Department of Tourism, Ionian University, 49100 Corfu, Greece
2. Department of Geography, University of the Aegean, 81100 Mytilene, Greece; akizos@aegean.gr
3. Department of Agricultural Economics & Rural Development, Agricultural University of Athens, 11855 Athens, Greece; koutsouris@aua.gr
* Correspondence: skarampela@ionio.gr
† Presented at the 17th International Conference of the Hellenic Association of Agricultural Economists, Thessaloniki, Greece, 2–3 November 2023.

Abstract: In this study, we aim to explore the possible relationships between innovation, social capital, and farm viability towards sustainability, using indicators from the literature and developing complex indexes for all examined concepts in the framework of an agriculture cooperative located on the Greek island of Samos. Data from the United Winemaking Agricultural Cooperative of Samos (UWC SAMOS) were collected through semi-structured questionnaires and further personal in-depth interviews. The findings revealed a highly complex relationship between these indexes that could not just be analyzed quantitatively. Instead, qualitative data explain the weak innovation and low level of social trust by identifying the "institutionalization of the members of the cooperative", emphasizing the importance of mixed methods approaches.

Keywords: pro-innovative behavior; trust; farm viability; cooperative; Samos Island

1. Introduction

Today, innovation is considered the key to success (and survival [1]) in all economic activities in an increasingly competitive world. The concept (and content) of innovation has been defined in many ways. In general, innovation is considered a novelty that either creates something objectively new or something that stakeholders who are involved in the innovation process perceive as qualitatively new, while other interpretations take innovation as progress or the synthesis of activities [2]. Different typologies have been developed, with one of the most well-known provided by [3], which distinguishes the following four complementary types of innovation: product innovation, process innovation, marketing innovation, and organizational innovation. This taxonomy has proved relevant to agriculture (see, inter alia, [4]). The literature also suggests that creativity is closely linked to innovation [5] and, therefore, despite the fact that its measurement is not easy, it is crucial to have some form of estimation [6]. Nevertheless, according to [7], creativity on its own is not enough to bring about innovation. Creativity is limited to idea generation alone; hence, every innovation requires creativity, but creativity does not necessarily lead to innovation [8]. In this respect, risk-taking has also been identified as a characteristic closely linked to innovation (see, inter alia, [9,10]). Being or becoming an innovator involves risk-taking in the sense that innovative individuals have to be willing to try and accept the possibility of failing [8]. Thus, several scales that measure a risk-taking propensity have been developed [11]. Furthermore, a proactive personality, able to take initiative, has been related to innovation [12]. The above dimensions, creativity, risk-taking, and proactive personality, for this study, synthesize the aspect of pro-innovation behavior similar to [13].

A concept that has been widely used recently in relation to innovation in agriculture and farming research is networks. They are considered to open wider 'windows of opportunity' with regard to innovation [14]; thus, they have, in general, become associated with many benefits in terms of agricultural and rural development (see [15]). Among others, in networks, the interactions between their members facilitate knowledge exchange and affect their behavior towards all types of innovation, including the adoption of innovations and the embeddedness of new knowledge, resulting, in the case of agriculture, in the viability of members' farms (see [15,16]). Among other types of interactions (e.g., family, circles of friendship and acquaintance, voluntary associations, etc.), cooperatives comprise a distinct type of professional and business network; according to [16], cooperatives are considered formalized forms of small firms collaboration or a specific form of social capital with significant benefits to their members and the respective communities. Social capital comprises features of social organization—networks, norms, and trust—that potentially connect and enable people to act together while also providing access to valuable resources. This concept has been used in studies related to agriculture (e.g., [17]) but not in studies of cooperatives as of yet.

Finally, farm viability is taken to be the ultimate farmers' objective (and one of the pillars of sustainability), in which the measurement of employment and satisfaction are also included. The viability of farms has been a concern of farm studies [18], and many different approaches have been proposed to examine it [19]. Some of these approaches favor economic reasoning only, considering the farm a business that has to maximize its output and/or profits [20]. Other approaches link viability, especially in family farms, to different considerations and decision-making models that include the long-term viability of the farm, the use of resources outside of the farm, and making use of other opportunities that may be available [21].

2. Materials and Methods

The assumptions behind the whole rationale of this work are that there is a relationship between innovation, social capital, and farm viability towards sustainability with respect to the members of networks. For the aspects of pro-innovative behavior, trust, and farm viability, a number of dimensions, variables, and indicators are used from the literature, and complex indexes are constructed. Data from the United Winemaking Agricultural Cooperative of Samos (UWC SAMOS), which is considered a network, were collected through semi-structured questionnaires and further personal in-depth interviews. The research was conducted during the high season of harvest, the period from June to September of the year 2021, to ensure the maximum participation of the respondents. In the final sample, 86 respondents were included, comprising members of the cooperative, employees, and selected participants—key representatives of the Board of Directors (for example, the President)—who played an important role in terms of rural development while also considering (a) vine growers, (b) those related to the innovation of farms and/or for the cooperative and (c) willingness to participate. A number of valid questionnaires were collected, and the data were analyzed with SPSS.

3. Results and Discussion

In this paper, we examine innovation, social capital, and farm viability with regard to the members of a network–agricultural cooperative based on the development of a theoretical framework and, consequently, variables and indexes to measure such aspects and their interconnections. A failure to document a statistically significant relationship between the dimensions of the examined concepts owes to various reasons, for example, the lack of time series data (not just for one year).

The results regarding the respondents' views on innovation are in line with [22], whose research rural Greece, underlining the extremely weak and fragmented nature of the Greek Agricultural Knowledge and Innovation System, which seems to be rather unique in the European Union. Also, in our case study, there is the human problem of managing attention, which is pointed out by [23] because the members of the cooperative largely focus on their harvest and preserving existing practices (as farmers argued through the history of their vineyards) rather than on developing new ideas. The more successful an organization is—in this case, a cooperative—the more difficult it is to trigger peoples' action thresholds to pay attention to new ideas, needs, and opportunities. The operation of the cooperative significantly determines the behavior of its members regarding innovation. It seems that "They can see the world only through the eyes of the cooperative as they would not imagine themselves out of it". Consequently, our findings reveal a highly complex relationship that cannot be analyzed through exclusively quantitative analysis Instead, qualitative data explain the weak innovation and low level of social trust by identifying the "institutionalization of the members of the cooperative", underlying the importance of mixed methods approaches.

4. Conclusions

This piece of work, with the goal to measure, operationalize, and understand the relationship between pro-innovative behavior, trust, and farm viability, combines and assesses the different dimensions of all the examined concepts, using variables and indicators from the literature and developing complex indexes. To the best of our knowledge, these concepts have not been examined together, let alone in the case of a cooperative. Thus, the United Winemaking Agricultural Cooperative of Samos was selected as a case study, as one of the oldest cooperatives in Greece and one of the biggest wineries nationally, where the authors tried to contribute to the deepening of knowledge for the participants and members of this network.

Author Contributions: Conceptualization, S.K., T.K., and A.K.; methodology, S.K., T.K., and A.K.; software, S.K. and T.K.; validation, S.K., T.K., and A.K.; formal analysis, S.K.; investigation, S.K.; resources, S.K.; data curation, S.K.; writing—original draft preparation, S.K.; writing—review and editing, S.K., T.K., and A.K.; supervision, T.K. All authors have read and agreed to the published version of the manuscript.

Funding: This research is co-financed by Greece and the European Union (European Social Fund-ESF) through the Operational Programme «Human Resources Development, Education and Lifelong Learning» in the context of the project "Reinforcement of Postdoctoral Researchers—2nd Cycle" (MIS-5033021), implemented by the State Scholarships Foundation (IKΥ).

Institutional Review Board Statement: Not applicable.

Informed Consent Statement: Informed consent was obtained from all subjects involved in the study.

Data Availability Statement: Data sharing is not applicable to this article.

Acknowledgments: The first author gives special thanks to the farmers and interviewees who graciously volunteered their time for the research presented in this article. This work includes a postdoctoral research fund to Sofia Karampela by the Greek State Scholarships Foundation.

Conflicts of Interest: The authors declare no conflicts of interest.

References

1. Buzan, A. Foreword. In *CATS: The Nine Lives of Innovation*; Lundin, S.C., Tan, J., Eds.; Management Press: Spring Hill, Queensland, 2007; pp. iv–viii.
2. Werynski, P. Resentment Barriers to Innovation Development of Small and Medium Enterprises in Upper Silesia. *Sustainability* **2022**, *14*, 15687. [CrossRef]
3. OECD/Eurostat. *Oslo Manual: Guidelines for Collecting and Interpreting Innovation Data*, 3rd ed.; The Measurement of Scientific, Technological and Innovation Activities; OECD Publishing: Paris, France, 2005.

4. Faure, G.; Knierim, A.; Koutsouris, A.; Ndah, H.T.; Audouin, S.; Zarokosta, E.; Wielinga, E.; Triomphe, B.; Mathé, S.; Temple, L.; et al. How to strengthen innovation support services in agriculture with regards to multi-stakeholders approaches. *J. Innov. Econ. Manag.* **2019**, *28*, 145–169.
5. Amabile, T.M. Stimulate creativity by fuelling passion. In *Handbook of Principle of Organizational Behavior*; Locke, E., Ed.; Blackwell: Malden, MA, USA, 2000; pp. 331–341.
6. Kaufman, J.C. Counting the muses: Development of the Kaufman Domains of Creativity Scale (K-DOCS). *Psychol. Aesthet. Creat. Arts* **2012**, *6*, 298–308. [CrossRef]
7. Miron, E.; Erez, M.; Naveh, E. Do personal characteristics and cultural values that promote innovation, quality, and efficiency compete or complement each other? *J. Organ. Behav.* **2004**, *25*, 175–199. [CrossRef]
8. Parzefall, M.-R.; Seeck, H.; Leppänen, A. Employee innovativeness in organizations: A review of the antecedents. *Finn. J. Bus. Econ.* **2008**, *2*, 165–182.
9. Amabile, T.M. *Creativity in Context: Update to the Social Psychology of Creativity*; Routledge: New York, NY, USA, 1996.
10. Cropley, A.J. Creativity and innovation: Men's business or women's work? *Balt. J. Psychol.* **2002**, *3*, 77–88.
11. Weber, E.U.; Blais, A.R.; Betz, N.E. A domain-specific risk-attitude scale: Measuring risk perceptions and risk behaviors. *J. Behav. Decis. Mak.* **2002**, *15*, 263–290. [CrossRef]
12. Seibert, S.E.; Kraimer, M.L.; Crant, J.M. What do proactive people do? A longitudinal model linking proactive personality and career success. *Pers. Psychol.* **2001**, *54*, 845–874. [CrossRef]
13. Baran, M.; Hazenberg, R.; Iwińska, K.; Kasianiuk, K.; Perifanos, I.; Ferreira Da Silva, J.M.; Vasconcelos, C. Between innovative and habitual behavior. Evidence from a study on sustainability in Greece, Poland, Portugal, Sweden, and the United Kingdom. *Front. Energy Res.* **2023**, *10*, 1030418. [CrossRef]
14. Corsaro, D.; Cantù, C.; Tunisini, A. Actors' heterogeneity in innovation networks. *Ind. Mark. Manag.* **2012**, *41*, 780–789.
15. Reed, G.; Hickey, G.M. Contrasting innovation networks in smallholder agricultural producer cooperatives: Insights from the Niayes Region of Senegal. *J. Co-Oper. Organ. Manag.* **2016**, *4*, 97–107.
16. Tregear, A.; Cooper, S. Embeddedness, social capital and learning in rural areas: The case of producer cooperatives. *J. Rural. Stud.* **2016**, *44*, 101–110.
17. Koutsouris, A.; Zarokosta, E. Farmers' Networks and the Quest for Reliable Advice: Innovating in Greece. *J. Agric. Educ. Ext.* **2022**, *28*, 625–629. [CrossRef]
18. Farrell, M.; Murtagh, A.; Weir, L.; Conway, S.F.; McDonagh, J.; Mahon, M. Irish organics, innovation and farm collaboration: A pathway to farm viability and generational renewal. *Sustainability* **2021**, *14*, 93. [CrossRef]
19. Wojewódzka-Wiewiórska, A.; Kłoczko-Gajewska, A.; Sulewski, P. Between the social and economic dimensions of sustainability in rural areas—In search of farmers' quality of life. *Sustainability* **2019**, *12*, 148.
20. Spicka, J.; Hlavsa, T.; Soukupova, K.; Stolbova, M. Approaches to estimation the farm-level economic viability and sustainability in agriculture: A literature review. *Agric. Econ.* **2019**, *65*, 289–297.
21. Saravia-Matus, S.; Amjath-Babu, T.S.; Aravindakshan, S.; Sieber, S.; Saravia, J.A.; Gomez y Paloma, S. Can enhancing efficiency promote the economic viability of smallholder farmers? A case of Sierra Leone. *Sustainability* **2021**, *13*, 4235. [CrossRef]
22. Koutsouris, A.; Zarokosta, E. Supporting bottom-up innovative initiatives throughout the spiral of innovations: Lessons from rural Greece. *J. Rural. Stud.* **2020**, *73*, 176–185. [CrossRef]
23. Van de Ven, A.H. Central problems in the management of innovation. *Manag. Sci.* **1986**, *32*, 590–607. [CrossRef]

Disclaimer/Publisher's Note: The statements, opinions and data contained in all publications are solely those of the individual author(s) and contributor(s) and not of MDPI and/or the editor(s). MDPI and/or the editor(s) disclaim responsibility for any injury to people or property resulting from any ideas, methods, instructions or products referred to in the content.

Proceeding Paper

Gastronomic Tourism and Festivals: Views from Potential Tourists in Greece and South Korea [†]

Aliki Dourountaki [1,*], Sofia Karampela [1,2] and Alex Koutsouris [1]

[1] Department of Agricultural Economics & Rural Development, Agricultural University of Athens, 11855 Athens, Greece; skarampela@ionio.gr (S.K.); koutsouris@aua.gr (A.K.)
[2] Department of Tourism, Ionian University, 49100 Corfu, Greece
* Correspondence: alice.dourountaki@gmail.com
[†] Presented at the 17th International Conference of the Hellenic Association of Agricultural Economists, Thessaloniki, Greece, 2–3 November 2023.

Abstract: Tourism is multifaceted and primarily encompasses cultural activities, aiming mainly to ensure visitors' relaxation and rejuvenation. Therefore, activities that involve exploring the culinary richness of a destination, which provide elements of the local culture and history, are important. The purpose of this study is to outline the profile of potential gastronomic tourists in order to identify the motivating factors for tasting local cuisine. Additionally, gastronomic festivals are raising the question of whether they serve as a means to attract potential tourists and what conditions they must meet in order to become an attractive activity. A survey was carried out targeting two nationalities with a deep gastronomic culture: Greeks and Koreans. The findings indicate that despite their common perspectives on food selection motives and that the existence of a gastronomic festival at the travel destination interests both nationalities, they differ in terms of choosing a gastronomic festival as the primary factor in visiting a destination. Overall, it is deemed useful to implement and promote gastronomic festivals with a focus on the particular needs of the respective potential tourists.

Keywords: gastronomic tourism; festival; primary research; Greece; Korea

Citation: Dourountaki, A.; Karampela, S.; Koutsouris, A. Gastronomic Tourism and Festivals: Views from Potential Tourists in Greece and South Korea. *Proceedings* **2024**, *94*, 63. https://doi.org/10.3390/proceedings2024094063

Academic Editor: Eleni Theodoropoulou

Published: 25 March 2024

Copyright: © 2024 by the authors. Licensee MDPI, Basel, Switzerland. This article is an open access article distributed under the terms and conditions of the Creative Commons Attribution (CC BY) license (https://creativecommons.org/licenses/by/4.0/).

1. Introduction

In this piece of work, an attempt was made to connect two concepts: the gastronomic tourism and gastronomic festivals. Therefore, it was deemed necessary to outline the profile of potential gastronomic tourists in order to identify their motivations concerning food choices at a tourism destination and their preferences regarding the content of a gastronomic festival. Starting with identifying the conceptual link between gastronomy and tourism, an effort was made to review the literature regarding the profile of gastronomic tourists. Based on the information gathered mainly from the models presented by [1] and [2], categories of food motivations, festival attendance criteria, and festival content, each referring to specific factors, were developed that could frame a gastronomic tourist's behavior during a gastronomic trip. Those were then included in the survey questionnaire, combining quantitative and qualitative elements.

2. Materials and Methods

An online questionnaire was created and distributed via Google Forms through social media in September 2021. The questionnaire comprised three categories that addressed each factor outlined in Table 1 by including closed- and open-ended questions. Greece and South Korea are selected as case study areas of particular interest due to their rich history and culinary culture, and in recent years, efforts to promote gastronomic tourism have been observed. The sample of research included Greek potential tourists, residents of Athens, Greece, Korean potential tourists, and residents of South Korea (221 in total: 168 Greeks and 53 Koreans). For this reason, the questionnaire was initially written in Greek and then translated to Korean. The statistical analysis and processing of the results were performed using SPSS.

Table 1. Main preferences for each factor between Greeks and Koreans.

Categories	Factor	Greeks	Koreans
Food choice motivations	Behavioral/psychological	Trying food (from different countries)	
	Taste, price, quality, social environment, religion	Influenced by food quality	Influenced by food taste
	Tradition/culture	When it relates to the history and culture of the area	
	Gastronomic activities	Trying the local cuisine	
	Gastronomic image of the destination	I am familiar with Greek/Korean local cuisine	I am familiar with Greek/Korean recipes
Attendance criteria and conditions of a culinary festival	Primary reason for travelling	I am willing to plan a trip to taste the local cuisine	
	Place, season	Gastronomic festival: willing to attend if it is located within a short distance	
	Fame/image (marketing)	Gastronomic festival: willing to attend if it was suggested to me	
Content of a gastronomic festival	Specialized cuisine	preferring it if it utilizes local ingredients	preferring it if it focuses on local traditional cuisine
	Activities	preferring it if it includes tastings	
	Events	preferring it if they are selling food	

3. Results and Discussion

The main preferences of Greeks and Koreans are provided in Table 1 and are derived from the analysis of dispersion measures. The strongest correlation (r Pearson factor ≥ 0.7) was identified for the questions "Would it encourage me to visit a food festival again?" and "Would I recommend it to others?". This is reasonable because a tourist who has already visited a festival that ultimately left them with a positive impression is followed by a positive assessment and, therefore, a recommendation to others [3,4]. Furthermore, 95.9% of those from the sample who desire to try food from different countries are associated with trying unknown food (strong correlation $r = 0.62$). In addition, the main preference under the behavioral subcategory (Table 1) was found to be "wanting to try food from different countries" for both nationalities. This indicates a general preference for consuming local cuisine during their travels, which may include foods that are unfamiliar to them which they would try. That particular subcategory was created with reference to the FNS scale (by [5]). Regarding specialized cuisine within the context of a food festival, Greeks prefer the use of local products, while Koreans focus on local traditional recipes. Specifically, the use of local products was found to be correlated with fusion cuisine. Therefore, they would seek to combine local traditional food with techniques or ingredients from other countries. Koreans, in a much higher percentage as compared to Greeks (81.1% versus 60.6%), were found to desire the food they consume to be connected to the history and culture of the region (see also [6]). A significant percentage of Koreans (64,1%) chose the answer "tasting the local cuisine" as the primary factor for travelling, whereas 48,9% of Greeks disagreed. Similarly, when it came to participating in gastronomic events or activities as a primary factor, while Koreans remained neutral, Greeks disagreed at a rate of 73.8%. These factors are correlated ($r = 0.69$), indicating that Greek tourists plan their trips primarily for other reasons, such as visiting museums and archaeological sites, a fact verified through open-ended questions (see also [7]). Ideally, both nationalities would prefer a food festival to take place in close proximity to their place of residence (see also [8]).

4. Conclusions

The present research falls short in exploring the contribution of a specific gastronomic festival to local tourism in a particular region in Greece and South Korea, as well as its impact on the perception of the gastronomic image of the place by tourists of other nationalities. The study was conducted during the global pandemic of COVID-19, which did not allow for the survey to address tourists at a festival. The collection of questionnaires should ideally have taken place during a gastronomic festival during a period of high tourist arrival rates, aiming to gather data from tourists of various nationalities. This would contribute to a more comprehensive outcome, especially concerning Section 1: gastronomic image of the area. However, the respondents were limited to those directly involved in the research, namely Greek and Korean potential tourists. Moreover, the strengths and weaknesses of this particular festival would be identified, which would serve to enhance its content and conditions.

However, it seems that the two nationalities participating in the research are interested in modern culinary variations, such as trying international or fusion cuisine, and they share common views regarding food consumption in a travel destination, focusing on taste, quality, and local culinary culture, particularly the use of local products in gastronomic festivals. A significant difference was observed in the visitation of a gastronomic festival, with only Koreans tending to view it positively. It would be interesting to further investigate the reasons as to why Greeks do not consider gastronomic festivals as primary factors, even though they have visited them in the past at a much higher rate compared to Koreans and had positive impressions. Nevertheless, it is notable that the majority would visit a gastronomic festival if they heard that it takes place in their travel destination, partially confirming tourists' need to find entertainment options during their trip but not contributing to an increase in visitor influx to the respective area. On the contrary, Koreans would plan their trip to specifically visit a gastronomic festival, which would generate a tourist influx even in remote areas.

Author Contributions: Conceptualization, A.D. and S.K.; methodology, A.D. and S.K.; software, A.D. and S.K.; validation, A.D., S.K. and A.K.; formal analysis, A.D. and S.K.; investigation, A.D.; resources, A.D. and S.K.; data curation, A.D. and S.K.; writing—original draft preparation, A.D. and S.K.; writing—review and editing, A.D., S.K. and A.K.; supervision, S.K. and A.K. All authors have read and agreed to the published version of the manuscript.

Funding: This research received no external funding.

Institutional Review Board Statement: Not applicable.

Informed Consent Statement: Informed consent was obtained from all subjects involved in the study.

Data Availability Statement: Data are contained within the article.

Acknowledgments: The authors give special thanks to the interviewees who graciously volunteered their time for the research presented in this article.

Conflicts of Interest: The authors declare no conflicts of interest.

References

1. Kim, G.Y.; Eves, A.; Scarles, C. Building a model of local food consumption on trips and holidays: A grounded theory approach. *Int. J. Hosp. Manag.* **2009**, *28*, 423–431. [CrossRef]
2. Mak, A.H.N.; Lumbers, M.; Eves, A.; Chang, R.C.Y. Factors influencing tourist food consumption. *Int. J. Hosp. Manag.* **2011**, *31*, 928–936. [CrossRef]
3. Rimmington, M.; Yuksel, A. Tourist satisfaction and food service experience: Results and implications of an empirical investigation. *Anatolia Int. J. Tour. Hosp. Res.* **1998**, *9*, 37–57. [CrossRef]
4. Lee, J.; Lee, H. 2020 Global Consumer Survey on Korean Food. Ministry of Agriculture, Food and Rural Affairs (MAFRA). Available online: https://www.mafra.go.kr/bbs/english/25/327010/artclView.do (accessed on 26 April 2021).
5. Pliner, P.; Hobden, K. Development of a scale to measure the trait of food neophobia in humans. *Appetite* **1992**, *19*, 105–120. [CrossRef] [PubMed]

6. Mak, A.H.N.; Lumbers, M.; Eves, A.; Chang, R.C.Y. The effects of food-related personality traits on tourist food consumption. *Asia Pac. J. Tour. Res.* **2017**, *22*, 1–20. [CrossRef]
7. Quan, S.; Wang, N. Towards a structural model of tourist experience: An illustration from food experiences in tourism. *Tour. Manag.* **2004**, *25*, 297–305. [CrossRef]
8. Kim, Y.H.; Duncan, J.L.; Jai, T.M. A case study of a southern food festival: Using a cluster analysis approach. *Anatolia* **2014**, *25*, 457–473. [CrossRef]

Disclaimer/Publisher's Note: The statements, opinions and data contained in all publications are solely those of the individual author(s) and contributor(s) and not of MDPI and/or the editor(s). MDPI and/or the editor(s) disclaim responsibility for any injury to people or property resulting from any ideas, methods, instructions or products referred to in the content.

Proceeding Paper

Development of A Non-Invasive System for the Automatic Detection of Cattle Lameness [†]

George Bellis [1],*, Paris Papaggelos [1], Evangeli Vlachogianni [1], Ilias Laleas [1], Stefanos Moustos [1], Thanos Patas [2], Sokratis Poulios [2], Nikos Tzioumakis [2], Giannis Giakas [3], Giorgos Tsiogkas [4], Christos Kokkotis [5] and Dimitrios Tsaopoulos [6]

1. Biomechanical Solutions (BME), 43100 Karditsa, Greece; ppapangel@gmail.com (P.P.); gb@bme.gr (E.V.); laleasilias@yahoo.gr (I.L.); stefmoy74@gmail.com (S.M.)
2. Polytech S.A., 41222 Larissa, Greece; gaspar611@gmail.com (T.P.); spoulios@polytech.com.gr (S.P.); ngt.peds@gmail.com (N.T.)
3. Department of Physical Education and Sport Science, University of Thessaly (UTH), 42100 Trikala, Greece; ggiakas@uth.gr
4. TSIOGKAS FARM, 42100 Trikala, Greece; goletsa@yahoo.gr
5. Department of Physical Education and Sport Science, Democritus University of Thrace (DUTH), 69100 Komotini, Greece; ckokkoti@affil.duth.gr
6. Institute of Bioeconomy and Agriculture (IBO-CERTH), 57001 Thessaloníki, Greece; d.tsaopoulos@certh.gr
* Correspondence: ge.mpellis@gmail.com; Tel.: +30-2441024575
† Presented at the 17th International Conference of the Hellenic Association of Agricultural Economists, Thessaloniki, Greece, 2–3 November 2023.

Abstract: Lameness is a crucial welfare issue in the modern dairy cattle industry, that if not identified and treated early causes losses in milk production and leads to early culling of animals. At present, the most common methods used for lameness detection and assessment are various visual locomotion scoring systems, which are labour-intensive, and the results may be subjective. The purpose of this project is to develop an integrated system for early detection of lameness in cattle, using force plate gait analysis and pattern recognition techniques to identify changes in gait which indicate the onset of lameness. The system will be tested on the natural onset of lameness in an organised farm environment.

Keywords: lameness; prevention; biomechanics; cattle; animal

1. Introduction

One of the most significant problems in cattle is lameness. The percentage of lameness prevalence among cattle that were exposed to the same environmental risk factors was 36.8% for U.K. [1] and relatively lower (18.7%) but also high for Greece [2]. Lameness is the declination from normal limb motion, usually with the presence of pain. Figure 1 presents some of the most frequent forms of infections in cattle hoofs that cause lameness.

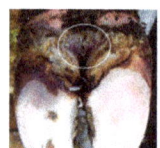   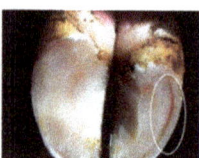

Figure 1. Usual types of hoof infections that cause lameness in cattle. From left to right: abscess, sole ulcer, abscess and ulcer, white line disease due to overtrimming or excessive wear.

Lameness impacts cattle welfare and health negatively; thus, a non-preventive attitude constitutes a violation of the five principles of cattle welfare [3]. In addition to this, lameness causes great economic loss to cattle farms, as it relates to lower milk production, milk quantity rejection, and extra medical expenses and labour costs [4] and is by far the greatest cause of death among cattle, with a rate of 20% [5].

Modern lameness detection methods are based on visual and clinical observation [6] and usually classify subjects on a scale from 1 to 5, according to the severity of observed lameness prevalence, where 1 is for light lameness and 5 is for severe lameness. In fact, these methods rely on human factors (lack of experience, subjectivity, and non-repeatability) and in most cases result in an understatement of the problem at an early stage. It has been mentioned that only one in four light lameness cases are detected in dairy cattle farms and only one in three cases of lameness is correctly classified as per the scale of severity [7].

The purpose of this project is the development of a prototype of a system that is objective, reliable, and automatic and will detect lameness in cattle at an early stage of prevalence. Early detection of the problem helps to establish early treatment, successively resulting in more effective and rapid treatment, improved welfare and health for cattle, and reduced financial loss for farmers. Meanwhile, existing lameness detection methods, that are based on visual observation, fail to assess the problem in an early stage, as is graphically presented in Figure 2.

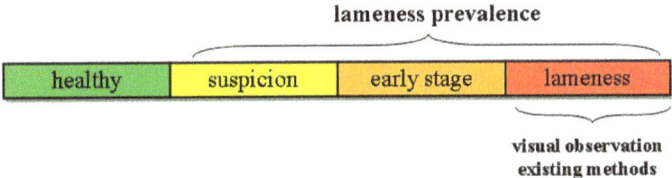

Figure 2. Stages of lameness progression.

2. Methods

The system prototype is simplistically presented in Figure 3 and generally consists of the following described basic parts, which are force plates in a walkway arrangement, an RFID ear tag system with a receiver and tags in all cattle ears, and a PC running purpose-built software that includes a user interface and a machine learning algorithm that supports the decision-making process.

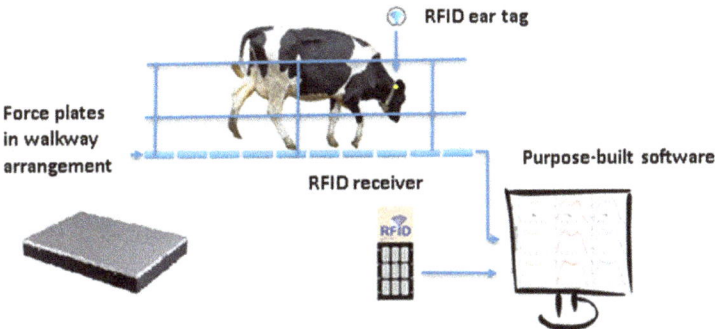

Figure 3. Prototype system main components.

The force plates acquire the ground reaction forces of the vertical axis Z of all limbs of cattle passing along the walkway one by one. The force plates were developed specifically to be suited to the dirty, dusty, and moist environment of where they would be installed, including the need to wash away mud from the top surface, the requirement of the levelling regulation of the system on top of non-flat surfaces such as soil, the constraints in surface materials that can be used due to injury hazards of cow hoofs, and the necessity of easy replacement and availability in the market of all walkway and force plate mechanical parts.

The electronic parts of the force plates were also designed and built specifically for the project. Figure 4 presents a typical load cell of nominal weight of 500 kg, like the ones that are used in force plate assembly, and the two types of custom PCBs—the analogue and the digital circuit, respectively. The PCBs were purpose-designed and -built in order to meet the main project challenges, which meant acquiring a number of analogue channels in a scale of magnitude of 100, with 24-bit precision, and a sampling rate of 1000 Hz, producing packages of synchronised data.

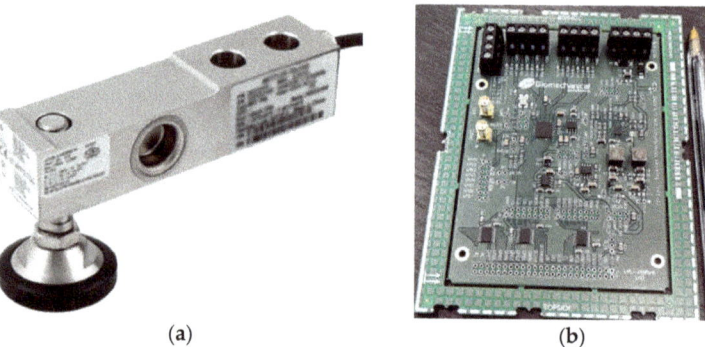

Figure 4. Electronic components of the system. (**a**) Shear beam load cell. (**b**) Custom-designed and -built PCBs for project needs.

The RFID ear tag system is a widespread commercial technology that covers the needs of identification of cattle that pass along a walkway one by one. The incorporation of the RFID system is in line with the guidance of the software, so that data are gathered separately for each distinct cow and any fluctuations in already-recorded motion patterns of limbs are also identified via the functioning machine learning algorithm.

3. Results and Discussion

Early visual detection of lameness is by default a difficult task. Thus, the pattern recognition algorithm training proved to be a challenging process, as many cattle that were tagged as completely healthy by visual observation were actually in a very early stage of lameness prevalence which could have been detected by the system, but not visually. This contrast had been creating a complication in lameness prevalence recognition from the very beginning.

In order to overcome this challenge, field data acquisition lasted much longer than was initially planned. Increased measured trials served the purpose of acquiring adequate gait patterns of healthy cattle, until meeting the aim of detecting slight gait pattern changes in some of them. This process allowed for safe segregation of cattle that were healthy and those that were in a very early stage of lameness prevalence, following the creation of a proper dataset for machine learning algorithm training.

An approximate outcome of the trials is that gait pattern changes that were recognised by the system resulted in 80% actual early-stage lameness prevalence.

4. Conclusions

The main goal of this research and development project was the development of a system that detects lameness in the very early stages, long before a visual observation can be effective. Initial field data acquisition showed encouraging results, as most of the cattle that slight gait pattern changes were detected in actually developed early-stage lameness that could be visually observed.

Author Contributions: Conceptualization, G.G., D.T.; methodology, G.B. and C.K.; software, C.K. and I.L.; practical implementation, P.P., E.V., S.M., T.P., N.T., S.P. and G.T.; writing—original draft preparation, G.B.; writing—review and editing, G.B.; All authors have read and agreed to the published version of the manuscript.

Funding: This project was funded by the Hellenic General Secretariat of Research and Innovation, project code T2EDK-00758.

Institutional Review Board Statement: Not applicable.

Informed Consent Statement: Not applicable.

Data Availability Statement: Data are contained within the article.

Conflicts of Interest: The authors George Bellis, Paris Papaggelos, Evangeli Vlachogianni, Ilias Laleas and Stefanos Moustos are employed by Biomechanical Solutions (BME). The author Thanos Patas, Sokratis Poulios and Nikos Tzioumakis are employed by Polytech S.A. George Tsiogkas is employed by TSIOGKAS FARM. The remaining authors declare that the research was conducted in the absence of any commercial or financial relationships that could be construed as potential conflicts of interest.

References

1. Barker, Z.E.; Leach, K.A.; Whay, H.R.; Bell, N.J.; Main, D.C. Assessment of lameness prevalence and associated risk factors in dairy herds in England and Wales. *J. Dairy Sci.* **2010**, *93*, 932–941. [CrossRef] [PubMed]
2. Katsoulos, P.D.; Christodoulopoulos, G. Prevalence of lameness and of associated claw disorders in Greek dairy cattle industry. *Livest. Sci.* **2008**, *122*, 354–358. [CrossRef]
3. Webster, A.J. Farm animal welfare: The five freedoms and the free market. *Vet. J.* **2001**, *161*, 229–237. [CrossRef] [PubMed]
4. Bruijnis, M.R.; Hogeveen, H.; Stassen, E.N. Measures to improve dairy cow foot health: Consequences for farmer income and dairy cow welfare. *J. Dairy Sci.* **2010**, *93*, 2419–2432. [CrossRef] [PubMed]
5. NAHMS. Part I: Reference of Dairy Cattle Health and Management Practices in the United States, 2007. In *USDA Animal Health Report*; USDA: Washington, DC, USA, 2007; p. 95.
6. Sprecher, D.J.; Hostetler, D.E.; Kaneene, J.B. A lameness scoring system that uses posture and gait to predict dairy cattle reproductive performance. *Theriogenology* **1997**, *47*, 1179–1187. [CrossRef] [PubMed]
7. Espejo, L.; Endres, M.I.; Salfer, J.A. J. Scientific report on the effects of farming systems on dairy cow welfare and disease. *Dairy Sci.* **2006**, *89*, 3052–3058. [CrossRef] [PubMed]

Disclaimer/Publisher's Note: The statements, opinions and data contained in all publications are solely those of the individual author(s) and contributor(s) and not of MDPI and/or the editor(s). MDPI and/or the editor(s) disclaim responsibility for any injury to people or property resulting from any ideas, methods, instructions or products referred to in the content.

Proceeding Paper

Pastoral Schools: Diffusing the Italian and Spanish Experience for Sustainable Mediterranean Pastoralism through Co-Creation [†]

Antonello Franca [1], Marta G. Ferre-Rivera [2], Feliu Lopez-i-Gelats [3], Giovanni M. Altana [4], Dimitrios Skordos [5], Marisol Dar Ali [5] and Athanasios Ragkos [5,*]

1. CNR ISPAAM, Institute for the Animal Production System in the Mediterranean Environment, National Research Council, 07100 Sassari, Italy; antonio.franca@cnr.it
2. Spain INGENIO (CSIC-UPV), Spanish National Research Council, 46022 Valencia, Spain; mgrivfer@ingenio.upv.es
3. Inclusive Societies, Policies and Communities, Spain University of Vic, 08500 Vic, Spain; feliu.lopez@uvic.cat
4. Italy ReteAPPIA, Italian Network of Pastoralism, 07100 Sassari, Italy; altanagiovannim@gmail.com
5. Agricultural Economics Research Institute, Hellenic Agricultural Organization–DIMITRA, 11528 Athns, Greece; dimitrisskor@gmail.com (D.S.); darali.marisol@gmail.com (M.D.A.)
* Correspondence: ragkos@elgo.gr
† Presented at the 17th International Conference of the Hellenic Association of Agricultural Economists, Thessaloniki, Greece, 2–3 November 2023.

Abstract: Pastoralism constitutes an extensive livestock system offering a feasible alternative toward agro-ecological transition. People who are engaged in the sector are expected to have a high level of skills related to knowledge and experience of nature and climate, management of resources, and other significant elements that comprise Traditional Ecological Knowledge. The purpose of this paper is to present the emergence and operation of "Pastoral Schools" in various Mediterranean countries, which offer training to people who wish to be professionally involved in pastoralism. In particular, the co-creation approach that takes place within the PASTINNOVA project is presented, which involves the establishment of Regional Living Labs bringing together actors from several Mediterranean countries who are interested in analyzing the operation of pastoral schools, exchange relevant experiences and knowledge, and deliver solutions that will upscale the performance of these schools and permit their operation to be expanded in other Mediterranean settings.

Keywords: Innovations and Business Models; Traditional Ecological Knowledge; extensive livestock production

1. Introduction

A pastoral system is a complex structure emerging from interactions of raising livestock and utilizing natural resources, in which breeders share production purposes, traditions, and cultural values [1]. Pastoral systems in the Mediterranean share three features: agro-ecological constraints; traditional socio-cultural roles; and the potential to foster sustainable entrepreneurship. A particular element at the nexus of these three components is the Traditional Ecological Knowledge (TEK) related to pastoral systems. According to [2], TEK "... consists of the body of knowledge, beliefs, traditions, practices, institutions, and worldviews developed and sustained by indigenous, peasant, and local communities in interaction with their biophysical environment". TEK involves knowledge about the environment that is based on practice and experience and is transmitted across generations [3]. This body of knowledge is used for the livelihoods of populations that evolved from it. Particular elements of TEK include biophysical observations, management practices, institutions, values, and beliefs [4]. Domains of particular interest for pastoral-related TEK are the management of natural resources (land, rangelands, and vegetation); traditional routes and practices related to flock mobilities; knowledge about weather and climate; flock management practices (feeding, grazing, milking, animal health); farm labor and the allocation of roles and task allocation among workers; typical products and transformation practices; and social norms, customs, and traditions [5].

Working on a pastoral farm is not an easy task and requires a high level of specific skills. In intensive systems, workers are expected to be familiar with modern technologies and automated systems, as well as innovative methods of monitoring animal health and product quality. In addition, farm managers must anticipate market conditions and maintain a high level of communication and information. In pastoral systems, such skills must also be complemented by TEK in order to be able to resolve unexpected situations and risks (e.g., predators and extreme weather phenomena). This employment implies very harsh living and working conditions, isolation, and limited access to basic services, while pastoralists must be aware of their role in managing public goods and providing ecosystem services. Over the centuries, TEK has been an asset used to manage these issues. Work with owners of pastoral farms showed that they appreciate the skills of salaried migrant workers related to TEK [6].

Acknowledging the specificities of pastoral systems compared to intensive ones, as well as the importance of TEK for pastoralism, "Pastoral Schools" (PS) have been established in several Mediterranean countries (Italy, Spain, France), while the need to introduce them in other countries is becoming more and more evident. This issue will be examined within the PASTINNOVA project along with other Innovations and Business Models (IBMs).

2. Methods

Within the framework of the PASTINNOVA project, a wide variety of IBMs will be considered, which have been grouped into four thematic clusters. The project proposes a co-creation approach based on the theory of Living Laboratories [7]. Because of their importance and relevance, pastoral schools have been selected as a priority IBM for Mediterranean pastoralism through an open approach among all PASTINNOVA partners and related actors, and the common network of PASTINNOVA will provide a forum to upscale their performance and expand their operation across the Mediterranean basin. The specific context that lies at the foundations of this IBM is the fact that generational renewal is one of the most relevant challenges for pastoralism at present; however, at the same time, there is an increasing will by urban dwellers to go back to nature and rural lifestyles. The co-creation process will involve actors from existing successful examples of PS operations in Spain (Catalonia) and Italy (Piedmont/Lombardy, Sardinia, and Tuscany), as well as Greece, Algeria, and potentially from other countries in order to examine ways to disseminate and transfer know-how to countries where this IBM presents lower levels of maturity.

3. Results and Discussion

The Catalan Shepherd School is a school to train new pastoralists. This PS is addressed to all people who wish to be involved in pastoralism. Although some of the students have family traditions in pastoralism, most of them do not. It has been running for the last 14 years, which demonstrates that there is already a significant level of maturity and accumulated experience. The rationale behind the operation of the Shepherd School is to understand the real needs of pastoralists and provide them with training that will deliver solutions to these needs. To achieve this, there is a team of people behind the whole organization who address the needs of trainees while also performing follow-up surveys and training even after the main training period. The main challenge being addressed by the Catalan Shepherd School is generational renewal, which is actually the major problem for pastoralism in Spain in general. Nevertheless, it remains a challenge for trained people to actually start a new business because of constraints, such as land access, machinery acquisition, and capital availability. In addition, organizational problems, such as low funding, remain.

The "National School of Pastoralism" in Italy (Scuola Nazionale di Pastorizia, SNAP) aspires to provide elements of "training, information, innovation and dialogue" and proposes itself as modular, itinerant, and interactive. The SNAP was born from the recognized need by both sector operators (farmers, technicians, and researchers) and the local communities of the internal areas to think of a new figure of pastoralists, no longer only as food producers but also as providers of ecosystem services for the communities themselves. With this aim, a working group comprising research institutions, universities, associations, and extension services from across the country was created to develop a shared training model. As in Spain, the key challenge was to tackle generational renewal problems combined with the objective to facilitate the adoption of the necessary technological, organizational, and social innovations to combine income objectives, good management of ecosystems, and preservation of cultural identity. The SNAP model was the initial reference for various initiatives in Italy that are and will be active in different geographical/socio-economic contexts, including one in Northern Italy (Lombardy and Piedmont regions), financed with private and public funds, one in Sardinia, totally financed from regional funds, and one in Tuscany, funded by a LIFE project. These schools are open to women and men interested in undertaking this activity who may be workers in other sectors, unemployed, or students, as well as to managers of extensive livestock systems, breeders, shepherds, and cheese makers. From the SNAP perspective, the school does not exhaust its activities in the educational field, it also wants to be a place of both technical and social innovation (including exchange of information, organization of events, initiatives of territorial animation), thus providing an opportunity to enhance the territories and convey knowledge, awareness, and value regarding pastoralism and, therefore, rights and resources for those who practice it. Attendants will be awarded a certificate of attendance valid for the purposes of professional placement and/or continuation of studies in national and international companies/institutions. The final objective will be the official recognition of the professional figure of the shepherd in the regional registers of professions.

In Greece, there are no schools dedicated to pastoralism-related vocational training. Farm vocational training is offered to young high school graduates who want to be employed in farms and businesses in the agri-food sector. Six (6) public Vocational Training Institutes (DIEKs) offering seven (7) curricula to interested students operate in several parts of Greece under the jurisdiction of the Ministry of Rural Development and Food and the operational responsibility of ELGO-DIMITRA. The content and supervision of the curricula are formulated by the Ministry of Education and Religious Affairs. Out of the seven curricula, two are focused on livestock production: "Dairy technician–Cheese maker" (DIEK Ioannina) and "Livestock systems administration" (DIEK Larissa). Although currently there are no specific domains targeted to pastoralism, the experience of other countries could provide an alternative to develop such training activities for people who could be interested in the profession.

4. Conclusions

During the co-creation approach in PASTINNOVA, important exchanges are expected to be developed among partners from Mediterranean countries, including the organization of field visits and exchange excursions of stakeholders. There is also a linkage with other activities and Organizations; for instance, the Greek Ministry of Culture is particularly interested in the integration of TEK in vocational training.

Author Contributions: Conceptualization and methodology, A.R., A.F., G.M.A., M.G.F.-R. and F.L.-i.-G.; investigation, all authors; writing—original draft preparation, D.S. and M.D.A.; writing—review and editing A.F., M.G.F.-R. and F.L.-i.-G.; supervision, A.R. All authors have read and agreed to the published version of the manuscript.

Funding: This research is part of the project "PASTINNOVA–Innovative Models for Sustainable Future of Mediterranean Pastoral Systems", funded by the PRIMA Foundation, which is supported by the European Union.

Institutional Review Board Statement: Not applicable.

Informed Consent Statement: Not applicable.

Data Availability Statement: No new data were created or analyzed in this study. Data sharing is not applicable to this article.

Conflicts of Interest: The authors declare no conflicts of interest. The funders had no role in the design of the study; in the collection, analyses, or interpretation of data; in the writing of the manuscript; or in the decision to publish the results.

References

1. Caballero, R.; Fernández-González, F.; Pérez-Badia, R.; Molle, G.; Roggero, P.P.; Bagella, S.; D'Ottavio, P.; Papanastasis, V.P.; Fotiadis, G.; Sidiropoulo, A.; et al. Grazing Systems and Biodiversity in Mediterranean Areas: Spain, Italy and Greece. *Rev. De La Soc. Española Para El Estud. De Los Pastos* **2009**, *39*, 9–152.
2. Gómez-Baggethun, E.; Corbera, E.; Reyes-García, V. Traditional ecological knowledge and global environmental change: Research findings and policy implications. *Ecol. Soc.* **2013**, *18*, 72. [CrossRef] [PubMed]
3. Ianni, E.; Geneletti, D.; Ciolli, M. Revitalizing traditional ecological knowledge: A study in an alpine rural community. *Environ. Manag.* **2015**, *56*, 144–156. [CrossRef]
4. Berkes, F. Context of Traditional Ecological Knowledge. In *Sacred Ecology: Traditional Ecological Knowledge and Resource Management*; Taylor & Francis: London, UK, 1999; pp. 3–14.
5. Fernández-Giménez, M.E.; El Aich, A.; El Aouni, O.; Adrane, I.; El Aayadi, S. Ilemchane Transhumant Pastoralists' Traditional Ecological Knowledge and Adaptive Strategies: Continuity and Change in Morocco's High Atlas Mountains. *Mt. Res. Dev.* **2021**, *41*, R61–R73. [CrossRef]
6. Ragkos, A.; Nori, M. Foreign workers in grazing small ruminants: Assessment of their practical knowledge and skills. In Proceedings of the 9th Conference on Rangeland Science, Larissa, Greece, 9–12 October 2018.
7. European Network of Living Labs. Available online: https://enoll.org (accessed on 20 June 2023).

Disclaimer/Publisher's Note: The statements, opinions and data contained in all publications are solely those of the individual author(s) and contributor(s) and not of MDPI and/or the editor(s). MDPI and/or the editor(s) disclaim responsibility for any injury to people or property resulting from any ideas, methods, instructions or products referred to in the content.

MDPI
St. Alban-Anlage 66
4052 Basel
Switzerland
www.mdpi.com

Proceedings Editorial Office
E-mail: proceedings@mdpi.com
www.mdpi.com/journal/proceedings

Disclaimer/Publisher's Note: The statements, opinions and data contained in all publications are solely those of the individual author(s) and contributor(s) and not of MDPI and/or the editor(s). MDPI and/or the editor(s) disclaim responsibility for any injury to people or property resulting from any ideas, methods, instructions or products referred to in the content.

www.ingramcontent.com/pod-product-compliance
Lightning Source LLC
LaVergne TN
LVHW070141100526
838202LV00015B/1871